# Mathematical Problems from Applied Logic II

# INTERNATIONAL MATHEMATICAL SERIES

Series Editor:  Tamara Rozhkovskaya
*Sobolev Institute of Mathematics of the Siberian Branch
of the Russian Academy of Sciences, Novosibirsk, Russia*

# Mathematical Problems from Applied Logic II

## Logics for the XXIst Century

Edited by

### Dov M. Gabbay

*King's College London*
*London, UK*

### Sergei S. Goncharov

*Sobolev Institute of Mathematics*
*SB Russian Academy of Sciences*
*Novosibirsk, Russia*

### Michael Zakharyaschev

*Birkbeck College*
*London, UK*

 Springer

Dov M. Gabbay
Department of Computer Science
King's College London
Strand, London WC2R 2LS
UK
dg@dcs.kcl.ac.uk

Sergei S. Goncharov
Sobolev Institute of Mathematics
SB Russian Academy of Sciences
Novosibirsk 630090
Russia
gonchar@math.nsc.ru

Michael Zakharyaschev
School of Computer Science and
    Information Systems
Birkbeck College
Malet Street
London WC1E 7HX
UK
michael@dcs.bbk.ac.uk

This series was founded by Kluwer/Plenum Publishers (now Springer) and the Russian publisher Tamara Rozhkovskaya (Novosibirsk, Russia, tamara@mathbooks.ru) in 2002. Each volume is simultaneously published in English and in Russian and presents contributions from volume editors and authors exclusively invited by the series editor. The English camera-ready manuscript was prepared by Tamara Rozhkovskaya.

e-ISBN-10: 0-387-69245-2
ISBN-13: 978-1-4419-2408-7          e-ISBN-13: 978-0-387-69245-6
ISSN 1571-5485

Printed on acid-free paper.

9 8 7 6 5 4 3 2 1

springer.com

*Mathematical Problems from Applied Logic I, II*
*Logics for the XXIst Century*

Two volumes of the *International Mathematical Series* present the most important thematic topics of logic confronting us in this century, including problems arising from successful applications areas such as computer science, AI language, etc. etc.

Invited authors—world-known specialists in the field of logic—were asked to write a chapter (in the form of a survey, a specific problem, or a point of view) basically outlining

## WHAT IS ON MY MIND AS MOST
## STRIKING/IMPORTANT/PRESSING
## NEED TO BE DONE?

# Main Topics

- Provability logic, intuitionistic provability logic, classification of provability logics, provability logics with additional operators, generalized provability predicates, interpretability and conservativity logics, Magari algebras and propositional second-order provability logic, logic of proofs, quantified logics of proofs, intuitionistic logic of proofs, the logic of single conclusion proofs, justification logic

  Sergei Artemov, Vol. II

- Nonstandard inferences in description logics; an overview of the modern state, open problems, and perspectives for future research

  Franz Baader and Ralf Küsters, Vol. I

- Logic of provability and a list of open problems in informal concepts of proof, intuitionistic arithmetic, bounded arithmetic, bimodal and polymodal logics, Magari algebras and Lindenbaum Heyting algebras, interpretability logic and its kin, graded provability algebras

  Lev Beklemishev and Albert Visser, Vol. I

- Logical dynamics: a survey of conceptual issues and open mathematical problems emanating from the recent development of various "dynamic-epistemic logics" for information update and belief revision. These systems put many-agent activities at the center stage of logic, such as speech acts, communication, and general interaction

  Johan van Benthem, Vol. I

- Computability theory, appication of computability theory to biology, psychology, physics, chemistry, economics, and other basic sciences, machine inductive inference and computability-theoretic learning, computability theory for computational complexity

  John Case, Vol. II

- The continuing relevance of Turing's approach to real-world computability and incomputability, and the mathematical modeling of emergent phenomena. Related open questions of a research interest in computability theory

  S. Barry Cooper, Vol. I

- What logics do we need? What are logical systems and what should they be? What is a proof? What foundations do we need?

  John N. Crossley, Vol. I

- Computability theory: bounds and complexity for computable models, isomorphism problem, classes of computable models and index sets, 30+ open questions in the theory of computable models

  Sergei S. Goncharov, Vol. II

- Two doors to open: Mathematical logic and cognitive science. Semantics of medieval Arab linguists

  Wilfrid A. Hodges, Vol. I

- Logic and spacetime geometry, first-order logic foundation of relativity theories, effect of gravitation on clocks

  Judit X. Madarász, István Németi, and Gergely Székely, Vol. II

- Applied logic: characterization and relation with other trends in logic, computer science, and mathematics

  Lawrence S. Moss, Vol. I

- Hybrid systems, digital programs, continuous plants and controllers, discretization, continualization

  Anil Nerode, Vol. II

- Region-based theory of space: algebras of regions, models, representation theory, contact algebras, region-based propositional modal logics of space

  Dimiter Vakarelov, Vol. II

# Editors

*Dov M. Gabbay*
King's College London
London, UK

*Sergei S. Goncharov*
Sobolev Institute of Mathematics
SB Russian Academy of Sciences

Novosibirsk State University
Novosibirsk, Russia

*Michael Zakharyaschev*
Birkbeck College
London, UK

# Dov M. Gabbay

Augustus De Morgan Professor of Logic
Department of Computer Science
King's College London
Strand, London WC2R 2LS
UK

dg@dcs.kcl.ac.uk
www.dcs.kcl.ac.uk/staff/dg

- Author of the books ○ *Temporal Logics, Vols. 1,2*, OUP, 1994, 2000 ○ *Fibred Semantics*, OUP, 1998 ○ *Elementary Logic*, Prentice-Hall, 1998, ○ *Fibring Logics*, OUP, 1998 ○ *Neural-Symbolic Learning Systems* (with A. Garcez and K. Broda), Springer, 2002 ○ *Agenda Relevance* (with J. Woods), Elsevier, 2003, etc.

- Editor-in-Chief of journals ○ *Journal of Logic and Computation* ○ *Logic Journal of the IGPL* ○ *Journal of Applied Logic* (with J. Siekmann and A. Jones) ○ *Journal of Language and Computation* (with T. Fernando, U. Reyle, and R. Kempson) ○ *Journal of Discrete Algorithms*

- Editor of journals and series ○ *Autonomous Agents* ○ *Studia Logica* ○ *Journal of Applied Non-Classical Logics* ○ *Journal of Logic, Language & Information* ○ *F News* ○ *Handbook of Logic in Computer Science* (with S. Abramsky and T. Maibaum), ○ *Handbook of Logic in AI and Logic Programming* (with C. Hogger and J.A. Robinson), ○ *Handbook of Philosophical Logic* (with F. Guenthner) ○ *Handbook of the History of Logic* ○ *Handbook of the Philosophy of Science* ○ *Studies in Logic and the Foundations of Mathematics*, etc.

Scientific interests: Logic and computation, dynamics of practical reasoning, proof theory and goal-directed theorem proving, non-classical logics and non-monotonic reasoning, labelled deductive systems, fibring logics, logical modelling of natural language.

# Sergei S. Goncharov

Sobolev Institute of Mathematics
SB Russian Academy of Sciences
4, Prospekt Koptyuga
Novosibirsk 630090
Russia

gonchar@math.nsc.ru
www.math.nsc.ru

- Head of Mathematical Logic Department and Laboratory of Computability and Applied Logic, Sobolev Institute of Mathematics SB Russian Academy of Sciences
- Council Member of the *Association for Symbolic Logic*
- Professor of Logic and Dean of Mathematical Department of the Novosibirsk State University
- Vice-Chairman of *Siberian Fund of Algebra and Logic*
- Author of the books ○ *Countable Boolean Algebras and Decidability*, Consultants Bureau, New York, 1997 ○ *Constructive Models* (with Yu.L. Ershov), Kluwer Academic/ Plenum Publishers, 2000
- Editor-in-Chief of ○ *Bulletin of the Novosibirsk State University. Ser. Mathematics, Mechanics, and Informatics*
- Editor of journals and series ○ *Algebra and Logic* (Associate Ed.) ○ *Siberian Mathematical Journal* ○ *Siberian School of Algebra and Logic* (monograph series), Kluwer Academic / Plenum Publishers ○ *Handbook of Recursive Mathematics* (with Yu.L. Ershov, A. Nerode, J.B. Remmel, and V.W. Marek), Vols. 1,2, Elsevier, 1998 ○ *Computability and Models (Perspectives East and West)* (with S.B. Cooper), Kluwer Academic/ Plenum Publishers, 2003

*Scientific interests*: Theory of computability, computable and decidable models, abstract data types, model theory and algebra, computer science and logic programming, applied logic.

# Michael Zakharyaschev

School of Computer Science
and Information Systems
Birkbeck College
Malet Street
London WC1E 7HX
UK
michael@dcs.bbk.ac.uk
www.dcs.bbk.ac.uk/~michael

- Professor of Computer Science, Birkbeck University of London
- Logic coordinator of *Group of Logic, Language and Computation* www.dcs.kcl.ac.uk/research/groups/gllc
- Member of the Steering Committee of *Advances in Modal Logic* (a bi-annual workshop and book series in Modal Logic), www.aiml.net
- Author of the books ○ *Modal Logic* (with A. Chagrov), Oxford Logic Guides: 35, Clarendon Press, Oxford, 1997 ○ *Many-Dimensional Modal Logics: Theory and Applications*, (with D. Gabbay, A. Kurucz, and F. Wolter), Series in Logic and the Foundation of Mathematics, 148, Elsevier, 2003
- Editor of journals and series ○ *Studia Logica* (Associate Ed.) ○ *Journal of Applied Logic* ○ *Journal of Logic and Computation* ○ *Advances in Modal Logic*

*Scientific interests*: Knowledge representation and reasoning, modal and temporal logics, description logics, spatial and temporal reasoning, automated theorem proving, intuitionistic and intermediate logics.

# Authors of Volume I†

*Franz Baader*

> Technische Universität Dresden
> Dresden, Germany

*Lev Beklemishev*

> Steklov Mathematical Institute RAS
> Moscow, Russia
>
> Universiteit Utrecht
> Utrecht, The Netherlands

*Johan van Benthem*

> University of Amsterdam
> Amsterdam, The Netherlands
>
> Stanford University
> Stanford, USA

*S. Barry Cooper*

> University of Leeds
> Leeds, UK

*John N. Crossley*

> Monash University
> Melbourne, Australia

---

† See www.springer.com ISBN 0-387-28688-8, ISSN 1571-5485.

*Wilfrid A. Hodges*
  Queen Mary University of London
  London, UK

*Ralf Küsters*
  Institut für Informatik
  und Praktische Mathematik
  Christian-Albrechts-Universität zu Kiel
  Kiel, Germany

*Lawrence S. Moss*
  Indiana University
  Bloomington, USA

*Albert Visser*
  Universiteit Utrecht
  Utrecht, The Netherlands

# Content of Volume I†

† See www.springer.com ISBN 0-387-28688-8, ISSN 1571-5485.

# Authors of Volume II

*Sergei Artemov*

   The City University of New York
   New York, USA
   Moscow State University
   Moscow, Russia

*John Case*

   University of Delaware
   Newark, USA

*Sergei S. Goncharov*

   Sobolev Institute of Mathematics
   SB Russian Academy of Sciences
   Novosibirsk State University
   Novosibirsk, Russia

*Judit X. Madarász*

   Alfréd Rényi Institute of Mathematics
   Hungarian Academy of Sciences
   Budapest, Hungary

*István Németi*

   Alfréd Rényi Institute of Mathematics
   Hungarian Academy of Sciences
   Budapest, Hungary

*Anil Nerode*
    Cornell University
    Ithaca, USA

*Gergely Székely*
    Eötvös Loránd University Budapest
    Alfréd Rényi Institute of Mathematics
    Hungarian Academy of Sciences
    Budapest, Hungary

*Dimiter Vakarelov*
    Sofia University
    Sofia, Bulgaria

# Sergei Artemov

The Graduate Center of
the City University of New York
365 Fifth Avenue, New York, NY 10016
USA

Moscow State University
Vorob'evy Gory, 119899 Moscow
Russia
sartemov@gc.cuny.edu
http://web.cs.gc.cuny.edu/~sartemov/

- Distinguished Professor, the Head of the Research Laboratory for Logic and Computation, CUNY
- Editor of Annals of Pure and Applied Logic, Moscow Mathematical Journal, the monograph series *Studied in Logic and Foundations of Mathematics*

*Scientific interests*: Logic in computer science, mathematical logic, proof theory, knowledge representation, artificial intelligence, automated deduction, verification, optimal control, hybrid systems.

# John Case

Department of Computer
and Information Sciences
University of Delaware
Newark, DE 19716
USA

case@cis.udel.edu
www.cis.udel.edu/~case

- Professor and the former Chair of current department and former Associate Dean, Faculty of Natural Science and Mathematics, University of Buffalo (USA)
- Author (with James Royer) of *Subrecursive Programming Systems: Complexity and Succinctness*, Birkhäuser, 1994
- A Founding Editor *J. Universal Computer Science*

*Scientific interests*: computational learning theory, machine learning, functional and comparative genomics, lattice-connected computer models of parallel processing.

# Judit X. Madarász

Department of Algebraic Logic
Alfréd Rényi Institute of Mathematics
Hungarian Academy of Sciences
Budapest, H-1053 Hungary

madarasz@renyi.hu
www.renyi.hu/~madarasz

- Author (with H. Andréka and I. Németi) of the Internet book *On the Logical Structure of Relativity Theories* [http://www.math-inst.hu/pub/algebraic-logic/olsort.html]

*Scientific interests*: Logic of spacetime, Gödel's rotating universe, geometry, algebraic logic.

# István Németi

Department of Algebraic Logic
Alfréd Rényi Institute of Mathematics
Hungarian Academy of Sciences
Budapest, H-1053 Hungary

nemeti@renyi.hu
www.logicart.hu/index_hu.html
www.renyi.hu/~nemeti

- Author of the books ○ *Cylindric Set Algebras*, Lect. Notes Math. **881**, Springer, 1981 (with L. Henkin, J. D. Monk, A. Tarski, and H. Andréka) ○ *Decision Problems for Equational Theories of Relation Algebras*, Mem. Am. Math. Soc., 1997 (with H. Andréka and S. R. Givant) ○ *Generalization of the Concept of Variety and Quasi-Variety to Partial Algebras through Category Theory*, Warsaw, 1983 (with H. Andréka)
- Editor of journals and series ○ *Journal of Applied Nonclassical Logic* ○ *Journal of Applied Logic* ○ *Advanced Studies in Mathematics and Logic book series* ○ *Logica Universalis*, etc.

*Scientific interests*: Logic as a unifying foundation of all branches of science, relativity theory, computer science and AI, universal algebra, model theory.

# Anil Nerode

Goldwin Smith Professor of Mathematics
Department of Mathematics
Cornell University
545 Malott Hall
Ithaca, NY 14853-4201
USA
anil@math.cornell.edu
www.math.cornell.edu/~anil

- Chairman (1982-1987) of Department of Mathematics and Director (1987-1996) of Mathematical Sciences Institute, Cornell University • Vice President (1991-1994), American Mathematical Society • Chair, advisor, member of centers on mathematics, information technologies, engineering, etc. of numerous organizations (NAS, NSF, AMS, MAA, IBM, etc.)
- Author of the books ○ *Automata Theory and Its Applications* (with B. Khoussainov), Birkhauser, 2001 ○ *Logic for Applications* (with R. A. Shore), Springer, 1993, 1997 ○ *Principles of Logic Programming. With the Cooperation of A. Sinachopoulos* (with G. Metakides), North-Holland, 1996, etc.
- Editor Emeritus ○ *Annals of Pure and Applied Logic* • Associate Editor ○ *Annals of Mathematics and Artificial Intelligence* • Editor of journals ○ *Computer Modelling and Simulation* ○ *Mathematics and Computer Modelling* ○ *Documenta Mathematica* ○ *Grammars* • Editor of *Handbook of Recursive Mathematics. Vols. 1, 2* (with Yu. L. Ershov, S.S. Goncharov, J.B. Remmel, and V.W. Marek) North-Holland, 1998 • Advisory Editorial Board ○ *Discrete Mathematics and Theoretical Computer Science Series* • Editor of journals and series of past years ○ *Proceedings of the American Mathematical Society* (1962-1965) ○ *Advances in Mathematics* (1968-1971) ○ *Journal of Symbolic Logic* (1968-1983) ○ *Annals of Pure and Applied Logic* (1983-1997) ○ *Future Generation Computing* (1984-1998) ○ *ORCA Journal of Computing* (1988-1990), ○ *Constraints* (1995-1999), etc.

Scientific interests: Mathematical logic, computability, complexity, automata, recursive algebra, recursive analysis, multiple agent distributed control, relaxed calculus of variations on manifolds, hybrid systems, non-monotonic logics, modal logics for computer science, constructive logics for computer science, distributed systems, history of mathematics.

# Gergely Székely

Eötvös Loránd University Budapest

Department of Algebraic Logic
Alfréd Rényi Institute of Mathematics
Hungarian Academy of Sciences
Budapest, H-1053 Hungary

turms@renyi.hu
http://www.renyi.hu/~turms

- PhD student

*Scientific interests*: Logic of spacetime, conceptual analysis of spacetime and relativity theories.

# Dimiter Vakarelov

Department
of Mathematical Logic
and Its Applications
Sofia University
15 Tsar Osvoboditel Blvd.
1504 Sofia, Bulgaria

dvak@fmi.uni-sofia.bg

- Professor of Mathematics, Faculty of Mathematics and Computer Science, Sofia University
- Editor of ○ *Journal of Applied Non-Classical Logic*

*Scientific interests*: Applied non-classical logic, modal logic, region-based theory of space.

# Content of Volume II

**Sergei Artemov**

**Judit X. Madarász,
István Németi,
and Gergely Székely**

First-Order Logic Foundation
of Relativity Theories ........................................ 217

## Anil Nerode

### Beyond Hybrid Systems ................................... 253

## Dimiter Vakarelov

### Region-Based Theory of Space: Algebras of Regions, Representation Theory, and Logics ................................................ 267

# On Two Models
# of Provability

## Sergei Artemov

*CUNY Graduate Center*
*New York, USA*

Gödel's modal logic approach to analyzing provability attracted a great deal of attention and eventually led to two distinct mathematical models. The first is the modal logic GL, also known as the Provability Logic, which was shown in 1979 by Solovay to be the logic of the formal provability predicate. The second is Gödel's original modal logic of provability S4, together with its explicit counterpart, the Logic of Proofs LP, which was shown in 1995 by Artemov to provide an exact provability semantics for S4. These two models complement each other and cover a wide range of applications, from traditional proof theory to $\lambda$-calculi and formal epistemology.

Mathematical Problems from Applied Logics. Logics for the XXIst Century. II. Edited by Dov M. Gabbay *et al*. / International Mathematical Series, 5, Springer, 2007

# 1. Introduction

In his 1933 paper [79], Gödel chose the language of propositional modal logic to describe the basic logical laws of provability. According to his approach, the classical logic is augmented by a new unary logical connective (modality) '□' where $\Box F$ should be interpreted as

*F is provable.*

> Gödel's treatment of provability as modality in [79] has an interesting prehistory. In his letter to Gödel [185] of January 12, 1931, John von Neumann actually used formal provability as a modal-like operator $B$ and gave a shorter, modal-style derivation of Gödel's second incompleteness theorem. Von Neumann freely used such modal logic features as the transitivity axiom $B(a) \rightarrow B(B(a))$, equivalent substitution, and the fact that the modality commutes with the conjunction '∧.'

Gödel's goal was to provide an exact interpretation of intuitionistic propositional logic within a classical logic with the provability operator, hence giving classical meaning to the basic intuitionistic logical system.

According to Brouwer, the founder of intuitionism, truth in intuitionistic mathematics means the existence of a proof. An axiom system for intuitionistic logic was suggested by Heyting in 1930; its full description may be found in the fundamental monographs [93, 106, 171]. By IPC, we infer Heyting's intuitionistic propositional calculus. In 1931–34, Heyting and Kolmogorov gave an informal description of the intended proof-based semantics for intuitionistic logic [91, 92, 93, 107], which is now referred to as the *Brouwer-Heyting-Kolmogorov (BHK) semantics*. According to the *BHK*-conditions, a formula is 'true' if it has a proof. Furthermore, a proof of a compound statement is connected to proofs of its parts in the following way:

- a proof of $A \wedge B$ consists of a proof of proposition $A$ and a proof of proposition $B$,

- a proof of $A \vee B$ is given by presenting either a proof of $A$ or a proof of $B$,
- a proof of $A \to B$ is a construction transforming proofs of $A$ into proofs of $B$,
- falsehood $\perp$ is a proposition which has no proof; $\neg A$ is shorthand for $A \to \perp$.

From a foundational point of view, it did not make much sense to understand the above 'proofs' as proofs in an intuitionistic system, which those conditions were supposed to specify. So in 1933 ([79]), Gödel took the first step towards developing an exact semantics for intuitionism based on **classical provability**. Gödel considered the classical modal logic S4 to be a calculus describing properties of provability in classical mathematics:

(i) *Axioms and rules of classical propositional logic,*

(ii) $\Box(F \to G) \to (\Box F \to \Box G)$,

(iii) $\Box F \to F$,

(iv) $\Box F \to \Box\Box F$,

(v) *Rule of necessitation:* $\dfrac{\vdash F}{\vdash \Box F}$ .

Based on Brouwer's understanding of logical truth as provability, Gödel defined a translation $tr(F)$ of the propositional formula $F$ in the intuitionistic language into the language of classical modal logic, i.e., $tr(F)$ was obtained by prefixing every subformula of $F$ with the provability modality $\Box$. Informally speaking, when the usual procedure of determining classical truth of a formula is applied to $tr(F)$, it will test the provability (not the truth) of each of $F$'s subformulas in agreement with Brouwer's ideas.

Even earlier, in 1928, Orlov published the paper [147] in Russian, in which he considered an informal modal-like operator of provability, introduced modal postulates (ii)–(v), and described the translation $tr(F)$ from propositional formulas to modal formulas. On the other hand, Orlov chose to base his modal system on a type of relevance logic; his system fell short of S4.

From Gödel's results in [**79**], and the McKinsey-Tarski work on topological semantics for modal logic [**130**], it follows that the translation $tr(F)$ provides a proper embedding of the intuitionistic logic IPC into S4, i.e., an embedding of IPC into classical logic extended by the provability operator.

**Theorem 1.1** (Gödel, McKinsey, Tarski). IPC *proves* $F$ $\Leftrightarrow$ S4 *proves* $tr(F)$.

Still, Gödel's original goal of defining IPC in terms of classical provability was not reached, since the connection of S4 to the usual mathematical notion of provability was not established. Moreover, Gödel noticed that the straightforward idea of interpreting modality $\Box F$ as $F$ *is provable in a given formal system* $T$ contradicted Gödel's second incompleteness theorem (cf. [**48, 51, 70, 89, 165**] for basic information concerning proof and provability predicates, as well as Gödel's incompleteness theorems).

Indeed, $\Box(\Box F \to F)$ can be derived in S4 by the rule of necessitation from the axiom $\Box F \to F$. On the other hand, interpreting modality $\Box$ as the predicate $\mathsf{Provable}_T(\cdot)$ of formal provability in theory $T$ and $F$ as contradiction, i.e., $0 = 1$, converts this formula into the false statement that the consistency of $T$ is internally provable in $T$:

$$\mathsf{Provable}_T(\lceil Consis(T) \rceil) \, .$$

To see this, it suffices to notice that the following formulas are provably equivalent in $T$:

$$\mathsf{Provable}_T(\lceil 0 = 1 \rceil) \to (0 = 1) \, ,$$
$$\neg \mathsf{Provable}_T(\lceil 0 = 1 \rceil) \, ,$$
$$Consis(T) \, .$$

Here $\lceil \varphi \rceil$ stands for the Gödel number of $\varphi$. Below we will omit Gödel number notation whenever it is safe, for example, we will write $\mathsf{Provable}(\varphi)$ and $\mathsf{Proof}(t, \varphi)$ instead of $\mathsf{Provable}(\lceil \varphi \rceil)$ and $\mathsf{Proof}(t, \lceil \varphi \rceil)$.

The situation after Gödel's paper [79] can be described by the following figure where '↪' denotes a proper embedding:

$$\text{IPC} \hookrightarrow \text{S4} \hookrightarrow ? \hookrightarrow \textit{CLASSICAL PROOFS} \, .$$

In a public lecture in Vienna in 1938 [80], Gödel suggested using the format of explicit proofs $t$ *is a proof of* $F$ for interpreting his provability calculus S4, though he did not give a complete set of principles of the resulting logic of proofs. Unfortunately, Gödel's work [80] remained unpublished until 1995, when the Gödelian logic of proofs had already been axiomatized and supplied with completeness theorems connecting it to both S4 and classical proofs.

The provability semantics of S4 was discussed in [48, 51, 56, 81, 108, 117, 121, 133, 138, 140, 141, 145, 157, 158] and other papers and books. These works constitute a remarkable contribution to this area, however, they neither found the Gödelian logic of proofs nor provided S4 with a provability interpretation capable of modeling the *BHK* semantics for intuitionistic logic. Comprehensive surveys of work on provability semantics for S4 may be found in [12, 17, 21].

The Logic of Proofs LP was first reported in 1994 at a seminar in Amsterdam and at a conference in Münster. Complete proofs of the main theorems of the realizability of S4 in LP, and about the completeness of LP with respect to the standard provability semantics, were published in the technical report [10] in 1995. The foundational picture now is

$$\text{IPC} \hookrightarrow \text{S4} \hookrightarrow \text{LP} \hookrightarrow \textit{CLASSICAL PROOFS} \, .$$

The correspondence between intuitionistic and modal logics induced by Gödel's translation $tr(F)$ has been studied by Blok, Dummett, Esakia, Flagg, Friedman, Grzegorczyk, Kuznetsov, Lemmon, Maksimova, McKinsey, Muravitsky, Rybakov, Shavrukov, Tarski, and many others. A detailed survey of modal companions of intermediate (or superintuitionistic) logics is given in [60]; a brief one is in [61], Sections 9.6 and 9.8.

Gödel's 1933 paper [79] on a modal logic of provability left two natural open problems:

(A) Find a modal logic of Gödel's predicate of formal provability Provable($x$), which appeared to be 'a provability semantics without a calculus.'

(B) Find a precise provability semantics for the modal logic S4, which appeared to be 'a provability calculus without a provability semantics.'

Problem (A) was solved in 1976 by Solovay, who showed that the modal logic GL (a.k.a. G, L, K4.W, PRL) axiomatized all propositional properties of the provability predicate Provable($F$) ([48, 51, 63, 166, 167]). The solution to problem (B) was obtained through the Logic of Proofs LP (see above and Section 3).

The provability logic GL is given by the following list of postulates:

(i) *Axioms and rules of classical propositional logic,*

(ii) $\Box(F{\to}G) \to (\Box F{\to}\Box G)$,

(iii) $\Box(\Box F{\to}F) \to \Box F$,

(iv) $\Box F{\to}\Box\Box F$,

(v) *Rule of necessitation:* $\dfrac{\vdash F}{\vdash \Box F}$.

Models (A) and (B) have quite different expressive capabilities. The logic GL formalizes Gödel's second incompleteness theorem $\neg\Box(\neg\Box\bot)$, Löb's theorem $\Box(\Box F{\to}F) \to \Box F$, and a number of other meaningful provability principles. However, proofs as objects are not present in this model. LP naturally extends typed $\lambda$-calculus, modal logic, and modal $\lambda$-calculus ([14, 15]). On the other hand, model (A) cannot express Gödel's incompleteness theorem.

Provability models (A) and (B) complement each other by addressing different areas of application. The provability logic GL finds applications in traditional proof theory (cf. Subsection 2.11).

The Logic of Proofs LP targets areas of typed theories and programming languages, foundations of verification, formal epistemology, etc. (cf. Subsection 3.8).

## 2. Provability Logic

A significant step towards finding a modal logic of formal provability was made by Löb who formulated in [125], on the basis of previous work by Hilbert and Bernays from 1939 (see [94]), a number of natural modal-style properties of the formal provability predicate and observed that these properties were sufficient to prove Gödel's second incompleteness theorem. These properties, known as the *Hilbert-Bernays-Löb derivability conditions*, essentially coincide with postulates (ii), (iv), and (v) of the above formulation of GL, i.e., with the modal logic K4. Moreover, Löb found an important strengthening of the Gödel theorem. He established the validity of the following *Löb Rule* about formal provability:

$$\frac{\vdash \Box F \rightarrow F}{\vdash F} .$$

It was later noticed in (cf. [127]) that this rule can be formalized in arithmetic, which gave a valid law of formal provability known as *Löb's principle*:

$$\Box(\Box F \rightarrow F) \rightarrow \Box F .$$

This principle provided the last axiom of the provability logic GL, named after Gödel and Löb. Neither Gödel nor Löb formulated the logic explicitly, though they established the validity of the underlying arithmetical principles. Presumably, it was Smiley, whose work [164] on the foundations of ethics was the first to consider GL a modal logic.

Significant progress in the general understanding of the formalization of metamathematics, particularly in [70], inspired Kripke, Boolos, de Jongh, and others to look into the problem of modal axiomatization of the logic of provability. More specifically, the effort was concentrated on establishing GL's completeness with respect to the formal provability interpretation. Independently, a similar problem in an algebraic context was considered by Magari and his school in Italy (see [129]). A comprehensive account of these early developments in provability logic can be found in [52].

H. Friedman formulated the question of decidability of the letterless fragment of provability logic as his Problem 35 in [74]. This question, which happened to be much easier than the general case, was immediately answered by a number of people including Boolos [46], van Benthem, Bernardi, and Montagna. This result was apparently known to von Neumann as early as 1931 [185].

## 2.1. Solovay's completeness theorem

The problem of finding a modal logic of Gödel's predicate of formal provability Provable($x$) was solved in 1976 by Solovay.

Let $*$ be a mapping from the set of propositional letters to the set of arithmetical sentences. We call such a mapping an (arithmetical) *interpretation*. Given a standard provability predicate Provable($x$) in PA, we can extend the interpretation $*$ to all modal formulas as follows:

- $\bot^* = \bot$; $\top^* = \top$;
- $*$ commutes with all Boolean connectives;
- $(\Box G)^* = $ Provable($G^*$) .

The Hilbert-Bernays-Löb derivability conditions, together with the validity of Löb's principle, essentially mean that GL is sound with respect to the arithmetical interpretation.

**Proposition 2.1.** *If* GL $\vdash X$, *then for all interpretations* $*$, PA $\vdash X^*$.

Solovay in [167] established that GL is also complete with respect to the arithmetical interpretation. Solovay also showed that the set of modal formulas expressing universally *true* principles of provability was axiomatized by a decidable extension of GL, which is usually denoted by S. The system S has the axioms

- all theorems of GL (a decidable set),
- $\Box X \to X$,

and *modus ponens* as the sole rule of inference.

**Theorem 2.1** (Solovay, [**167**]).

(1) GL $\vdash X$ *iff for all interpretations* $*$, PA $\vdash X^*$,

(2) S $\vdash X$ *iff for all interpretations* $*$, $X^*$ *is true.*

For the proof of this theorem in [**167**], Solovay invented an elegant technique of embedding Kripke models into arithmetic. Variants and generalizations of this construction have been applied to obtain arithmetical completeness results for various logics with provability and interpretability semantics. An inspection of Solovay's construction shows that it works for all natural formal theories containing a rather weak *elementary arithmetic* EA. Such robustness allows us to claim that GL is indeed a universal propositional logic of formal provability.

Whether or not Solovay's theorem can be extended to bounded arithmetic theories such as $S_2^1$ or $S_2$ remains an intriguing open question. Interesting partial results here were obtained by Berarducci and Verbrugge in [**43**].

Solovay's results and methods opened a new page in the development of provability logic. Several groups of researchers in the USA (Solovay, Boolos, Smoryński), the Netherlands (D. de Jongh, Visser), Italy (Magari, Montagna, Sambin, Valentini), and the former USSR (Artemov and his students), have started to work intensively in this area. An early textbook by Boolos [**48**], followed by Smoryński's [**166**], played an important educational role.

The following uniform version of Solovay's Theorem 2.1.1 was established independently by Artemov, Avron, Boolos, Montagna, and Visser [**3, 4, 49, 135, 175**]:

*There is an arithmetical interpretation* $*$ *such that for each modal formula* $X$, PA $\vdash X^*$ *iff* GL $\vdash X$ .

The main thrust of the research efforts in the wake of Solovay's theorem was in the direction of generalizing Solovay's results to more expressive languages. Some of the problems that have received prominent attention are covered below.

## 2.2. Fixed point theorem

As an important early result on the application of modal logic to the study of the concept of provability in formal systems, a theorem stands out that was found independently by de Jongh and Sambin, who established that GL has the fixed point property (see [48, 51, 165, 166]). The de Jongh-Sambin fixed point theorem is a striking reproduction of Gödel's fixed point lemma in a propositional language free of coding, self-substitution functions, etc.

A modal formula $F(p)$ is said to be *modalized in* $p$ if every occurrence of the sentence letter $p$ in $F(p)$ is within the scope of $\Box$.

**Theorem 2.2** (de Jongh, Sambin). *For every modal formula $F(p)$ modalized in the sentence letter $p$, there is a modal formula $H$ containing only sentence letters from $F$, not containing $p$, and such that GL proves*

$$H \leftrightarrow F(H) .$$

*Moreover, any two solutions to this fixed-point equation with respect to $F$ are provably equivalent in GL.*

The uniqueness segment was also established by Bernardi in [44].

The proof actually provided an efficient algorithm that, given $F$, calculates its fixed point $H$. Here are some examples of $F$'s and their fixed points $H$.

| Modal formula $F(p)$ | Its fixed point $H$ |
|---|---|
| $\Box p$ | $\top$ |
| $\Box \neg p$ | $\Box \bot$ |
| $\neg \Box p$ | $\neg \Box \bot$ |

$$\neg\Box\neg p \qquad\qquad\qquad\qquad\qquad \bot$$
$$q \wedge \Box p \qquad\qquad\qquad\qquad q \wedge \Box q$$

Perhaps the most famous fixed point of the above sort is given by the second Gödel incompleteness theorem. Indeed, consider $\neg\Box p$ as $F(p)$. By the above table, the corresponding fixed point $H$ is $\neg\Box\bot$. Hence GL proves

$$\neg\Box\bot \rightarrow \neg\Box(\neg\Box\bot). \tag{1}$$

Since the arithmetical interpretation of $\neg\Box\bot$ for a given theory $T$ is the consistency formula $Consis(T)$, this yields that (1) represents the formalized second Gödel incompleteness theorem:

*If $T$ is consistent, then $T$ does not prove its consistency*

and that this theorem is provable in $T$.

The fixed point theorem for GL allowed van Benthem [173] and then Visser [184] to interpret the modal $\mu$-calculus in GL. Together with van Benthem's observation that GL is faithfully interpretable in $\mu$-calculus [173], this relates two originally disjoint research areas.

## 2.3. First-order provability logics

The natural problem of axiomatizing first-order provability logic was first introduced by Boolos in [48, 50] as the major open question in this area. A straightforward conjecture that the first-order version of GL axiomatizes first-order provability logic was shown to be false by Montagna [137]. A final negative solution was given in papers by Artemov [5] and Vardanyan [174].

**Theorem 2.3** (Artemov, Vardanyan). *First-order provability logic is not recursively axiomatizable.*

In particular, Artemov showed that the set of the first-order modal formulas that are true under any arithmetical interpretation is not arithmetical. This proof used Tennenbaum's well-known theorem about the uniqueness of the recursive model of Peano

Arithmetic. Vardanyan showed that the set of first-order modal formulas that are provable in PA under any interpretation is $\Pi_2^0$-complete, thus not effectively axiomatizable. Independently but somewhat later, similar results were obtained by McGee in his Ph.D. thesis; they were never published.

Even more dramatically, [7] showed that first-order provability logics are sensitive to a particular formalization of the provability predicate and thus are not robustly defined.

The material on first-order provability logic is extensively covered in a textbook [51] and in a survey [63].

## 2.4. Intuitionistic provability logic

The question of generalizing Solovay's results from classical theories to intuitionistic ones, such as Heyting arithmetic HA, proved to be remarkably difficult. Visser, in [175], found a number of nontrivial principles of the provability logic of HA. Similar observations were independently made by Gargov and Gavrilenko. In [177], a characterization and a decision algorithm for the letterless fragment of the provability logic of HA were obtained, thus solving an intuitionistic analog of Friedman's 35th problem.

**Theorem 2.4** (Visser, [177]). *The letterless fragment of the provability logic of HA is decidable.*

Some significant further results were obtained in [65, 95, 96, 97, 177, 180, 182, 183], but the general problem of axiomatizing the provability logic of HA remains a major open question.

## 2.5. Classification of provability logics

Solovay's theorems naturally led to the notion of *provability logic for a given theory T relative to a metatheory U*, which was suggested by Artemov in [3, 4] and Visser in [175]. This logic, denoted $PL_T(U)$, is defined as the set of all propositional principles of provability in $T$ that can be established by means of $U$. In particular, GL is the provability logic $PL_T(U)$ with $U = T = \mathsf{PA}$, and Solovay's provability logic S from Theorem 2.1.2 corresponds to $T = \mathsf{PA}$ and $U$'s being the set of all true sentences of arithmetic. The problem of describing all provability logics for a given theory $T$ relative to a metatheory $U$, where $T$ and $U$ range over extensions of Peano arithmetic, has become known as the *classification problem for provability logics*. Each of these logics extends GL and hence can be represented in the form GLX which is GL with additional axioms $X$ and modus ponens as the sole rule of inference. Within this notational convention, $\mathsf{S=GL}\{\Box p \rightarrow p\}$. Consider sentences $F_n = \Box^{n+1}\bot \rightarrow \Box^n\bot$, for $n \in \omega$. In [**4, 6, 176**], the following three families of provability logics were found:

$$\mathsf{GL}_\alpha = \mathsf{GL}\{F_n \mid n \in \alpha\}, \text{ where } \alpha \subseteq \omega \; ;$$

$$\mathsf{GL}_\beta^- = \mathsf{GL}\Big\{ \bigvee_{n \notin \beta} \neg F_n \Big\}, \text{ where } \beta \text{ is a confinite subset of } \omega \; ;$$

$$\mathsf{S}_\beta = \mathsf{S} \cap \mathsf{GL}_\beta^-, \text{ where } \beta \text{ is a confinite subset of } \omega \; .$$

The families $\mathsf{GL}_\alpha$, $\mathsf{GL}_\beta^-$ and $\mathsf{S}_\beta$ are ordered by inclusion of their indices, and $\mathsf{GL}_\beta \subset \mathsf{S}_\beta \subset \mathsf{GL}_\beta^-$ for cofinite $\beta$.

In [**6**], the classification problem was reduced to finding all provability logics in the interval between $\mathsf{GL}_\omega$ and S. In [**101**], Japaridze found a new provability logic D in this interval,

$$\mathsf{D} = \mathsf{GL}\{\neg\Box\bot, \Box(\Box p \vee \Box q) \rightarrow (\Box p \vee \Box q)\} \; .$$

He showed that D is the provability logic of PA relative to PA+ *formalized $\omega$-consistency of* PA. This discovery produced one more provability logic series,

$$\mathsf{D}_\beta = \mathsf{D} \cap \mathsf{GL}_\beta^-, \text{ where } \beta \text{ is a confinite subset of } \omega \; ,$$

with $\mathsf{GL}_\beta \subset \mathsf{D}_\beta \subset \mathsf{S}_\beta \subset \mathsf{GL}_\beta^-$ for cofinite $\beta$.

The classification was completed by Beklemishev who showed in [33] that no more provability logics exist.

**Theorem 2.5** (Beklemishev, [33]). *All provability logics occur in* $GL_\alpha$, $GL_\beta^-$, $S_\beta$, *and* $D_\beta$, *for* $\alpha, \beta \subseteq \omega$, *and* $\beta$ *cofinite.*

The proof of Theorem 2.5 produced yet another provability interpretation of D which was shown to be the provability logic of any $\Sigma_1$-sound-but-not-sound theory relative to the set of all true sentences of arithmetic. For more details, see [21, 33, 41].

## 2.6. Provability logics with additional operators

Solovay's theorems have been generalized to various extensions of the propositional language by additional operators having arithmetical interpretations.

One straightforward generalization is obtained by simultaneously considering several provability operators corresponding to different theories. Already in the simplest case of *bimodal provability logic*, the axiomatization of such logics turns out to be very difficult. The bimodal logics for many natural pairs of theories have been characterized in [34, 35, 59, 101, 166]. However, the general classification problem for bimodal provability logics for pairs of recursively enumerable extensions of PA remains a major open question.

Bimodal logic has been used to study relationships between provability and interesting related concepts such as the Mostowski operator, and Rosser, Feferman, and Parikh provabilities (see [124, 160, 161, 166, 178]). In a number of cases, Solovay-style arithmetical completeness theorems have been obtained. These results have their origin in an important paper by Guaspari and Solovay [86] (see also [166]). They consider an extension of the propositional modal language by a *witness comparison* operator, thus allowing the formalization of Rosser-style arguments from his well-known proof of the incompleteness theorem [153]. Similar logics have since been used in [57, 58, 64], for example, in the study of the speed-up of proofs.

## 2.7. Generalized provability predicates

A natural generalization of the provability predicate is given by the notion of $n$-provability which is, by definition, a provability predicate in the set of all true arithmetical $\Pi_n$-sentences. For $n = 0$, this concept coincides with the usual notion of provability. As was observed in [166], the logic of each individual $n$-provability predicate coincides with GL. A joint logic of $n$-provability predicates for $n = 0, 1, 2, \ldots$ contains the modalities [0], [1], [2], etc. The arithmetical interpretation of a formula in this language is defined as usual, except that we now require, for each $n \in \omega$, that $[n]$ be interpreted as $n$-provability.

The system GLP introduced by Japaridze [101, 102] is given by the following axioms and rules of inference.

  (i)  *Axioms of* GL *for each operator* $[n]$,
 (ii)  $[m]\varphi \rightarrow [n]\varphi$, *for* $m \leq n$,
(iii)  $\langle m \rangle \varphi \rightarrow [n]\langle m \rangle \varphi$, *for* $m < n$,
 (iv)  *Rule modus ponens*,
  (v)  *Rule* $\varphi \vdash [n]\varphi$.

**Theorem 2.6** (Japaridze). GLP *is sound and complete with respect to the $n$-provability interpretation.*

Originally, Japaridze established in [101, 102] the completeness of GLP for an interpretation of modalities $[n]$ as the provability in arithmetic using not more than $n$ nested applications of the $\omega$-rule. Later, Ignatiev in [99] observed that Japaridze's theorem holds for the $n$-provability interpretation. Ignatiev also found normal forms for letterless formulas in GLP which play a significant role in Section 2.11 (where only the soundness of GLP is essential).

## 2.8. Interpretability and conservativity logics

*Interpretability* is one of the central concepts of mathematics and logic. A theory $X$ is interpretable in $Y$ iff the language of $X$

can be translated into the language of $Y$ in such a way that $Y$ proves the translation of every theorem of $X$. For example, Peano Arithmetic PA is interpretable in Zermelo-Fraenkel set theory ZF. The importance of this concept lies in its ability to compare theories of different mathematical character in different languages, for example, set theory and arithmetic. The notion of interpretability was given a mathematical shape by Tarski in 1953 in [**170**]. There is not much known about interpretability in general. The modal logic approach provides insights into the structure of interpretability in some special situations when $X$ and $Y$ are finite propositional-style extensions of a base theory containing a certain sufficient amount of arithmetic.

Visser, following Švejdar [**168**], introduced a binary modality $A \triangleright B$ to stand for the arithmetization of the statement

$$\textit{the theory } T + A \textit{ interprets } T + B,$$

where $T$ contains a sufficient amount of arithmetic, and $A$'s and $B$'s are propositional formulas in the language with '$\triangleright$.' This new modality emulates provability $\Box F$ by $\neg F \triangleright \bot$ and thus is more expressive than the ordinary $\Box$. The resulting *interpretability logic* substantially depends on the basis theory $T$.

The following logic IL is the collection of some basic interpretability principles valid in all reasonable theories: axioms and rules of GL plus

- $\Box(A \to B) \to A \triangleright B$,
- $(A \triangleright B \land B \triangleright C) \to A \triangleright C$,
- $(A \triangleright C \land B \triangleright C) \to (A \lor B) \triangleright C$,
- $A \triangleright B \to (\Diamond A \to \Diamond B)$,
- $\Diamond A \triangleright A$.

(We assume here that the interpretability modality '$\triangleright$' binds stronger than the Boolean connectives.)

For two important classes of theories $T$, the interpretability logic has been characterized axiomatically.

Let ILP be IL augmented by the principle

$$A \triangleright B \to \Box(A \triangleright B) \,.$$

**Theorem 2.7** (Visser, [**179**]). *The interpretability logic of a finitely axiomatizable theory satisfying some natural conditions is* ILP.

In particular, the class of theories covered by this theorem includes the arithmetical theories $I\Sigma_n$ for all $n = 1, 2, 3, \ldots$, the second-order arithmetic $ACA_0$, and the von Neumann-Gödel-Bernays theory GB of sets and classes.

Let ILM be IL augmented by Montagna's principle

$$A \rhd B \to (A \wedge \Box C) \rhd (B \wedge \Box C).$$

The following theorem was established independently in [**159**] and [**42**].

**Theorem 2.8** (Shavrkurov, Berarducci). *The interpretability logic of essentially reflexive theories satisfying some natural conditions is* ILM.

In particular, this theorem states that ILM is the interpretability logic for Peano arithmetic PA and Zermelo-Fraenkel set theory ZF.

An axiomatization of the minimal interpretability logic, i.e., of the set of interpretability principles that hold over all reasonable arithmetical theories, is not known. Important progress in this area has been made by Goris and Joosten, who have found new universal interpretability principles (cf. [**84**, **105**]). Yet more new interpretability principles have been found recently by Goris; they were discovered using Kripke semantics and later shown sound for arithmetic.

The $\rhd$ modality has a related *conservativity* interpretation, which leads to the conservativity logics studied in [**87**, **88**, **98**]. Logics of *interpolability* and of *tolerance*, introduced by Ignatiev and Japaridze [**66**, **67**, **100**], have a related arithmetical interpretation, but a format which is different from that of interpretability logics; see [**63**] for an overview.

An excellent survey of interpretability logic is given in [**181**]; see also [**63**].

## 2.9. Magari algebras and propositional second-order provability logic

An algebraic approach to provability logic was initiated by Magari and his students [128, 129, 135, 136]. The *provability algebra* of a theory $T$, also called the *Magari algebra of $T$*, is defined as the set of $T$-sentences factorized modulo provable equivalence in $T$ and equipped with the usual Boolean operations together with the provability operator mapping a sentence $F$ to $\mathsf{Provable}_T(F)$.

Using the notion of provability algebra, one can impart a provability semantics to a representative subclass of propositional second-order modal formulas, i.e., modal formulas with quantifiers over arithmetical sentences. These are just first-order formulas over the provability algebra. For several years, the questions of decidability of the propositional second-order provability logic and of the first-order theory of the provability algebra of $\mathsf{PA}$ remained open (cf. [20]). Shavrukov in [162] provided a negative solution to both of these questions.

**Theorem 2.9** (Shavrukov, [162]). *The first-order theory of the provability algebra of $\mathsf{PA}$ is mutually interpretable with the set of all true arithmetical formulas $\mathsf{TA}$.*

This result was proved by one of the most ingenious extensions of Solovay's techniques.

## 2.10. 'True and Provable' modality

A gap between the provability logic $\mathsf{GL}$ and $\mathsf{S4}$ can be bridged to some extent by using the *strong provability* modality $\Box F$ which is interpreted as

$$(\Box F)^* = F^* \wedge \mathsf{Provable}(F^*) \, .$$

The reflexivity principle

$$\Box F \to F$$

is then vacuously provable, hence the strong provability modality is $\mathsf{S4}$-compliant.

This approach has been explored in [47, 81, 118], where it was shown independently that the arithmetically complete modal logic of strong provability coincides with Grzegorczyk's logic Grz, which is the extension of S4 by the axiom

$$\Box(\Box(F\to\Box F)\to F)\to F \ .$$

The modality of strong provability has been further studied in [142, 143]; it played a significant role in introducing justification into formal epistemology (cf. [26, 28, 27]), as well as in the topological semantics for modal logic (cf. surveys [68, 76]).

Strong provability also plays a certain foundational role: it provides an exact provability-based model for intuitionistic logic IPC. Indeed, by Grzegorczyk's result from [85], Gödel's translation $tr$ specifies an exact embedding of IPC into Grz (cf. Theorem 1.1):

$$\text{IPC } \textit{proves } F \quad \Leftrightarrow \quad \text{Grz } \textit{proves } tr(F) \ .$$

However, the foundational significance of this reduction for intuitionistic logic is somewhat limited by a nonconstructive meaning of strong provability as 'classically true and formally provable,' which is incompatible with the intended intuitionistic semantics. The aforementioned embedding does not bring us closer to the *BHK* semantics for IPC either. For more discussion on these matters, see [8, 12, 119].

## 2.11. Applications

The methods of modal provability logic are applicable to the study of fragments of Peano arithmetic.

Using provability logic methods, Beklemishev in [36] answered a well-known question: what kind of computable functions could be proved to be total in the fragment of PA where induction is restricted to $\Pi_2$-formulas without parameters? He showed that these functions coincide with those that are primitive recursive. In general, provability logic analysis substantially clarified the behavior of parameter-free induction schemata.

Later results [37, 39] revealed a deeper connection between provability logic and traditional proof-theoretic questions, such as consistency proofs, ordinal analysis, and independent combinatorial principles. In [39], Beklemishev gave an alternative proof of Gentzen's famous theorem on the proof of the consistency of PA by transfinite induction up to the ordinal $\epsilon_0$.

In [38] (cf. also surveys [21, 40]), Beklemishev suggested a simple PA-independent combinatorial principle called *the Worm Principle*, which is derived from Japaridze's polymodal extension GLP of provability logic (cf. Section 2.7). Finite words in the alphabet of natural numbers will be called *worms*. The Worm Principle asserts the termination of any sequence $w_0, w_1, w_2, \ldots$ of worms inductively constructed according to the following two rules. Suppose $w_m = x_0 \ldots x_n$, then

(i) if $x_n = 0$, then $w_{m+1} := x_0 \ldots x_{n-1}$ (the head of the worm is cut away);

(ii) if $x_n > 0$, set $k := \max\{i < n : x_i < x_n\}$ and let $w_{m+1} = x_0 \ldots x_k (x_{k+1} \ldots x_{n-1} (x_n - 1))^{m+1}$ (the head of the worm decreases by one, and the part after position $k$ is appended to the worm $m$ times).

Clearly, the emerging sequence of worms is fully determined by the initial worm $w_0$. For example, consider a worm $w_0 = 2031$. Then the sequence looks as follows:

$$
\begin{aligned}
w_0 &= 2031 \\
w_1 &= 203030 \\
w_2 &= 20303 \\
w_3 &= 20302222 \\
w_4 &= 2030222122212221222212221 \\
w_5 &= 2030(22212221222122212220)^6 \\
&\ldots
\end{aligned}
$$

**Theorem 2.10** (Beklemishev, [38]).

(1) *For any initial worm $w_0$, there is an $m$ such that $w_m$ is empty.*

(2) *The previous statement is unprovable in Peano arithmetic*
PA. *In fact, Statement 1 is equivalent to the 1-consistency of*
PA.

For other PA-independent principles, cf. [169].

Japaridze used a technique from the area of Provability Logic
to investigate fundamental connections between provability, computability, and truth in his work on Computability Logic [103, 104].

Artemov's Logic of Proofs (Section 3) with its applications also
emerged from studies in Provability Logic.

## 3. Logic of Proofs

The source of difficulties in the provability interpretation of modality lies in the implicit nature of the existential quantifier $\exists$. Consider, for instance, the reflection principle in PA, i.e., all formulas of type $\text{Provable}(F) \to F$. By Gödel's second incompleteness theorem, this principle is not provable in PA, since the consistency formula $\text{Con(PA)}$ coincides with a special case of the reflection principle, namely $\text{Provable}(\bot) \to \bot$. The formula $\text{Provable}(F)$ is $\exists x \text{Proof}(x, F)$ where $\text{Proof}(x, y)$ is Gödel's *proof predicate*

$x$ *is (a code of) a proof of a formula (having code)* $y$.

Assuming $\text{Provable}(F)$ does not yield pointing to any specific
proof of $F$, since this $x$ may be a nonstandard natural number
which is not a code of any actual derivation in PA.

For proofs represented by explicit terms, the picture is very
different. The principle of *explicit reflection* $\text{Proof}(p, F) \to F$ is
provable in PA for each specific derivation $p$. Indeed, if $\text{Proof}(p, F)$
holds, then $F$ is evidently provable in PA, and so is the formula
$\text{Proof}(p, F) \to F$. Otherwise, if $\text{Proof}(p, F)$ does not hold, then
$\neg\text{Proof}(p, F)$ is true and provable, therefore $\text{Proof}(p, F) \to F$ is
also provable.

This observation suggests a remedy: representing proofs by terms $t$ in the proof formula $\mathsf{Proof}(t, F)$ instead of implicit representation of proofs by existential quantifiers in the provability formula $\exists x \mathsf{Proof}(x, F)$. As we have already mentioned, Gödel suggested using the format of explicit proof terms for the interpretation of S4 as early as 1938, but that paper remained unpublished until 1995 ([**80**]). Independently, the study of explicit modal logics was initiated in [**10, 29, 30, 31, 172**]. In modern terminology, the Logic of Proofs is an instance of Gabbay's Labelled Deductive Systems (cf. [**75**]).

*Proof polynomials* are terms built from *proof variables* $x, y, z, \ldots$ and *proof constants* $a, b, c, \ldots$ by means of three operations: *application* '$\cdot$' (binary), *union* '$+$' (binary), and *proof checker* '$!$' (unary). The language of *Logic of Proofs* LP is the language of classical propositional logic supplemented by a new rule for building formulas, namely for each proof polynomial $p$ and formula $F$, there is a new formula $p{:}F$ denoting '$p$ is a proof of $F$.' It is also possible to read this language type-theoretically: formulas become types, and $p{:}F$ denotes 'term $p$ has type $F$.' We assume also that '$t{:}$' and '$\neg$' bind stronger than '$\wedge, \vee$' which, in turn, bind stronger than '$\rightarrow$.'

Axioms and inference rules of LP:

(i) *Axioms of classical propositional logic*

(ii) $t{:}(F \rightarrow G) \;\rightarrow\; (s{:}F \rightarrow (t \cdot s){:}G)$          (*application*)

(iii) $t{:}F \rightarrow F$                               (*reflection*)

(iv) $t{:}F \;\rightarrow\; !t{:}(t{:}F)$                     (*proof checker*)

(v) $s{:}F \rightarrow (s+t){:}F, \quad t{:}F \rightarrow (s+t){:}F$       (*sum*)

(vi) *Rule modus ponens*

(vii) $\vdash c{:}A$, *where* $A$ *is from* (i)–(v), *and* $c$ *is a proof constant* (*Rule of constant specification*)

As one can see from the principles of LP, constants denote proofs of axioms. The application operation corresponds to the internalized *modus ponens* rule: for each $s$ and $t$, a proof $s \cdot t$ is a proof of all formulas $G$ such that $s$ is a proof of $F \rightarrow G$ and $t$ is a proof of $F$ for some $F$. The union '$s + t$' of proofs $s$ and $t$ is a proof which proves everything that either $s$ or $t$ does. Finally, '!' is interpreted as a universal program for checking the correctness of proofs, which given a proof $t$, produces a proof that $t$ proves $F$ ([10, 12]). In [13], it was noted that proof polynomials represent the whole set of possible operations on proofs for a propositional language. It was shown that any operation on proofs which is invariant with respect to a choice of a normal proof system and which can be specified in a propositional language can be realized by a proof polynomial.

In what follows, '⊢' denotes derivability in LP unless stated otherwise. By a *constant specification CS*, we mean a set of formulas $\{c_1{:}A_1, c_2{:}A_2, \ldots\}$ where each $A_i$ is an axiom from (i)–(v) of LP, and each $c_i$ is a proof constant. By default, with each derivation in LP, we associate a constant specification $CS$ introduced in this derivation by the use of the rule of constant specification.

One of the basic properties of LP is its capability of internalizing its own derivations. The weak form of this property yields the following admissible rule for LP ([10, 12]):

*if* ⊢ $F$, *then* ⊢ $p{:}F$ *for some proof polynomial $p$* .

This rule is a translation of the well-known necessitation rule of modal logic

$$\frac{\vdash F}{\vdash \Box F}$$

into the language of explicit proofs. The following more general *internalization rule* holds for LP: *if*

$$A_1, \ldots, A_n \vdash B ,$$

*then it is possible to construct a proof polynomial $t(x_1, \ldots, x_n)$ such that*

$$x_1{:}A_1, \ldots, x_n{:}A_n \vdash t(x_1, \ldots, x_n){:}B .$$

One might notice that the Curry-Howard isomorphism covers only a simple instance of the proof internalization property where all of $A_1, \ldots, A_n, B$ are purely propositional formulas containing no proof terms. For the Curry-Howard isomorphism basics, see, for example, [78].

The decidability of LP was established in [134]. Kuznets in [115] obtained an upper bound $\Sigma_2^p$ on the satisfiability problem for LP-formulas in $M$-models. This bound was lower than the known upper bound $PSPACE$ on the satisfiability problem in S4 (under the assumption that $\Sigma_2^p \neq PSPACE$). A possible explanation of why LP wins in complexity over S4 is that the satisfiability test for LP is somewhat similar to type checking, i.e., checking the correctness of assigning types (formulas) to terms (proofs), which is known to be relatively easy in classical cases. Milnikel in [132] established $\Pi_2^p$-completeness of LP for some natural classes of constant specifications, including so-called injective ones, when each constant denotes a proof of not more than one axiom. $\Pi_2^p$-hardness for the whole LP remains an open problem.

N. Krupski in [109, 110] considered a representative subsystem of LP, rLP, consisting of formulas $t{:}F$ derivable in LP. The system rLP is as expressible as LP itself, since every $F$ derivable in LP is represented in rLP by $t{:}F$ for an appropriate proof term $t$. A better upper bound ($NP$) for the decision procedure in rLP was found. In addition, the disjunctive property for the original logic of proofs LP was also established:

*if* LP $\vdash s{:}F \vee t{:}G$, *then* LP $\vdash s{:}F$ *or* LP $\vdash t{:}G$.

## 3.1. Arithmetical Completeness

The Logic of Proofs LP is sound and complete with respect to the natural provability semantics. By *proof system* we mean

1. provably in PA decidable predicate $\mathsf{Proof}(x, y)$ that enumerates all theorems of PA, i.e.,

$$\mathsf{PA} \vdash \varphi \quad \text{iff} \quad \mathsf{Proof}(n, \varphi) \text{ holds for some } n,$$

2. computable functions $\mathbf{m}(x, y)$, $\mathbf{a}(x, y)$ and $\mathbf{c}(x)$ such that, for all arithmetical formulas $\varphi, \psi$ and all natural numbers $k, n$ the following holds:

$$\mathsf{Proof}(k, \varphi \to \psi) \wedge \mathsf{Proof}(n, \varphi) \to \mathsf{Prf}(\mathbf{m}(k, n), \psi)$$

$$\mathsf{Proof}(k,\varphi) \rightarrow \mathsf{Proof}(\mathbf{a}(k,n),\varphi)$$
$$\mathsf{Proof}(n,\varphi) \rightarrow \mathsf{Proof}(\mathbf{a}(k,n),\varphi)$$
$$\mathsf{Proof}(k,\varphi) \rightarrow \mathsf{Proof}(\mathbf{c}(k),\mathsf{Proof}(k,\varphi)) .$$

The class of proof systems includes the Gödelian proof predicate in PA

> $x$ is a Gödel number of a derivation in PA that
> contains a formula with a Gödel number $y$

with the obvious choice of operations $\mathbf{m}(x,y)$, $\mathbf{a}(x,y)$, and $\mathbf{c}(x)$. In particular, $\mathbf{a}(k,n)$ is the concatenation of proofs $k$ and $n$, and $\mathbf{c}$ is a computable function that given a Gödel number of a proof $k$, returns the Gödel number $\mathbf{c}(k)$ of a proof, containing formulas $\mathsf{Proof}(k,\varphi)$ for all $\varphi$'s such that $\mathsf{Proof}(k,\varphi)$ holds.

An arithmetical interpretation $*$ is determined by a choice of proof system as well as an interpretation of proof variables and constants by numerals (denoting proofs), and propositional variables by arithmetical sentences. Boolean connectives are understood in the same way in both LP and PA, and a formula $p{:}F$ is interpreted as an arithmetical formula $\mathsf{Proof}(p^*, F^*)$.

> This kind of provability semantics is referred to as *call-by-value* semantics; it was introduced in [11] and used in [12, 14, 24, 83, 189]. A more sophisticated *call-by-name* semantics of the language of LP was introduced in [10] and used in [112, 113, 163, 186]. Under the call-by-name semantics, proof polynomials are interpreted as Gödel numbers of definable provably recursive arithmetical terms. Call-by-value interpretations may be regarded as a special case of call-by-name interpretations since numerals are definable provably recursive arithmetical terms.

For a given constant specification $CS$, an interpretation $*$ is called a $CS$-*interpretation* if all formulas from $CS$ are true under a given $*$. The following arithmetical completeness theorem has been established in [10] for the call-by-name semantics, and in [11] for the call-by-value semantics (see also articles [12, 14]):

**Theorem 3.1** (Artemov, [**10, 11**]). *A formula F is derivable in* LP *with a given constant specification* $CS$ *iff* PA $\vdash$ $F^*$, *for any* $CS$-*interpretation* $*$.

This theorem stands if one replaces 'PA $\vdash$ $F^*$' by '$F^*$ holds in the standard model of arithmetic.'

## 3.2. Realization Theorem

Another major feature of the Logic of Proofs is its ability to realize all S4-derivable formulas by restoring corresponding proof polynomials inside all occurrences of modality. This fact may be expressed by the following realization theorem ([**10, 12**]). We understand *forgetful projection* of an LP-formula $F$ to be a modal formula obtained by replacing all occurrences of $t{:}(\cdot)$ in $F$ by $\Box(\cdot)$.

**Theorem 3.2** (Artemov, [**10**]). S4 *is the forgetful projection of* LP.

That the forgetful projection of LP is S4-compliant is a straightforward observation. The converse has been established in [**10, 12**] by presenting an algorithm which substitutes proof polynomials for all occurrences of modalities in a given cut-free Gentzen-style S4-derivation of a formula $F$, thereby producing a formula $F^r$ derivable in LP. The original realization algorithms from [**10, 12**] were exponential. Brezhnev and Kuznets in [**55**] offered a realization algorithm of S4 into LP which is polynomial in the size of a cut-free derivation in S4. The lengths of realizing proof polynomials can be kept quadratic in the length of the original cut-free S4-derivation.

Here is an example of an S4-derivation realized as an LP-derivation in the style of the realization theorem 3.2. There are two columns in the table below. The first is a Hilbert-style S4-derivation of a modal formula $\Box A \vee \Box B \rightarrow \Box(\Box A \vee B)$. The second column displays corresponding steps of an LP-derivation of a formula

$$x{:}A \vee y{:}B \rightarrow (a{\cdot}!x + b{\cdot}y){:}(x{:}A \vee B)$$

with constant specification

$$\{ \; a{:}(x{:}A \to x{:}A \vee B), \quad b{:}(B \to x{:}A \vee B) \; \} \; .$$

| Derivation in S4 | Derivation in LP |
|---|---|
| 1. $\Box A \to \Box A \vee B$ | $x{:}A \to x{:}A \vee B$ |
| 2. $\Box(\Box A \to \Box A \vee B)$ | $a{:}(x{:}A \to x{:}A \vee B)$ |
| 3. $\Box\Box A \to \Box(\Box A \vee B)$ | $!x{:}x{:}A \to (a{\cdot}!x){:}(x{:}A \vee B)$ |
| 4. $\Box A \to \Box\Box A$ | $x{:}A \to !x{:}x{:}A$ |
| 5. $\Box A \to \Box(\Box A \vee B)$ | $x{:}A \to (a{\cdot}!x){:}(x{:}A \vee B)$ |
| 5′. | $(a{\cdot}!x){:}(x{:}A \vee B) \to (a{\cdot}!x + b{\cdot}y){:}(x{:}A \vee B)$ |
| 5″. | $x{:}A \to (a{\cdot}!x + b{\cdot}y){:}(x{:}A \vee B)$ |
| 6. $B \to \Box A \vee B$ | $B \to x{:}A \vee B$ |
| 7. $\Box(B \to \Box A \vee B)$ | $b{:}(B \to x{:}A \vee B)$ |
| 8. $\Box B \to \Box(\Box A \vee B)$ | $y{:}B \to (b{\cdot}y){:}(x{:}A \vee B)$ |
| 8′. | $(b{\cdot}y){:}(x{:}A \vee B) \to (a{\cdot}!x + b{\cdot}y){:}(x{:}A \vee B)$ |
| 8″. | $y{:}B \to (a{\cdot}!x + b{\cdot}y){:}(x{:}A \vee B)$ |
| 9. $\Box A \vee \Box B \to \Box(\Box A \vee B)$ | $x{:}A \vee y{:}B \to (a{\cdot}!x + b{\cdot}y){:}(x{:}A \vee B)$ |

Extra steps 5′, 5″, 8′, and 8″ are needed in the LP case to reconcile different internalized proofs of the same formula: $(a{\cdot}!x){:}(x{:}A \vee B)$ and $(b{\cdot}y){:}(x{:}A \vee B)$. The resulting realization respects Skolem's idea that negative occurrences of existential quantifiers (here over proofs hidden in the modality of provability) are realized by free variables whereas positive occurrences are realized by functions of those variables.

Switching from the provability format to the language of specific witnesses reveals hidden self-referentiality of modal logic, i.e., the necessity of using proof assertions of the form $t{:}F(t)$, where $t$ occurs in the very formula $F(t)$ of which it is a proof. A recent result by Kuznets in [55] shows that self-referentiality is an intrinsic feature of the modal logic approach to provability in general.

**Theorem 3.3** (Kuznets, [55]). *Self-referential constant specifications of the sort $c{:}A(c)$ are necessary for realization of the modal logic S4 in the Logic of Proofs LP.*

In particular, the S4-theorem

$$\neg\Box\neg(S\to\Box S)$$

cannot be realized in LP without self-referential constant specifications of the sort $c{:}A(c)$.

Systems of proof polynomials for other classical modal logics K, K4, D, D4, T were described in [53, 54]. The case of S5 = S4 + $(\neg\Box F \to \Box\neg\Box F)$ was special because of the presence of negative information about proofs and its connections to formal epistemology. The paper by Artemov, Kazakov, and Shapiro [24] introduced a system of proof terms for S5, and established realizability of the logic S5 by these terms, decidability, and completeness of the resulting logic of proofs. An alternative approach, not connected to the arithmetical provability, to representing negative information in the logic of proofs was considered in [126].

## 3.3. Fitting models

The original idea of epistemic semantics for LP can be traced back to Mkrtychev and Fitting. It consists of augmenting Boolean or Kripke models with an *evidence function*, which assigns 'admissible evidence' terms to a statement before deciding its truth value.

Fitting models are defined as follows. A *frame* is a structure $(W, R)$, where $W$ is a non-empty set of *possible worlds* and $R$ is a binary reflexive and transitive *evidence accessibility* relation on $W$. Given a frame $(W, R)$, a *possible evidence* function $\mathcal{E}$ is a mapping from worlds and proof polynomials to sets of formulas. We can read $F \in \mathcal{E}(u, t)$ as

'F is one of the formulas for which

t serves as possible evidence in world u.'

An evidence function respects the intended meanings of the operations on proof polynomials, i.e., for all proof polynomials $s$ and $t$, for all formulas $F$ and $G$, and for all $u, v \in W$, each of the following hold:

(i) *Monotonicity*: $uRv$ implies $\mathcal{E}(u, t) \subseteq \mathcal{E}(v, t)$;

(ii) *Closure*

- *Application*: $F \to G \in \mathcal{E}(u, s)$ and $F \in \mathcal{E}(u, t)$ implies $G \in \mathcal{E}(u, s \cdot t)$;
- *Inspection*: $F \in \mathcal{E}(u, t)$ implies $t{:}F \in \mathcal{E}(u, !t)$;
- *Sum*: $\mathcal{E}(u, s) \cup \mathcal{E}(u, t) \subseteq \mathcal{E}(u, s + t)$.

A model is a structure $\mathcal{M} = (W, R, \mathcal{E}, \Vdash)$ where $(W, R)$ is a frame, $\mathcal{E}$ is an evidence function on $(W, R)$, and $\Vdash$ is an arbitrary mapping from sentence variables to subsets of $W$. Given a model $\mathcal{M} = (W, R, \mathcal{E}, \Vdash)$, the forcing relation $\Vdash$ is extended from sentence variables to all formulas by the following rules. For each $u \in W$:

(i) $\Vdash$ respects Boolean connectives ($u \Vdash F \wedge G$ iff $u \Vdash F$ and $u \Vdash G$, $u \Vdash \neg F$ iff $u \nVdash F$, etc.);

(ii) $u \Vdash t{:}F$ iff $F \in \mathcal{E}(u, t)$ and $v \Vdash F$ for every $v \in W$ with $uRv$.

We consider the modality $\square$, associated with the evidence accessibility relation $R$. In this terms, the last item of the above definition can be recast as

(ii') $u \Vdash t{:}F$ iff $u \Vdash \square F$ and $t$ is an admissible evidence for $F$ at $u$.

Mkrtychev models are Fitting models with singleton $W$'s. LP was shown to be sound and complete with respect to both Mkrtychev models ([**134**]) and Fitting models ([**72, 73**]). Fitting models were adapted for a multi-agent epistemic setting in [**16, 26, 27, 71**] and became the standard semantics for epistemic modal logics with justification.

In his recent paper [**83**], Goris showed that LP is sound and complete with respect to the call-by-value semantics of proofs in Buss's weak arithmetic $S_2^1$, thus showing that explicit knowledge can be realized by *PTIME*-computable operations on proofs in a natural mathematical system. Note that the corresponding question for the Provability Logic GL remains a major open problem (cf. Subsection 2.1).

### 3.4. Joint logics of proofs and provability

The problem of finding a joint logic of proofs and provability
has been a natural next step, since there are important principles formulated in a mixed language of formal provability and
explicit proofs. For example, the modal principle of negative introspection $\neg\Box F \to \Box\neg\Box F$ is not valid in the provability semantics; neither is a purely explicit version of negative introspection
$\neg(x{:}F) \to t(x){:}\neg(x{:}F)$. However, a mixed explicit-implicit principle
$\neg(t{:}F) \to \Box\neg(t{:}F)$ is valid in the standard provability semantics.

The joint system of provability and explicit proofs without
operations on proof terms, system B, was found in [9]. This system
describes those principles that have a pure logical character and
do not depend on any specific operations of proofs.

The postulates of B consist of those of GL together with the
following new principles:

A1. $t{:}F \to F$,

A2. $t{:}F \to \Box t{:}F$,

A3. $\neg t{:}F \to \Box\neg t{:}F$,

RR. *Rule of reflection:* $\dfrac{\vdash \Box F}{\vdash F}$ .

**Theorem 3.4** (Artemov, [9]). B *is sound and complete with
respect to the semantics of proofs and provability in Peano arithmetic.*

The problem of joining two models of provability, GL and LP,
into one model can be specified as that of finding an arithmetically
complete logic containing postulates of both GL and LP and closed
under internalization.

The first solution to this problem was offered by Yavorskaya
(Sidon) in [163, 186] who found an arithmetically complete system of provability and explicit proofs, LPP, containing both GL
and LP. Along with natural extensions of principles and operations from GL and LP, LPP contains additional operations '⇑' and
'⇓' which were used to secure the internalization property of LPP.

The operation '⇑' given a proof $t$ of $F$, returns a proof $⇑t$ of Provable($F$). The operation '⇓' takes a proof $t$ of Provable($F$) and returns a proof $⇓t$ of $F$. The set of postulates of LPP consists of those of GL and LP together with A2, A3, and RR from B, plus two new principles:

A4. $t{:}F \rightarrow (⇑t){:}\Box F$,

A5. $t{:}\Box F \rightarrow (⇓t){:}F$.

Finally, Nogina in [26, 144] noticed that each specific instance of operations '⇑' and '⇓' can be eliminated, and introduced an arithmetically complete logic GLA joining GL and LP in their original languages. The system GLA is presented in [26, 144] by the set of postulates of GL and LP augmented by the principles:

- $t{:}F \rightarrow \Box F$,
- $\neg t{:}F \rightarrow \Box \neg t{:}F$,
- $t{:}\Box F \rightarrow F$.

and *Rule of reflection* RR.

**Theorem 3.5.**

(1) (Yavorskaya (Sidon), [163, 186]) LPP *is sound and complete with respect to the semantics of proofs and provability in Peano arithmetic.*

(2) (Nogina, [26, 144]) GLA *is sound and complete with respect to the semantics of proofs and provability in Peano arithmetic.*

It was the system GLA which served in [26, 27] as a prototype of basic logic of knowledge with justification (cf. Subsection 3.8).

## 3.5. Quantified logics of proofs

The arithmetical provability semantics for the logic of proofs may be naturally generalized to first-order language and to the language of LP with quantifiers over proofs. Both possibilities of enhancing the expressive power of LP were investigated and in both cases, axiomatizability questions have been answered negatively.

**Theorem 3.6.**

(1) (Artemov, Yavorskaya (Sidon), [**32**]) *The first-order logic of proofs is not recursively enumerable.*

(2) (Yavorsky, [**190**]) *The logic of proofs with quantifiers over proofs is not recursively enumerable.*

An interesting decidable fragment of the first-order logic of the standard proof predicate was found in [**189**].

## 3.6. Intuitionistic logic of proofs

As in the case of Provability Logic, a natural question is that of efficient axiomatization of the logic of proofs for Heyting Arithmetic HA. However, unlike the Provability Logic case, the first layer of problems here has a definite resolution. Let us consider so-called intuitionistic basic logic of proofs iBLP where no specific operations on proofs are in the langauge.

The first thing to notice is that in addition to the principles borrowed from the classical Logic of Proofs, there is a principle of decidability of proof assertions

$$t{:}F \lor \neg t{:}F \ .$$

Another source of new principles is the set of admissible propositional rules in HA. As was noticed by Iemhoff, for each admissible rule $F/G$ in HA there is a logic of proofs principle

$$x{:}F \to G \ .$$

A complete decidable axiomatization iBLP was found by Artemov and Iemhoff in [**22, 23**] with the use of the ideas and technique of Ghilardi. The next natural goal in this direction is the establishment of the arithmetical completeness of intuitionistic logic of proofs with operations corresponding to all admissible rules in HA, cf. [**22, 23**] for precise formulations.

## 3.7. The logic of single conclusion proofs

By definition, each single conclusion proof, also known as *functional proofs*, proves a unique formula. In the functional logic of proofs, a formula $t{:}F$ still has the meaning '$t$ is a proof of formula $F$,' but the class of its interpretations is limited to functional proof systems only. It is easy to see that single conclusion proofs lead to modal identities inconsistent with any normal modal logic, e.g., $x{:}\top \rightarrow \neg x{:}(\top \wedge \top)$ is a valid principle of the functional proofs which, however, has the forgetful projection $\Box\top \rightarrow \neg\Box(\top\wedge\top)$ which is incompatible with any normal modal logic.

The mathematical problem here was to give a full axiomatization of all resulting tautologies in the language of LP (without the operation '+,' which does not work on functional proofs); this problem was solved by V. Krupski in [112].

The functionality property of proofs, which states that if $p{:}$ $F \wedge p{:}G$, then $F$ and $G$ must coincide syntactically, does not look like a propositional condition, since it operates with the strong notion of syntactic coincidence. An adequate propositional description of this property was found in [29] using so-called *conditional unification*. It was then generalized in [112, 113] to the full language of the logic of proofs.

Each formula $C$ of type $t_1{:}F_1 \wedge \ldots \wedge t_n{:}F_n$ generates a set of quasi-equations of type $S_C := \{\, t_i = t_j \Rightarrow F_i = F_j \mid 1 \le i, j \le n \,\}$. A *unifier* $\sigma$ of $S_C$ is a substitution $\sigma$ such that either $t_i\sigma \not\equiv t_j\sigma$ or $F_i\sigma \equiv F_j\sigma$ holds for any $i, j$. Here and below '$X \equiv Y$' denotes the syntactic equality of $X$ and $Y$. $A = B\,(mod\,S)$ means that for each unifier $\sigma$ of system $S$, the property $A\sigma \equiv B\sigma$ holds. This conditional unification was shown to be decidable in the cases under consideration (cf. [29, 112, 113]). By *Unification Axiom* we understand the schema

$$t_1{:}F_1 \wedge \ldots \wedge t_n{:}F_n \rightarrow (A \leftrightarrow B)$$

for each condition $C$ of type $t_1{:}F_1 \wedge \ldots \wedge t_n{:}F_n$ and each $A$, $B$ such that $A = B\,(mod\,S_C)$.

The Logic of Functional Proofs FLP was introduced by V. Krupski in [112]. The language of FLP is the language of LP without the operation "+" and without proof constants. The axioms and rules of FLP are:

A0. *Axiom and rules of classical propositional logic*

A1. $t{:}(F \rightarrow G) \rightarrow (s{:}F \rightarrow (t{\cdot}s){:}G)$

A2. $t{:}F \rightarrow F$

A3. $t{:}F \rightarrow !t{:}t{:}F$

A4. *Unification axiom.*

**Theorem 3.7** (V. Krupski, [112, 113]). *The logic* FLP *is decidable, sound, and complete with respect to the arithmetical provability interpretation based on functional proof predicates.*

The logic of functional proofs was further developed in [114], where its extension with references $\mathsf{FLP}_{ref}$ was introduced. System $\mathsf{FLP}_{ref}$ extends FLP with second-order variables which denote the operation of reconstructing an object from its reference, e.g., determining a formula proven by a given derivation. $\mathsf{FLP}_{ref}$ may be also viewed as a natural formal system for admissible inference rules in arithmetic. See also follow-up articles [156, 187].

## 3.8. Applications

Here we will list some conceptual applications of the Logic of Proofs.

1. *Existential semantics for modal logic.* Proof polynomials and LP represent an exact *existential semantics* for mainstream modal logic. Initially, Gödel regarded the modality $\Box F$ as the provability assertion, i.e.,

there exists a proof for $F$.

Thus, according to Gödel, modality is a $\Sigma_1$-sentence, i.e., the one which consists of an existential quantifier (here over proofs) followed by a decidable condition. Such an understanding of modality is typical of 'naive' semantics for a wide range of epistemic and provability logics. Nonetheless, before LP was discovered, major modal logics lacked a mathematical semantics of an existential character. The exception to the rule is the arithmetical provability interpretation for the Provability Logic GL, which still cannot be extended to the major modal logics S4 and S5.

Almost 30 years after the first work by Gödel on the subject, a semantics of a *universal* character was discovered for modal logic, namely Kripke semantics. Modality in that semantics is read informally as the sentence:

*In each possible situation, F holds.*

Such a reading of modality naturally appears in dynamic and temporal logics aimed at describing computational processes, states of which usually form a (possibly branching) Kripke structure. Universal semantics has been playing a prominent role in modal logic. However, it is not the only possible semantical tool in the study and application of modality. The existential semantics of realizability by proof polynomials can also be useful for foundations and application of modal logic. For more discussion on the existential semantics for modal logic, see [18].

2. *Justification Logic.* A major area of application of the Logic of Proofs is epistemology. Books [69, 131] serve as excellent introductions to the mathematical logic of knowledge.

Plato's celebrated tripartite definition of knowledge as *justified true belief* is generally regarded in mainstream epistemology as a set of necessary conditions for the possession of knowledge. Due to Hintikka, the 'true belief' components have been fairly formalized by means of modal logic and its possible worlds semantics. The remaining 'justification' condition has received much attention in epistemology (cf., for example, [45, 77, 82, 90, 120, 122, 123, 146]), but lacked formal representation. The issue of finding a formal epistemic logic with justification has also been discussed in [172]. Such a logic should contain assertions of the form $\Box F$ (*F is known*), along with those of the form $t{:}F$ (*t is a justification for F*). Justification was introduced into formal epistemology in [16, 26, 27, 28] by combining Hintikka-style epistemic modal logic with justification calculi arising from the Logic of Proofs LP. Epistemic logic with justification was used in [16, 19] to offer a new approach to *common knowledge*. A new modal operator $J\varphi$ for *justified knowledge* introduced in [16, 19] is defined as a forgetful projection of justification assertions $t:\varphi$ in a multi-agent epistemic logic with common justification. Justified knowledge was shown to be a lighter, constructive version of common knowledge. In particular, in [2] it was shown that for a typical epistemic problem, common knowledge systems are conservative over those of justified knowledge, hence whenever the former work, the latter can be used, too. This line of research is picking up rapidly, cf. also [71, 116, 148, 151, 152, 154, 155, 188].

3. *Tackling the logical omniscience problem.* The traditional Hintikka-style modal logic approach to knowledge has the well-known defect of *logical omniscience*, which is the unrealistic feature that an agent knows all logical consequences of his/her assumptions ([69, 139, 149, 150]). Epistemic systems with justification address the issue of logical omniscience in a natural way. A justified knowledge $t:F$ cannot be asserted without presenting an explicit justification $t$ for $F$, hence justified knowledge does not lead to logical omniscience. This property was formally established in [25], where it was shown that Justification Logic is logically omniscient w.r.t. the usual knowledge represented by Hintikka-style epistemic modalities '$F$ *is known*' (modulo common complexity assumptions), and is not logically omniscient w.r.t. the evidence-based knowledge '$t$ *is a justification for* $F$.'

4. *Reflection in typed combinatory logic and λ-calculus.* Typed λ-calculi and Combinatory Logic are mathematical prototypes of functional programming languages with types (cf., for example, [62]). There are reasons to believe that this area would benefit from extending λ-calculi and Combinatory Logic by self-referential capacities which enable systems to simultaneously operate with related objects of different abstraction level: functions, their high level programs, their low level codes, etc. Reflexive Combinatory Logic RCL ([17]) was invented to meet these kinds of expectations. RCL introduces a reflexivity mechanism into Combinatory Logic,

hence to $\lambda$-calculus. RCL has the implicative intuitionistic (minimal) logic as a type system, a rigid typing. Reflexive combinatory terms are built from variables, 'old' combinators **k** and **s**, and new combinators **d**, **o**, and **c**. The principles of RCL are

A1. $t{:}A \to A$
A2. $\mathbf{k}{:}(A \to (B \to A))$
A3. $\mathbf{s}{:}[(A \to (B \to C)) \to ((A \to B) \to (A \to C))]$
A4. $\mathbf{d}{:}(t{:}A \to A)$
A5. $\mathbf{o}{:}[u{:}(A \to B) \to (v{:}A \to (u \cdot v){:}B)]$
A6. $\mathbf{c}{:}(t{:}A \to !t{:}t{:}A)$

*Rule modus ponens,*

$$\frac{A \to B \quad A}{B} \ .$$

RCL has a natural provability semantics inherited from LP. Combinatory terms stand for proofs in PA or in intuitionistic arithmetic HA. Formulas $t{:}F$ are interpreted as arithmetical statements about provability, $\mathsf{Proof}(t, F)$, combinators **k**, **s**, **d**, **o**, and **c** denote terms corresponding to proofs of arithmetical translations of axioms A2–A6.

RCL evidently contains implicative intuitionistic logic, ordinary Combinatory Logic CL$_\to$, and is closed under the combinatory application rule

$$\frac{u{:}(A \to B) \quad v{:}A}{(u \cdot v){:}B} \ .$$

Furthermore, RCL enjoys the internalization property ([**17**]): if $A_1, \ldots, A_n \vdash B$ then for any set of variables $x_1, \ldots, x_n$ of respective types, it is possible to construct a term $t(x_1, \ldots, x_n)$ such that

$$x_1{:}A_1, \ldots, x_n{:}A_n \vdash t(x_1, \ldots, x_n){:}B \ .$$

It is interesting to consider the following natural (though so far informal) computational semantics for combinators of RCL. This semantics is based on the standard set-theoretic semantics of types, i.e., a type is a set and the implication type $U \to V$ is a set of functions from $U$ to $V$. Some elements of a given type may be constructive objects which have *names*, i.e., computational programs. Terms of RCL are names of constructive objects, some of them specific, e.g., combinators $k$, $s$, $d$, $o$, or $c$). The type $t{:}F$ is interpreted as a set consisting of the object corresponding to term $t$.

Basic combinators of RCL are understood as follows. Combinators $k$ and $s$ have the same meaning as in the combinatory logic $CL_\to$. For example, $k$ maps an element $x \in A$ into the constant function $\lambda y.x$ with $y$ ranging over $B$. The *denotate* combinator $d : [t : F \to F]$ realizes the function which maps a name (program) into the object with the given name. A primary example is the correspondence between indexes of computable functions and functions themselves. The *interpreter* combinator $o : [u : (F \to G) \to (v : F \to (u \cdot v) : G)]$ realizes the program which maps program $u$ and input $v$ into the result of applying $u$ to $v$. The *coding* combinator $c{:}[t{:}F \to !t{:}(t{:}F)]$ maps program $t$ into its code $!t$ (alias, specific key in a database, etc.).

In the followup papers [109, 111], N. Krupski established that typability and type restoration can be done in polynomial time and that the derivability relation for RCL is decidable and *PSPACE-complete*.

In [1], some version of reflexive $\lambda$-calculus was considered that has an unrestricted internalization property.

## 4. Acknowledgements

The real authors of this survey are those who contributed to this wonderful field scientifically. The author is especially indebted to his student, close collaborator, and friend Lev Beklemishev who strongly influenced this text, especially its Provability Logic portion. Special thanks to Karen Kletter for proofreading and editing this paper.

# References

1. J. Alt and S. Artemov, *Reflective λ-calculus*, In: Proceedings of the Dagstuhl-Seminar on Proof Theory in Computer Science, Lect. Notes Comput. Sci. **2183**, Springer, 2001, pp. 22–37.

2. E. Antonakos, *Justified knowledge is sufficient*, Technical Report TR-2006004, CUNY Ph.D. Program in Computer Science (2006).

3. S. Artemov, *Extensions of Arithmetic and Modal Logics* (in Russian), Ph.D. Thesis, Moscow State University - Steklov Mathematical Insitute (1979).

4. S. Artemov, *Arithmetically complete modal theories* (in Russian), In: Semiotika Informatika **14**, VINITI, Moscow, 1980, pp. 115–133; English transl.: S. Artemov, et al., *Six Papers in Logic*, Am. Math. Soc., Translations (2), **135**, 1987.

5. S. Artemov, *Nonarithmeticity of truth predicate logics of provability* (in Russian), Dokl. Akad. Nauk SSSR **284** (1985), 270–271; English transl.: Sov. Math. Dokl. **32** (1985), 403–405.

6. S. Artemov, *On modal logics axiomatizing provability* (in Russian), Izv. Dokl. Akad. Nauk SSSR Ser. Mat. **49** (1985), 1123–1154; English transl.: Math. USSR Izv. **27** (1986), 401–429.

7. S. Artemov, *Numerically correct provability logics* (in Russian), Dokl. Akad. Nauk SSSR **290** (1986), 1289–1292; English transl. Sov. Math. Dokl. **34** (1987), 384–387.

8. S. Artemov, *Kolmogorov logic of problems and a provability interpretation of intuitionistic logic*, In: Theoretical Aspects of Reasoning about Knowledge - III Proceedings (1990), pp. 257–272.

9. S. Artemov, *Logic of proofs*, Ann. Pure Appl. Logic **67** (1994), 29–59.

10. S. Artemov, *Operational modal logic*, Technical Report MSI 95-29, Cornell University (1995).

11. S. Artemov, *Logic of proofs: a unified semantics for modality and λ-terms*, Technical Report CFIS 98-06, Cornell University (1998).

12. S. Artemov, *Explicit provability and constructive semantics*, Bull. Symb. Log. **7** (2001), 1–36.

13. S. Artemov, *Operations on proofs that can be specified by means of modal logic*, In: Advances in Modal Logic, Vol. 2, CSLI Publications, Stanford University, 2001, pp. 59–72.

14. S. Artemov, *Unified semantics for modality and λ-terms via proof polynomials*, In: Algebras, Diagrams and Decisions in Language, Logic and Computation, K. Vermeulen and A. Copestake (Eds.), CSLI Publications, Stanford University, 2002, pp. 89–119.

15. S. Artemov, *Embedding of modal lambda-calculus into the logic of proofs*, Proc. Steklov Math. Inst. **242** (2003), 36–49.

16. S. Artemov, *Evidence-based common knowledge*, Technical Report TR-2004018, CUNY Ph.D. Program in Computer Science (2004), revised version of 2005.

17. S. Artemov, *Kolmogorov and Gödel's approach to intuitionistic logic: current developments* (in Russian), Usp. Mat. Nauk **59** (2003), No.2, 9–36; English transl.: Russ. Math. Surv. **59** (2004), 203–229.

18. S. Artemov, *Existential semantics for modal logic*, In: We Will Show Them: Essays in Honour of Dov Gabbay, Vol. 1, H. Barringer, A. d'Avila Garcez, L. Lamb, and J. Woods (Eds.), College Publications, London, 2005, pp. 19–30.

19. S. Artemov, *Justified common knowledge*, Theor. Comput. Sci. **357** (2006), 4–22.

20. S. Artemov and L. Beklemishev, *On propositional quantifiers in provability logic*, Notre Dame J. Formal Logic **34** (1993), 401–419.

21. S. Artemov and L. Beklemishev, *Provability logic*, In: Handbook of Philosophical Logic, 2nd edition, D. Gabbay and F. Guenthner (Eds.), Kluwer, 2004, pp. 229–403.

22. S. Artemov and R. Iemhoff, *The basic intuitionistic logic of proofs*, Technical Report TR-2005002, CUNY Ph.D. Program in Computer Science (2005).

23. S. Artemov and R. Iemhoff, *The basic intuitionistic logic of proofs*, J. Symb. Log. (2006). [To appear]

24. S. Artemov, E. Kazakov, and D. Shapiro, *Epistemic logic with justifications*, Technical Report CFIS 99-12, Cornell University (1999).

25. S. Artemov and R. Kuznets, *Logical omniscience via proof complexity*, accepted to Computer Science Logic '06.

26. S. Artemov and E. Nogina, *Logic of knowledge with justifications from the provability perspective*, Technical Report TR-2004011, CUNY Ph.D. Program in Computer Science (2004).

27. S. Artemov and E. Nogina, *Introducing justification into epistemic logic*, J. Log. Comput. **15** (2005), 1059–1073.

28. S. Artemov and E. Nogina, *On epistemic logic with justification*, In: Theoretical Aspects of Rationality and Knowledge. Proceedings of the Tenth Conference (TARK 2005), June 10–12, 2005, R. van der Meyden (Ed.), Singapore. 2005, pp. 279–294.

29. S. Artemov and T. Strassen, *The basic logic of proofs*, In: Computer Science Logic. 6th Workshop, CSL '92. San Miniato, Italy, September/October 1992. Selected Papers, E. Börger, G. Jäger, H. K. Büning, S. Martini, and M. Richter (Eds.), Lect. Notes Comput. Sci. **702**, Springer, 1992, pp. 14–28.

30. S. Artemov and T. Strassen, *Functionality in the basic logic of proofs*, Technical Report IAM 93-004, Department of Computer Science, University of Bern, Switzerland (1993).

31. S. Artemov and T. Strassen, *The logic of the Gödel proof predicate*, In: Computational Logic and Proof Theory. Third Kurt Gödel Colloquium, KGC '93. Brno, Chech Republic, August 1993. Proceedings, G. Gottlob, A. Leitsch, and D. Mundici (Eds.), Lect. Notes Comput. Sci. **713**, Springer, 1993, pp. 71–82.

32. S. Artemov and T. Yavorskaya (Sidon), *On the first order logic of proofs*, Moscow Math. J. **1** (2001), 475–490.

33. L. Beklemishev, *On the classification of propositional provability logics* (in Russian), Izv. Dokl. Akad. Nauk SSSR Ser. Mat. **53** (1989), 915–943; English transl.: Math. USSR Izv. **35** (1990), 247–275.

34. L. Beklemishev, *On bimodal logics of provability*, Ann. Pure Appl. Logic **68** (1994), 115–160.

35. L. Beklemishev, *Bimodal logics for extensions of arithmetical theories*, J. Symb. Log. **61** (1996), 91–124.

36. L. Beklemishev, *Parameter free induction and provably total computable functions*, Theor. Comput. Sci. **224** (1999), 13–33.

37. L. Beklemishev, *Proof-theoretic analysis by iterated reflection*, Arch. Math. Logic **42** (2003), 515–552.

38. L. Beklemishev, *The Worm principle*, Logic Group Preprint Series 219, University of Utrecht (2003).

39. L. Beklemishev, *On the idea of formalisation in logic and law* (2004), Logic and Law, 6th Augustus De Morgan Workshop, King's College London.

40. L. Beklemishev, *Reflection principles and provability algebras in formal arithmetic* (in Russian), Usp. Mat. Nauk **60** (2005), 3–78; English transl.: Russ. Math. Surv. **60** (2005), 197–268.

41. L. Beklemishev, M. Pentus, and N. Vereshchagin, *Provability, complexity, grammars*, Am. Math. Soc., Translations (2), **192** (1999).

42. A. Berarducci, *The interpretability logic of Peano Arithmetic*, J. Symb. Log. **55** (1990), 1059–1089.

43. A. Berarducci and R. Verbrugge, *On the provability logic of bounded arithmetic*, Ann. Pure Appl. Logic **61** (1993), 75–93.

44. C. Bernardi, *The uniqueness of the fixed point in every diagonalizable algebra*, Stud. Log. **35** (1976), 335–343.

45. L. Bonjour, *The coherence theory of empirical knowledge*, Philos. Stud. **30** (1976), 281–312. [Reprinted in: Contemporary Readings in Epistemology, M. F. Goodman and R.A. Snyder (Eds), Prentice Hall, 1993, pp. 70–89.]

46. G. Boolos, *On deciding the truth of certain statements involving the notion of consistency*, J. Symb. Log. **41** (1976), 779–781.

47. G. Boolos, *Reflection principles and iterated consistency assertions*, J. Symb. Log. **44** (1979), 33–35.

48. G. Boolos, *The Unprovability of Consistency: An Essay in Modal Logic,* Cambridge Univ. Press, 1979.

49. G. Boolos, *Extremely undecidable sentences*, J. Symb. Log. **47** (1982), 191–196.

50. G. Boolos, *The logic of provability*, Am. Math. Mon. **91** (1984), 470–480.

51. G. Boolos, *The Logic of Provability,* Cambridge Univ. Press, 1993.

52. G. Boolos and G. Sambin, *Provability: the emergence of a mathematical modality*, Stud. Log. **50** (1991), 1–23.

53. V. Brezhnev, *On explicit counterparts of modal logics*, Technical Report CFIS 2000-05, Cornell University (2000).

54. V. Brezhnev, *On the logic of proofs*, In: Proceedings of the Sixth ESSLLI Student Session, Helsinki, 2001, pp. 35–46.

55. V. Brezhnev and R. Kuznets, *Making knowledge explicit: How hard it is*, Theor. Comput. Sci. **357** (2006), 23–34.

56. S. Buss, *The modal logic of pure provability*, Notre Dame J. Formal Logic **31** (1990), 225–231.

57. A. Carbone and F. Montagna, *Rosser orderings in bimodal logics*, Z. Math. Logik Grundlagen Math. **35** (1989), 343–358.

58. A. Carbone and F. Montagna, *Much shorter proofs: A bimodal investigation*, Z. Math. Logik Grundlagen Math. **36** (1990), 47–66.

59. T. Carlson, *Modal logics with several operators and provability interpretations*, Isr. J. Math. **54** (1986), 14–24.

60. A. Chagrov and M. Zakharyaschev, *Modal companions of intermediate propositional logics*, Stud. Log. **51** (1992), 49–82.

61. A. Chagrov and M. Zakharyaschev, *Modal Logic,* Oxford Science Publications, 1997.

62. R. Constable, *Types in logic, mathematics and programming*, In: Handbook of Proof Theory, S. Buss (Ed.), Elsevier, 1998, pp. 683–786.

63. D. de Jongh and G. Japaridze, *Logic of provability*, In: Handbook of Proof Theory, S. Buss (Ed.), Elsevier, 1998, pp. 475–546.

64. D. de Jongh and F. Montagna, *Much shorter proofs*, Z. Math. Logik Grundlagen Math. **35** (1989), 247–260.

65. D. de Jongh and A. Visser, *Embeddings of Heyting algebras*, In: Logic: From Foundations to Applications. European Logic Colloquium, Keele, UK, July 20–29, 1993, W. Hodges, M. Hyland, C. Steinhorn, and J. Truss (Eds.), Clarendon Press, Oxford, 1996, pp. 187–213.

66. G. Dzhaparidze (Japaridze), *The logic of linear tolerance*, Stud. Log. **51** (1992), 249–277.

67. G. Dzhaparidze (Japaridze), *A generalized notion of weak interpretability and the corresponding modal logic*, Ann. Pure Appl. Logic **61** (1993), 113–160.

68. L. Esakia, *Intuitionistic logic and modality via topology*, Ann. Pure Appl. Logic **127** (2004), 155–170. [Provinces of logic determined. Essays in the memory of Alfred Tarski. Parts IV, V and VI, Z. Adamowicz, S. Artemov, D. Niwinski, E. Orlowska, A. Romanowska, and J. Wolenski (Eds.)]

69. R. Fagin, J. Halpern, Y. Moses, and M. Vardi, *Reasoning About Knowledge,* The MIT Press, 1995.

70. S. Feferman, *Arithmetization of metamathematics in a general setting*, Fundam. Math. **49** (1960), 35–92.

71. M. Fitting, *Semantics and tableaus for* LPS4, Technical Report TR-2004016, CUNY Ph.D. Program in Computer Science (2004).

72. M. Fitting, *A logic of explicit knowledge*, In: The Logica Yearbook 2004, L. Behounek and M. Bilkova (Eds.), Filosofia, 2005, pp. 11–22.

73. M. Fitting, *The logic of proofs, semantically*, Ann. Pure Appl. Logic **132** (2005), 1–25.

74. H. Friedman, *102 problems in mathematical logic*, J. Symb. Log. **40** (1975), 113–129.

75. D. Gabbay, *Labelled Deductive Systems*, Oxford Univ. Press, 1994.

76. D. Gabelaia, *Modal Definability in Topology* (2001), ILLC Publications, Master of Logic Thesis (MoL) Series MoL-2001-11.

77. E. Gettier, *Is Justified True Belief Knowledge?*, Analysis **23** (1963), 121–123.

78. J. Girard, Y. Lafont, and P. Taylor, *Proofs and Types*, Cambridge Univ. Press, 1989.

79. K. Gödel, *Eine Interpretation des intuitionistischen Aussagenkalkuls*, Ergebnisse Math. Kolloq. **4** (1933), 39–40; English transl. in: Kurt Gödel Collected Works, Vol. 1,, S. Feferman et al. (Eds.), Oxford Univ. Press, Oxford, Clarendon Press, New York, 1986, pp. 301–303.

80. K. Gödel, *Vortrag bei Zilsel, 1938*, In: Kurt Gödel Collected Works. Volume III, S. Feferman (Ed.), Oxford Univ. Press, 1995, pp. 86–113.

81. R. Goldblatt, *Arithmetical necessity, provability and intuitionistic logic*, Theoria **44** (1978), 38–46.

82. A. Goldman, *A causal theory of knowing*, J. Philos. **64** (1967), 335–372.

83. E. Goris, *Logic of proofs for bounded arithmetic*, In: Computer Science - Theory and Application, D. Grigoriev, J. Harrison, and E. Hirsch (Eds.), Lect. Notes Comput. Sci. **3967**, Springer, 2006, pp. 191–201.

84. E. Goris and J. Joosten, *Modal matters in interpretability logics*, Technical report, Utrecht University. Institute of Philosophy (2004), Logic Group preprint series; 226.

85. A. Grzegorczyk, *Some relational systems and the associated topological spaces*, Fundam. Math. **60** (1967), 223–231.

86. D. Guaspari and R. Solovay, *Rosser sentences*, Ann. Pure Appl. Logic **16** (1979), 81–99.

87. P. Hájek and F. Montagna, *The logic of $\Pi_1$-conservativity*, Arch. Math. Logic **30** (1990), 113–123.

88. P. Hájek and F. Montagna, *The logic of $\Pi_1$-conservativity continued*, Arch. Math. Logic **32** (1992), 57–63.

89. P. Hájek and P. Pudlák, *Metamathematics of First Order Arithmetic*, Springer-Verlag, Berlin, Heidelberg, New York, 1993.

90. V. Hendricks, *Active agents*, J. Logic Lang. Inf. **12** (2003), no. 4, 469–495.

91. A. Heyting, *Die intuitionistische grundlegung der mathematik*, Erkenntnis **2** (1931), 106–115.

92. A. Heyting, *Mathematische Grundlagenforschung. Intuitionismus. Beweistheorie*, Springer, 1934.

93. A. Heyting, *Intuitionism: An Introduction*, North-Holland, 1956.

94. D. Hilbert and P. Bernays, *Grundlagen der Mathematik, Vols. I and II, 2d ed.* Springer, 1968.

95. R. Iemhoff, *A modal analysis of some principles of the provability logic of Heyting arithmetic*, In: Advances in Modal Logic, Vol. 2, CSLI, M. Zakharyaschev, K. Segerberg, M. de Rijke, and H. Wansing (Eds.), Lect. Notes **119**, CSLI Publications, Stanford, 2001, pp. 301–336.

96. R. Iemhoff, *On the admissible rules of intuitionistic propositional logic*, J. Symb. Log. **66** (2001), 281–294.

97. R. Iemhoff, *Provability Logic and Admissible Rules,* Ph.D. Thesis, University of Amsterdam (2001).

98. K. Ignatiev, *Partial conservativity and modal logics*, ITLI Prepublication Series X–91–04, University of Amsterdam (1991).

99. K. Ignatiev, *On strong provability predicates and the associated modal logics*, J. Symb. Log. **58** (1993), 249–290.

100. K. Ignatiev, *The provability logic for $\Sigma_1$-interpolability*, Ann. Pure Appl. Logic **64** (1993), 1–25.

101. G. Japaridze, *The Modal Logical Means of Investigation of Provability* (in Russian), Ph.D. Thesis, Moscow State University (1986).

102. G. Japaridze, *The polymodal logic of provability* (in Russian), In: Intensional Logics and Logical Structure of Theories: Material from

the fourth Soviet-Finnish Symposium on Logic, Telavi, May 20–24, 1985, Metsniereba, Tbilisi, 1988, pp. 16–48.

103. G. Japaridze, *Introduction to computability logic*, Ann. Pure Appl. Logic **123** (2003), 1–99.

104. G. Japaridze, *From truth to computability I*, Theor. Comput. Sci. **357** (2006), 100–135.

105. J. Joosten, *Interpretability Formalized*, Ph.D. Thesis, University of Utrecht (2004).

106. S. Kleene, *Introduction to Metamathematics*, Van Norstrand, 1952.

107. A. Kolmogoroff, *Zur Deutung der intuitionistischen logik* (in German), Math. Z. **35** (1932), 58–65; English transl.: Selected works of A.N. Kolmogorov. Vol. I: Mathematics and Mechanics, V. M. Tikhomirov (Ed.), Kluwer, 1991 pp. 151–158.

108. S. Kripke, *Semantical considerations on modal logic*, Acta Philos. Fenn. **16** (1963), 83–94.

109. N. Krupski, *Some Algorithmic Questions in Formal Systems with Internalization* (in Russian), Ph.D. Thesis, Moscow State University (2006).

110. N. Krupski, *On the complexity of the reflected logic of proofs*, Theor. Comput. Sci. **357** (2006), 136–142.

111. N. Krupski, *Typing in reflective combinatory logic*, Ann. Pure Appl. Logic **141** (2006), 243–256.

112. V. Krupski, *Operational logic of proofs with functionality condition on proof predicate*, In: Logical Foundations of Computer Science '97, Yaroslavl', S. Adian and A. Nerode (Eds.), Lect. Notes Comput. Sci. **1234**, Springer, 1997, pp. 167–177.

113. V. Krupski, *The single-conclusion proof logic and inference rules specification*, Ann. Pure Appl. Log. **113** (2001), 181–206.

114. V. Krupski, *Referential logic of proofs*, Theor. Comput. Sci. **357** (2006), 143–166.

115. R. Kuznets, *On the complexity of explicit modal logics*, In: Computer Science Logic 2000, Lect. Notes Comput. Sci. **1862**, Springer, 2000, pp. 371–383.

116. R. Kuznets, *Complexity of Evidence-Based Knowledge*, In: Proceedings of the Rationality and Knowledge Workshop, ESSLLI, 2006.

117. A. Kuznetsov and A. Muravitsky, *The logic of provability* (in Russian), In: Abstracts of the 4th All-Union Conference on Mathematical Logic, Kishinev, 1976, p. 73.

118. A. Kuznetsov and A. Muravitsky, *Magari algebras* (in Russian), In: Fourteenth All-Union Algebra Conference, Abstracts, Part 2: Rings, Algebraic Structures, 1977, pp. 105–106.

119. A. Kuznetsov and A. Muravitsky, *On superintuitionistic logics as fragments of proof logic*, Stud. Log. **45** (1986), 77–99.

120. K. Lehrer and T. Paxson, *Knowledge: undefeated justified true belief*, J. Philos. **66** (1969), 1–22.

121. E. Lemmon, *New foundations for Lewis's modal systems*, J. Symb. Log. **22** (1957), 176–186.

122. W. Lenzen, *Knowledge, belief and subjective probability*, In: Knowledge Contributors, K. J. V. Hendricks and S. Pedersen, (Eds.), Kluwer, 2003.

123. D. Lewis, *Elusive knowledge*, Australian J. Philos. **7** (1996), 549–567.

124. P. Lindstrm, *Provability logic – a short introduction*, Theoria **62** (1996), 19–61.

125. M. Löb, *Solution of a problem of Leon Henkin*, J. Symb. Log. **20** (1955), 115–118.

126. D. Luchi and F. Montagna, *An operational logic of proofs with positive and negative information*, Stud. Log. **63** (1999), no.1, 7–25.

127. A. Macintyre and H. Simmons, *Gödel's diagonalization technique and related properties of theories*, Colloquium Mathematicum **28** (1973).

128. R. Magari, *The diagonalizable algebras (the algebraization of the theories which express Theor.:II)*, Bollettino della Unione Matematica Italiana, Serie 4 **12** (1975), suppl. fasc. 3, 117–125.

129. R. Magari, *Representation and duality theory for diagonalizable algebras (the algebraization of theories which express Theor.:IV)*, Stud. Log. **34** (1975), 305–313.

130. J. McKinsey and A. Tarski, *Some theorems about the sentential calculi of Lewis and Heyting*, J. Symb. Log. **13** (1948), 1–15.

131. Meyer, J.-J. Ch. and W. van der Hoek, *Epistemic Logic for AI and Computer Science*, Cambridge Univ. Press, 1995.

132. R. Milnikel, *Derivability in certain subsystems of the logic of proofs is $\Pi_2^p$-complete*, Ann. Pure Appl. Logic (2006). [To appear]

133. G. Mints, *Lewis' systems and system T (a survey 1965–1973)* (in Russian), In: Modal Logic, Nauka, Moscow, 1974, pp. 422–509; English transl.: G. Mints, *Selected Papers in Proof Theory*, Bibliopolis, Napoli, 1992.

134. A. Mkrtychev, *Models for the logic of proofs*, In: Logical Foundations of Computer Science '97, Yaroslavl', S. Adian and A. Nerode (Eds.), Lect. Notes Comput. Sci. **1234**, Springer, 1997, pp. 266–275.

135. F. Montagna, *On the diagonalizable algebra of Peano arithmetic*, Boll. Unione Mat. Ital. B (5) **16** (1979), 795–812.

136. F. Montagna, *Undecidability of the first-order theory of diagonalizable algebras*, Stud. Log. **39** (1980), 347–354.

137. F. Montagna, *The predicate modal logic of provability*, Notre Dame J. Formal Logic **25** (1987), 179–189.

138. R. Montague, *Syntactical treatments of modality with corollaries on reflection principles and finite axiomatizability*, Acta Philos. Fenn. **16** (1963), 153–168.

139. Y. Moses, *Resource-bounded knowledge*, In: Proceedings of the Second Conference on Theoretical Aspects of Reasoning about Knowledge, held in Pacific Grove, California, USA, March 7–9, 1988, M. Vardi (Ed.), Morgan Kaufmann, 1988, pp. 261–276.

140. J. Myhill, *Some remarks on the notion of proof*, J. Philos. **57** (1960), 461–471.

141. J. Myhill, *Intensional set theory*, In: Intensional Mathematics, S. Shapiro (Ed.), North-Holland, 1985, pp. 47–61.

142. E. Nogina, *Logic of proofs with the strong provability operator*, Technical Report ILLC Prepublication Series ML-94-10, Institute for Logic, Language and Computation, University of Amsterdam (1994).

143. E. Nogina, *Grzegorczyk logic with arithmetical proof operators* (in Russian), Fundam. Prikl. Mat. **2** (1996), no. 2, 483–499.

144. E. Nogina, *On logic of proofs and provability*, Bull. Symb. Log. **12** (2006), no. 2, 356.

145. P. Novikov, *Constructive Mathematical Logic from the Viewpoint of the Classical One* (in Russian), Nauka, 1977.

146. Nozick, R., *Philosophical Explanations,* Harvard Univ. Press, 1981.

147. I. Orlov, *The calculus of compatibility of propositions* (in Russian), Mat. Sb. **35** (1928), 263–286.

148. E. Pacuit, *A note on some explicit modal logics* (2005), 5th Panhellenic Logic Symposium, Athens.

149. R. Parikh, *Knowledge and the problem of logical omniscience,* In: ISMIS-8 (International Symposium on Methodolody for Intellectual Systems) 1987, Z. Ras and M. Zemankova (Eds.), pp. 432–439.

150. R. Parikh, *Logical omniscience,* In: Logic and Computational Complexity, International Workshop LCC '94, Indianapolis, Indiana, USA, 13–16 October 1994, D. Leivant (Ed.), Lect. Notes Comput. Sci. **960**, Springer, 1995, pp. 22–29.

151. B. Renne, *Bisimulation and public announcements in logics of explicit knowledge,* In: Proceedings of the Rationality and Knowledge Workshop, 2006.

152. B. Renne, *Semantic cut-elimination for an explicit modal logic,* In: Proceedings of the ESSLLI 2006 Student Session, Malaga, 2006.

153. J. Rosser, *Extensions of some theorems of Gödel and Church,* J. Symb. Log. **1** (1936), 87–91.

154. N. Rubtsova, *Semantics for Logic of Explicit Knowledge Corresponding to* S5, In: Proceedings of the Rationality and Knowledge Workshop, ESSLLI, 2006.

155. N. Rubtsova, *Evidence Reconstruction of Epistemic Modal Logic* S5, In: Computer Science - Theory and Application, D. Grigoriev, J. Harrison, and E. Hirsch (Eds.), Lect. Notes Comput. Sci. **3967**, Springer, 2006, pp. 313–321.

156. N. Rubtsova. and T. Yavorskaya-Sidon, *Operations on proofs and labels,* J. Appl. Non-Classical Logics. [To appear]

157. S. Shapiro, *Epistemic and intuitionistic arithmetic,* In: Intensional Mathematics, S. Shapiro (Ed.), North-Holland, 1985, pp. 11–46.

158. S. Shapiro, *Intensional mathematics and constructive mathematics,* In: Intensional Mathematics, S. Shapiro (Ed.), North-Holland, 1985, pp. 1–10.

159. V. Shavrukov, *The logic of relative interpretability over Peano arithmetic,* Preprint, Steklov Mathematical Institute, Moscow (1988), in Russian.

160. V. Shavrukov, *On Rosser's provability predicate*, Z. Math. Logik Grundlagen Math. **37** (1991), 317–330.

161. V. Shavrukov, *A smart child of Peano's*, N Notre Dame J. Formal Logic **35** (1994), 161–185.

162. V. Shavrukov, *Isomorphisms of diagonalizable algebras*, Theoria **63** (1997), 210–221.

163. T. Sidon, *Provability logic with operations on proofs*, In: Logical Foundations of Computer Science '97, Yaroslavl', S. Adian and A. Nerode (Eds.), Lect. Notes Comput. Sci. **1234**, Springer, 1997, pp. 342–353.

164. T. Smiley, *The logical basis of ethics*, Acta Philos. Fenn. **16** (1963), 237–246.

165. C. Smoryński, *The incompleteness theorems*, In: Handbook of Mathematical Logic, J. Barwise (Ed.), North Holland, 1977, pp. 821–865.

166. C. Smorỳnski, *Self-Reference and Modal Logic,* Springer, 1985.

167. R. Solovay, *Provability interpretations of modal logic*, Isr. J. Math. **25** (1976), 287–304.

168. V. Švejdar, *Modal analysis of generalized Rosser sentences*, J. Symb. Log. **48** (1983), 986–999.

169. G. Takeuti, *Proof Theory,* Elsevier, 1987.

170. A. Tarski, A. Mostovski, and R. Robinson, *Undecidable Theories,* North-Holland, 1953.

171. A. Troelstra and D. van Dalen, *Constructivism in Mathematics, Vols 1, 2,* North-Holland, 1988.

172. J. van Benthem, *Reflections on epistemic logic*, Logique Anal. Nouv. Ser. **133-134** (1993), 5–14.

173. J. van Benthem, *Modal frame correspondence generalized*, Technical Report PP-2005-08, Institute for Logic, Language, and Computation, Amsterdam (2005).

174. V. Vardanyan, *Arithmetic comlexity of predicate logicsof provability and their fragments* (in Russian), Dokl. Akad. Nauk SSSR **288** (1986), 11–14; English transl.: Sov. Math. Dokl. **33** (1986), 569–572.

175. A. Visser, *Aspects of Diagonalization and Provability,* Ph.D. Thesis, University of Utrecht (1981).

176. A. Visser, *The provability logics of recursively enumerable theories extending Peano Arithmetic at arbitrary theories extending Peano Arithmetic*, J. Philos. Logic **13** (1984), 97–113.

177. A. Visser, *Evaluation, provably deductive equivalence in Heyting arithmetic of substitution instances of propositional formulas*, Logic Group Preprint Series 4, Department of Philosophy, University of Utrecht (1985).

178. A. Visser, *Peano's smart children. A provability logical study of systems with built-in consistency*, Notre Dame J. Formal Logic **30** (1989), 161–196.

179. A. Visser, *Interpretability logic*, In: Mathematical Logic, P. Petkov (Ed.), Plenum Press, 1990, pp. 175–208.

180. A. Visser, *Propositional combinations of $\Sigma_1$-sentences in Heyting's arithmetic*, Logic Group Preprint Series 117, Department of Philosophy, University of Utrecht (1994).

181. A. Visser, *An overview of interpretability logic*, In: Advances in Modal Logic, Vol. 1, M. Kracht, M. de Rijke, and H. Wansing (Eds.), CSLI Publications, Stanford University, 1998, pp. 307–360.

182. A. Visser, *Rules and arithmetics*, Notre Dame J. Formal Logic **40** (1999), 116–140.

183. A. Visser, *Substitutions of $\Sigma_1^0$-sentences: Explorations between intuitionistic propositional logic and intuitionistic arithmetic*, Ann. Pure Appl. Logic **114** (2002), 227–271.

184. A. Visser, *Löb's Logic meets the $\mu$-Calculus*, In: Processes, Terms and Cycles: Steps on the Road to Infinity. Essays Dedicated to Jan Willem Klop on the Occasion of his 60th Birthday, A. Middeldorp, V. van Oostrom, F. van Raamsdonk, and R. de Vrijer (Eds.), Lect. Notes Comput. Sci. **3838**, Springer, 2005, pp. 14–25.

185. J. von Neumann, *A letter to Gödel on January 12, 1931*, In: Kurt Gödel Collected Works, Vol. V, S. Feferman, J. Dawson, W. Goldfarb, C. Parsons, and W. Sieg (Eds.), Oxford Univ. Press, 2003, pp. 341–345.

186. T. Yavorskaya (Sidon), *Logic of proofs and provability*, Ann. Pure Appl. Logic **113** (2001), 345–372.

187. T. Yavorskaya (Sidon), *Negative operations on proofs and labels*, J. Log. Comput. **15** (2005), 517–537.

188. T. Yavorskaya (Sidon), *Logic of proofs with two proof predicates*, In: Computer Science - Theory and Application, D. Grigoriev,

J. Harrison, and E. Hirsch (Eds.), Lect. Notes Comput. Sci. **3967**, Springer, 2006, pp. 369–380.

189. R. Yavorsky, *On the logic of the standard proof predicate*, In: Computer Science Logic 2000, Lect. Notes Comput. Sci. **1862**, Springer, 2000, pp. 527–541.

190. R. Yavorsky, *Provability logics with quantifiers on proofs*, Ann. Pure Appl. Logic **113** (2001), 373–387.

# Directions
# for Computability Theory
# Beyond Pure Mathematical

## John Case [†]

*University of Delaware*
*Newark, USA*

This paper begins by briefly indicating the principal, non-standard motivations of the author for his decades of work in Computability Theory (CT), a.k.a. Recursive Function Theory.

Then it discusses its proposed, general directions beyond those from pure mathematics for CT. These directions are as follows.

(1) Apply CT to basic *sciences*, for example, biology, psychology, physics, chemistry, and economics.

———————
[†] The author was supported in part by the National Science Foundation (NSF) grant CCR 0208616.

Mathematical Problems from Applied Logic. Logics for the XXIst Century. II. Edited by Dov M. Gabbay *et al.* / International Mathematical Series, 5, Springer, 2007

(2) Apply the resultant insights from (1) to philosophy and, more generally, apply CT to areas of philosophy *in addition to* the philosophy and foundations of mathematics.

(3) Apply CT for insights into engineering and other professional fields.

Lastly, this paper provides a progress report on the above non-pure mathematical directions for CT, including examples for biology, cognitive science and learning theory, philosophy of science, physics, applied machine learning, and computational complexity. Interweaved with the report are occasional remarks about the future.

## 1. Motivations

Ted Slaman [159] has nicely mentioned the central theme of his intellectual motivation (deriving from the influence of Sacks) for working in Computability Theory (CT), a.k.a. Recursive Function Theory, namely, *definability*. I like this very much; however, my own strongest intellectual motivations for devoting much of my research to CT have a very different nature and origin.

Since these latter motivations are directly or indirectly relevant to some of the directions I propose below, and are, I believe, fairly atypical among CT researchers, I describe them in some detail.

Before my undergraduate experiences, I was intellectually motivated by what I would now describe as philosophically oriented scientific curiosity. I naively hoped to discover the fundamental nature of the universe. It is likely I did not consider what that actually meant. I knew about Einstein, but not about Gödel and Turing. I considered studying physics, astronomy, or psychology. I remained flexible about fields in the future, expecting I should see what I liked, etc. I would now say I was a naive scientific reductionist and, so, chose physics (over psychology) for my

UG experience. My UG minors became mathematics and philosophy, and I took but one psychology course. Psychology seemed more easily learned on my own.

I considered physics, mathematics, or philosophy for graduate school. For various reasons, including generally better fit of cognitive style between myself and the field, I selected mathematics.

When I first learned CT the two major, intellectual ideas that captivated me were[1]

- *that the definition of computable is an absolute,* and
- *my realization that, very likely, the universe, above some level at least, is computable*[2].

*In a sense*, re the second bullet just above, with CT I felt that I was dealing with physics again in the form of a very abstract mechanics. Also, since biological organisms including humans are components of the physical universe, with CT I also had a very abstract handle on biology, psychology ... . Later I found [121] in which Myhill says on P. 149:

> ...in the author's view, the theorems of Church and Gödel are psychological laws. Mr. E. H. Galanter of the Department of Psychology, University of Pennsylvania, described them in conversation with the author as "the only known psychological laws comparable in exactitude with the laws of physics."[3]

---

[1] Many, for me, less major things also helped with the captivation, for example, the aesthetics of (some of) CT's tools, for example, recursion theorems [23, 24, 143].

[2] N.B. In a discrete, random universe but with *computable probability distributions* for its behaviors (for example, a discrete, quantum mechanical universe, perhaps, as I believe, ours is), the statistically *expected* behavior will still be computable [60] (and constructively so [72, 73]).

[3] However, when, many years later, I got to know Myhill and discussed with him the idea that people are essentially algorithmic mechanisms, he was, at that time, no longer in favor of the idea.

It is sometimes argued that Gödel's theorems imply people are not algorithmic. It is, I believe, never argued that there are people who can list or decide the set of truths of (first or second order)[4] arithmetic, or who solve the halting problem, or .... . Anyhow, regarding the arguments that are presented: they suffer not only from the usual problems of confusing "**T** being consistent" with "**T** being known to be consistent," but also, *and more basically*, with confusion about productive sets [148]. Many times these arguments are essentially, "I know an algorithm witnessing the set of truths of some arithmetic is productive; i.e., I know an algorithm which provides counterexamples to alleged complete recursive axiomatizations [113]; therefore, I am not a machine." More simply, "I know an algorithm; therefore, I am not a machine." In this form, these arguments are seen as absurd. See also [50, 51].[5]

Beginning with the next paragraph I list my suggested directions. After I present the directions, I present examples of progress to date and suggest future work.

## 2. Directions

**Direction 2.1.** Apply CT to the basic sciences.[6]

This leads to a second

---

[4] I mean second order in the sense of [148].

[5] I plan to write a philosophical paper in which I present some new arguments in this area and which tend a bit in the opposite direction of supporting a mechanistic world, or at least one with mechanistic expected behavior.

[6] By basic sciences, I have in mind biology, psychology, physics, chemistry, economics, etc., sciences which, each to varying degrees of predictive success, apply scientific method.

I do not consider set theory to be one of the sciences in the above sense, and it is not clear platonists would consider it to be anyhow — even if they think sets are a component of the universe independent of human invention. I am not a platonist. For me, set theory requires deciding what is useful and interesting to mean by sets, and I personally expect that that will have no absolute answer.

**Direction 2.2.** Apply the understandings from successes re Direction 2.1 to philosophy *and, more generally,* apply CT to more areas of philosophy than at present.[7]

*At least* for my personally captivating intellectual motivations for CT bulleted above, I believe CT *deserves* to survive! It is not so clear that it will.[8] This *partly* motivates

**Direction 2.3.** Apply CT to engineering and other professional or applied fields more generally.

## 3. Progress So Far And How One Might Go From Here

In this section, I proceed approximately chronologically with respect to my own first associations with the general subject headings. I interweave with the progress report occasional remarks about the future. I also make occasional remarks about proof techniques employed thus far[9], and prove one sample theorem (Theorem 3.1 in Section 3.2.1 below). I intend the progress report material partly as evidence that progress is possible, not as

---

[7] I like very much the idea of people *continuing* to apply CT to foundations of *mathematics* and associated philosophy. Directions 2.1 and 2.2 are, I believe, a needed expansion to areas *outside* of mathematics itself, i.e., moves away from math-centricism.

[8] For U.S. mathematical science departments, NSF commissioned [125], and in Appendix 2 **Assessment of Subfields**, under **Foundations** we see, among other things, the following.

> **Recursion (or computability) theory** is quiescent, with a substantial body of completed work. Barring a major breakthrough, or the further exploitation of connections with computational mathematics and computer science, the next decade is not expected to be very active. England plays a leading role, with the United States as a contributor, but the aging research population is not being replenished.

[9] I have not seen a need for $n$-jump priority arguments yet. There may eventually need to be new, complex methods created to obtain results in my proposed directions.

how one must necessarily proceed next. I will not be proposing particular and well-defined problems. Instead *and in general* I propose the creation of interesting, *insightful, and interpretable* definitions, problems, theorems for the sciences, philosophy, and applied fields.

Standard computability-theoretic notation will be from [148]. For example, $N$ will denote the set of natural numbers, $\{0, 1, 2, \ldots\}$. $\varphi$ will denote a fixed acceptable programming system (or numbering) for the class of partially computable functions: $N \to N$, where $\varphi_p$ is the partially computable function computed by program (or index) $p$ in the system. For a partial function $\psi$, $\delta\psi$ and $\rho\psi$ denote the domain and range of $\psi$, respectively. We write $W_p$ for the r.e. set accepted or enumerated by $\varphi$-program $p$, where, formally, $W_p \overset{\text{def}}{=} \delta\varphi_p$. We also write $\downarrow$ for converges or is defined and $\uparrow$ for diverges or is not defined.

## 3.1. Biology

Kleene [95, p. 59][10] was apparently the first to notice the connection between his second recursion theorem [148, p. 214] and Von Neumann's self-reproducing automata [124, 17]. I recall that Kleene told me (perhaps at the Kleene Symposium) that he had used his recursion theorem to understand Von Neumann's construction. This amazed me, since, by contrast, I had used von Neumann's construction to understand Kleene's proof of his recursion theorem [21].[11] Myhill's [123] seems to be the first *published* account featuring a connection between CT and self-reproducing automata although it does not employ a recursion theorem. This

---

[10] This paper is worth a read more generally.

[11] Wonderfully, the top level logic/refinement of Von Neumanns's construction is essentially identical to the top level logic/refinement of biological single-celled organisms' self-reproductive procedure. http://www.cis.udel.edu/~case/papers/krt-self-repo.pdf explicitly lays out the correspondence between Kleene's proof of his recursion theorem and (the top level of) a single-celled organism's self-reproductive procedure. This expands on the discussion of same in [21].

paper provided me with my first hint of how to connect CT to modeling the physical world, and it directly motivated my [21] (and the earlier [20]).

In [123], Myhill considers variants of machines which build strict copies of themselves, for example, machines which build distortions such as mirror images of themselves and machines which deterministically evolve with each generation a better automatic theorem prover.

Herein, in our brief discussion of [21], I omit most technical details. Suffice it to say one has a sequence of constructor machines $\mathcal{M}_0, \mathcal{M}_1, \mathcal{M}_2, \ldots$, and, besides their constructing capabilities, they have arbitrary effectively pre-assigned respective computing capabilities.

For these constructor machines, we write $\mathcal{M}_p \to \mathcal{M}_q$ to mean $\mathcal{M}_p$ constructs or begets $\mathcal{M}_q$.

One of the emphases in [21] was periodicity in generations of machines which construct offspring. For example, given any $n \in N$, we can obtain pairwise *distinct* constructor machines $\mathcal{M}_{e_0}, \mathcal{M}_{e_1}, \ldots, \mathcal{M}_{e_n}$ such that $\mathcal{M}_{e_0} \to \mathcal{M}_{e_1} \to \ldots \to \mathcal{M}_{e_n} \to \mathcal{M}_{e_0}$.[12] This sequence replicates with period $n + 1$. Self-reproduction is the $n + 1 = 1$ case. In nature we also see period two, the $n + 1 = 2$ case: the metagenic cœlenterates *Aurelia* and *Obelia* [84, p. 246] alternate between an attached polyp generation and a free swimming medusa generation with each looking and behaving different from the other. We see period three, the $n + 1 = 3$ case, in some parasites which occupy a succession of very different hosts. I could not find evidence of organisms reproducing with periodicity in generations greater than three. The obvious problem for biologists suggested by the existence of constructor machines which replicate every $n + 1$ generations for arbitrary large $n$ is, Why do we not see larger periods in nature?

---

[12] The proof can be done by a padded $n + 1$-ary recursion theorem. To handle elegantly the cases of *a*periodicity in generations (with no sterile descendent), I invented my Operator Recursion Theorem (ORT) [21], an infinitary self-reference principle [24]. My ORT is called the Functional Recursion Theorem in [127].

Another example from [21], given any $n \in N$, we can constructively obtain *different* pairwise distinct constructor machines $\mathcal{M}_{e_0}, \mathcal{M}_{e_1}, \ldots, \mathcal{M}_{e_n}$ such that $\mathcal{M}_{e_0} \rightarrow \mathcal{M}_{e_1} \rightarrow \ldots \rightarrow \mathcal{M}_{e_n}$ *and* $\mathcal{M}_{e_n}$ is sterile, and, additionally, where each $\mathcal{M}_{e_i}$, for $i \leqslant n$, has the *same* arbitrary pre-assigned computing capabilities. The $n$th descendant of $\mathcal{M}_{e_0}$ exists and is sterile, where $\mathcal{M}_{e_0}$ is considered to be the zeroth descendant of itself. In drafting this chapter, I thought of an engineering application. Many times humans have introduced a new organism into an environment to control a pre-existing pest organism only to discover that the new organism is a pest itself. The application is to employ the trick for producing the $\mathcal{M}_{e_0}, \mathcal{M}_{e_1}, \ldots, \mathcal{M}_{e_n}$ of this paragraph to engineer genetically a proposed new organism to introduce for pest control so as to have its $n$th descendant exist and be sterile (for whatever $n$ is desired), but to have its non-reproductive functions not be altered. Then, the altered new organism, if it turns out to be a pest itself, will die out anyway. If it does not seem to be a pest itself, and it is still needed to control the original pest, it can be reintroduced.

## 3.2. Machine inductive inference and computability-theoretic learning

CT applied to these areas first appeared in [141, 76]. Associated textbook material appears in [131, 86, 127].

One of the endorsement sentences I composed for MIT Press regarding the then upcoming [86] reads as follows.[13]

> Just as a conservation assumption from physics provides boundaries on and insight into the physically possible so too the computability assumption on learning provides herein boundaries on and insight into the cognitively possible.

---

[13] Another sentence with very different content was actually used in my endorsement on the book jacket.

The material in this section features Philosophy of Science (Section 3.2.1), Cognitive Science and Language Learning (Section 3.2.2), and Applied Machine Learning (Section 3.2.3).

Some CT work applied to computational complexity aspects of learning appears in Section 3.4 below.

### 3.2.1. Philosophy of Science. On [5, p. 125] it says the following.

> Consider the physicist who looks for a law to explain a growing body of physical data. His data consist of a set of pairs $(x, y)$, where $x$ describes a particular experiment, for example, a high-energy physics experiment, and $y$ describes the results obtained, for example, the particles produced and their respective properties. The law he seeks is essentially an algorithm for computing the function $f(x) = y$.

Here is another example from [35]: $x$ codes a particle diffraction experiment and $f(x)$ the resultant probable distribution (or fringe pattern) on the other side of the diffraction grating. Quantum theory provides *algorithmic* extraction of $f(x)$ from $x$. A program for $f$ is, then, a *predictive* explanation or law for the set of such diffraction experiments.

If, in our universe, people, including scientists (and collections thereof, including over historical time), are essentially algorithmic (as I believe), one can use CT to get theorems in philosophy of science. This realization occurred to me in the mid 70s and, for me, it was extremely intellectually exciting.

Some computability-theoretic inductive inference publications with something to say for philosophy of science are [141, 76, 4, 5, 93, 184, 46, 47, 37, 10, 35, 66, 105, 38, 1].

Some philosophy of science publications influenced by computability-theoretic inductive inference are [75, 91, 92, 101, 89, 102, 153, 154, 90].

In the rest of this section, I present a few sample results from computability-theoretic inductive inference with corresponding indications of their philosophical meaning.

In the following we will model inductive inference machines $\mathbf{M}$ extensionally as partially computable functions which take for their inputs finite initial segments of functions $f : N \to N$ and which either go undefined or return programs in the $\varphi$-system. Intuitively, for a finite initial segment $\sigma$, if $\mathbf{M}(\sigma)$ is defined (written: $\downarrow$) $= p$, then $p$ represents $\mathbf{M}$'s conjecture or hypothesis as to a program for $f$ based on the data points about $f$ contained in $\sigma$. We write $\sigma \subset f$ to mean $\sigma$ is a finite initial segment of $f$, i.e., $\sigma$ is a finite initial segment of a function, and its graph is a proper subset of that of $f$. We write $\overset{\infty}{\forall}$ to mean for all but finitely many. $\mathcal{R}$ denotes the class of computable functions mapping $N$ into $N$.

We next consider a criterion of successful inductive inference.

**Definition 3.1.** $\mathbf{M}$ *Ex-learns* $f \Leftrightarrow [(\forall \sigma \subset f)[\mathbf{M}(\sigma)\downarrow] \wedge (\exists p)(\overset{\infty}{\forall}\sigma \subset f)[\mathbf{M}(\sigma) = p \wedge \varphi_p = f]].$

Intuitively, $\mathbf{M}$ **Ex**-learns $f$ means that, $\mathbf{M}$ fed successively more data about $f$, outputs a corresponding succession of conjectures and *eventually* begins to output the same correct $\varphi$-program $p$ for $f$ over and over. For **Ex**-learning we think of $\mathbf{M}$ as eventually finding a predictive *ex*planation $p$ for $f$ [47].

For $p$ satisfying the right hand side of Definition 3.1 just above, we write $\mathbf{M}(f)\downarrow = p$.

$\mathbf{Ex} \overset{\text{def}}{=} \{\mathcal{S} \subseteq \mathcal{R} \mid (\exists \mathbf{M})[\mathbf{M} \ \mathbf{Ex}\text{-learns each } f \in \mathcal{S}]\}$. For example, the class of one-argument primitive recursive functions *is* in **Ex**, but $\mathcal{R}$ is *not* [76, 5]. Hence while some single $\mathbf{M}$ is "clever" enough to **Ex**-learn every primitive recursive function, no single $\mathbf{M}$ is clever enough to **Ex**-learn each computable function.

Next we begin to consider alternative criteria of success.

**Definition 3.2.** $\mathbf{M}$ *Conf-learns* $f \Leftrightarrow [\mathbf{M} \ \mathbf{Ex}$-learns $f \wedge (\forall \sigma \subset f)[\mathbf{M}(\sigma)\downarrow \wedge (\varphi_{\mathbf{M}(\sigma)} \cup \sigma) \ is \ single\text{-}valued]].$

Let $f[n] \overset{\text{def}}{=} f(0), \ldots, f(n-1)$.

We see that an **M** which **Conf**-learns a function $f$ **Ex**-learns $f$ *and* produces, on each input $f[n]$, a corresponding conjecture, $\mathbf{M}(f[n])$, based on the data in $f[n]$, *and* this conjecture does not explicitly output something *convergently contradicting* any data in $f[n]$. A program $\mathbf{M}(f[n])$ *may* go undefined on some inputs $< n$, but *on such inputs*, it, then, does *not converge* to anything different from what $f$ does. This *seems* like a very reasonable, common sense restriction. In fact, the stronger looking restriction for the related **Cons**-*learning* criterion may seem reasonable too: each program $\mathbf{M}(f[n])$ must converge to $f$ *on any inputs* $< n$. This restriction requires that each conjecture of **M** on $f$ has to be correct on the data about $f$ on which that conjecture is based.[14]

**Conf** $\overset{\text{def}}{=} \{\mathcal{S} \subseteq \mathcal{R} \mid (\exists \mathbf{M})[\mathbf{M} \ \mathbf{Conf}\text{-learns each } f \in \mathcal{S}]\}$.
Also, **Cons** $\overset{\text{def}}{=} \{\mathcal{S} \subseteq \mathcal{R} \mid (\exists \mathbf{M})[\mathbf{M} \ \mathbf{Cons}\text{-learns each } f \in \mathcal{S}]\}$.

Surprisingly, these natural, common sense restrictions on inductive inference (for **Conf** and **Cons**) strictly limit learning or inductive inference power (as measured by **Ex**-learning).

**Theorem 3.1** (Wiehagen [184]). **Cons** $\subset$ **Conf** $\subset$ **Ex**.

For example, the second, more surprising non-containment in Theorem 3.1 just above entails that, in some cases for explanatory inductive inference, employing conjectures convergently contradicting known data gives one strictly greater inferring power than not contradicting known data! This result is, I believe, of great interest for philosophy of science.

PROOF OF THEOREM 3.1. Of course **Cons** $\subseteq$ **Conf** $\subseteq$ **Ex**. Since, offhand, I know of no easily available proof of the second, somewhat harder, more surprising non-containment, and to provide herein at least one *illustrative, short, and sweet* proof

---

[14] **Conf** is short for *Conformal*, and **Cons** is short for *Consistent*. I sometimes like referring to Conformal as *postdictively-consistent* (not explicitly contradicting known data points in one's conjectures based on them) and Consistent as *postdictively-complete* (not missing any known data points in one's conjectures based on them).

re inductive inference, I prove the more surprising of the non-containments.

Let

$$S = \{f \in \mathcal{R} \mid \rho f \text{ is finite } \wedge \varphi_{\max(\rho f)} = f\}, \tag{3.1}$$

a self-referential class. We will show that

$$S \in (\mathbf{Ex} - \mathbf{Conf}).^{15} \tag{3.2}$$

Trivially, $S \in \mathbf{Ex}$ — as witnessed by a machine that, on any $f$, outputs the largest thing, if any[16], it has seen so far in the range of $f$.

Suppose for contradiction $\mathbf{M}$ witnesses that $S \in \mathbf{Conf}$. Hence

$$(\forall f \in S)(\forall \sigma \subset f)[\mathbf{M}(\sigma)\!\downarrow \wedge (\varphi_{\mathbf{M}(\sigma)} \cup \sigma) \text{ is single-valued}]. \tag{3.3}$$

**Claim 3.1.**
$$(\forall \sigma)(\exists f \in S)[f \supset \sigma]. \tag{3.4}$$

*Hence, by* (3.4), (3.3),

$$(\forall \sigma)[\mathbf{M}(\sigma)\!\downarrow \wedge (\varphi_{\mathbf{M}(\sigma)} \cup \sigma) \text{ is single-valued}]. \tag{3.5}$$

PROOF. We let $\max(\emptyset) \stackrel{\text{def}}{=} 0$.

Suppose $\sigma$ is given. By a padded version of Kleene's Recursion Theorem, there is program $e$ such that $e > \max(\rho\sigma))$, *and*, on input $x$, $e$ looks in a mirror to see which program it is[17], and, then,

$$\varphi_e(x) = \begin{cases} \sigma(x) & \text{if } x \in \delta\sigma, \\ e & \text{otherwise.} \end{cases} \tag{3.6}$$

Let $f = \varphi_e$. Then $f \supset \sigma$, and $f \in S$.                                    $\square$

---

[15] This and other such examples for witnessing the rest of the theorem as well as for other, related results are stated, but not proven correct, in [**40**]. For $(\mathbf{Ex} - \mathbf{Cons}) \neq \emptyset$, see [**4, 5, 183**].

[16] If nothing has yet appeared in the range of $f$, the machine can output any program.

[17] See [**24, 143**] for more about this way of understanding recursion theorems and program or machine self-reference in terms of mirrors. It is also discussed in Section 3.3 below.

We continue with the proof of the theorem.

We write $\sigma \cdot i$ for the finite sequence consisting of $\sigma$ followed by $i$.

By a padded version of Kleene's Recursion Theorem, there is a *different* program $e > 0, 1$ such that this $e$ looks in a mirror to see which program it is, and, then, the rest of this $e$'s behavior is described informally below.

> **begin** *Program e*;
>    set $\varphi_e(0) = e$;
>    let $\varphi_e^s$ be the finite *sequence* of successive values of $\varphi_e$ defined *before* stage $s$ below[18];
>    **do** stage $s$ **for** $s = 0$ to $\infty$;
>       **begin** stage $s$;
>          **if** (i) $\mathbf{M}(\varphi_e^s \cdot 0) = \mathbf{M}(\varphi_e^s \cdot 1)$ ($\ast$ i.e., $\mathbf{M}$ is *in*sensitive $\ast$)
>          **then**
>             set $\varphi_e^{s+1} = \varphi_e \cdot 0$ ($\ast$ passive ploy $\ast$)
>          **else** (ii) ($\ast$ i.e., $\mathbf{M}$ is *sensitive* $\ast$)
>             set $\varphi_e^{s+1} = \varphi_e \cdot \min(\{i \leqslant 1 \mid \mathbf{M}(\varphi_e^s) \neq \mathbf{M}(\varphi_e^s \cdot i)\})$ ($\ast$ aggressive ploy[19] $\ast$)
>          **endif**
>       **end** stage $s$
>    **enddo**
> **end** *Program e*.

Clearly, by (3.5), $\varphi_e$ is total and, then, is $\in \mathcal{S}$. Therefore, $(\exists p)[\mathbf{M}(\varphi_e)\!\downarrow = p \wedge \varphi_p - \varphi_e]$.

Hence $(\overset{\infty}{\forall} s)[(\text{i})$ holds at stage $s]$ — since each stage in which (ii) holds forces $\mathbf{M}$ to make another mind change.

Pick $s_0$ so large that $[(\text{i})$ holds at stage $s_0 \wedge \mathbf{M}(\varphi_e^{s_0+1}) = \mathbf{M}(\varphi_e^{s_0}) = p]$.

For each $i \leqslant 1$, let $\sigma_i = \varphi_e^{s_0} \cdot i$.

---

[18] $\varphi_e^s$ is an initial segment of a function; hence, it has domain $\{0, \ldots, n-1\}$ for some $n \geqslant 0$.

[19] This strategy forces $\mathbf{M}$ to make a change of conjecture, a "mind" change, on $\varphi_e$.

Then, $\sigma_0 = \varphi_e^{s_0+1} \neq \sigma_1 \wedge \delta\sigma_0 = \delta\sigma_1$.

Let $x^{s_0} = \max(\delta\sigma_0)$, which $= \max(\delta\sigma_1)$, and is $> 0$.

Since **M** is *insensitive* at stage $s_0$, $\mathbf{M}(\sigma_0) = p = \mathbf{M}(\sigma_1)$.

By (3.5),

$$[(\varphi_{\mathbf{M}(\sigma_1)} \cup \sigma_1) \text{ is single-valued}]. \qquad (3.7)$$

$x^{s_0} \in \delta\sigma_0 = \delta\sigma_1 \wedge \sigma_0(x^{s_0}) = 0 \neq 1 = \sigma_1(x^{s_0})$.

$\varphi_{\mathbf{M}(\sigma_0)} = \varphi_{\mathbf{M}(\sigma_1)} = \varphi_p = \varphi_e$.

$\delta\varphi_e = N$. Therefore, $x^{s_0} \in \delta\varphi_e$ too.

Since $\varphi_{\mathbf{M}(\sigma_0)} = \varphi_e$, $(x^{s_0}, 0) \in (\varphi_{\mathbf{M}(\sigma_0)} \cap \sigma_0)$.

Since $\varphi_{\mathbf{M}(\sigma_0)} = \varphi_{\mathbf{M}(\sigma_1)}$, $(x^{s_0}, 0) \in \varphi_{\mathbf{M}(\sigma_1)}$. Also, $(x^{s_0}, 1) \in \sigma_1$.

Hence

$$(x^{s_0}, 0), (x^{s_0}, 1) \in (\varphi_{\mathbf{M}(\sigma_1)} \cup \sigma_1), \qquad (3.8)$$

a contradiction to (3.7). $\qquad\qquad\qquad\qquad\qquad\qquad\qquad\qquad\square$

Next is another theorem I like very much.

**Theorem 3.2** (Bārzdiņš [4], Blum and Blum [5]). **Ex** *is not closed under union, i.e., there are classes* $\mathcal{S}_0, \mathcal{S}_1 \in$ **Ex** *such that* $(\mathcal{S}_0 \cup \mathcal{S}_1) \notin$ **Ex**.

Here are example such $\mathcal{S}_0, \mathcal{S}_1$ — similar to those from [5]. Let $\mathcal{S}_0 = \{f \in \mathcal{R} \mid \varphi_{f(0)} = f\}$ and $\mathcal{S}_1 = \{f \in \mathcal{R} \mid (\overset{\infty}{\forall} x)[f(x) = 0]\}$.

Theorem 3.2 essentially suggests that for success in inductive/scientific inference, one needs the diversity of incomparable "cognitive" styles of scientists: scientists $\mathbf{M}_0$ which **Ex**-learns $\mathcal{S}_0$ and $\mathbf{M}_1$ which **Ex**-learns $\mathcal{S}_1$ cannot be combined into a third scientist $\mathbf{M}_2$ which **Ex**-learns all that $\mathbf{M}_0$ does together with all that $\mathbf{M}_1$ does.

In physical optics there is a phenomenon called *anomalous dispersion*: the classical, quantitative explanation for different frequencies of light and, more generally, electromagnetic radiation being differentially bent according to frequency when passing through a prism does not work for X-rays.[20] Physicists used this model nonetheless until quantum mechanics provided a better model. In the mid 70s, I began to consider whether I could prove a theorem re machine inductive inference that would suggest a vindication of physicists' employing slightly faulty predictive explanations. For each $n \in N$, for partial functions $\eta, \theta$, we write $\eta =^n \theta$ to mean that there are at most $n$ counterexamples to $\eta = \theta$. We write $\eta =^* \theta$ to mean that there are at most finitely many counterexamples to $\eta = \theta$. For $a \in N \cup \{*\}$, we define a variant of **Ex**-learning, called **Ex**$^a$-*learning*, in which the eventual final programs $p$ are allowed to be mistaken on up to $a$ inputs in computing the input function $f$, i.e., success requires only that $\varphi_p =^a f$. A theorem indicating an increase in inferring or learning power comes with tolerance of some few mistakes in one's predictive explanations follows.

**Theorem 3.3** (Case and Smith [46, 47]). **Ex** = **Ex**$^0$ $\subset$ **Ex**$^1$ $\subset$ **Ex**$^2$ $\subset \ldots$ **Ex**$^*$.

Hence we see that tolerating up to just one single anomaly or mistake in one's final program provides a strict increase in inferring power, tolerating $n + 1$ anomalies provides an increase over tolerating no more than $n$, and tolerating a finite number provides an increase over tolerating a bounded number.[21] Of course, **Ex**$^*$ is not so "practical" a criterion as **Ex**$^n$, for small $n$, since, for the former, the finite number of anomalies in a final program $p$ may include all the data points for which the predictive explanation $p$ will ever be used. **Ex**$^*$ is mathematically interesting though.

In [47], we pointed out that Popper's Refutability Principle, the principle that purported scientific explanations ought to be subject to refutation by suitable experiments, needs some revision. The anomalies providing the hierarchy of Theorem 3.3 above

---

[20] Of course it is the model or predictive explanation which is anomalous, not the physical phenomenon itself.

[21] [5] announced the case of **Ex** $\subset$ **Ex**$^*$.

are *and must be* mistakes of *omission* [47], but this kind of mistake cannot be algorithmically detected in general. Explanations ought to be refutable when they make predictions, but may not be refutable when they fail to make a prediction at all — even if they should have.

Let minprogram$(f) \overset{\text{def}}{=} \min \{p \mid \varphi_p = f\}$.

**Definition 3.3** (Freivalds [68]). For $\mathcal{S} \subseteq \mathcal{R}$, $\mathcal{S} \in$ **Mex** as witnessed by **M** $\Leftrightarrow (\exists$ computable $h)(\forall f \in \mathcal{S})[$**M** **Ex**-learns $f \wedge (\forall p)[\mathbf{M}(f)\!\downarrow = p \Rightarrow p \leqslant h(\text{minprogram}(f))]]$.

For $\mathbf{M}, \mathcal{S}, h$ as in the just above definition, **M**'s final programs on $f \in \mathcal{S}$ are within "factor" $h$ of minprogram$(f)$. **Mex**-learning is intended as *a* model of inductive inference obeying a form of Occam's Razor. It is common in philosophy of science and in the applied part of artificial intelligence called *machine learning* to assume one's models for fitting and predicting data *should* obey some form of Occam's Razor. Yet we have the following theorem which shows that a simple, easily inferred subclass of $\mathcal{R}$ is *not* in **Mex**.

**Theorem 3.4** (Kinber [93]). $\mathcal{S}_1 = \{f \in \mathcal{R} \mid f$ *is the characteristic function of a finite set*$\} \in (\mathbf{Ex} - \mathbf{Mex})$.

Hence, at least some forms of another common sense principle, Occam's Razor, *restrict* one's inferring or learning power! Theorem 3.4 just above can be proved by a recursion theorem argument together with a finitary cancellation (zero-injury priority) scheme.[22] For more on **Mex** and variants, see [32, 33, 1].[23]

There are costly criteria providing inferring or learning power beyond that of **Ex**[*]. **M** **Bc**-*learns* $f \in \mathcal{R}$ means that **M** on $f$ outputs an infinite sequence of programs $p_0, p_1, p_2, \ldots$, and $(\overset{\infty}{\forall} i)[\varphi_{p_i} = $

---

[22] Chen in [32, 33] showed that, by contrast, $\mathcal{S}_0 = \{f \in \mathcal{R} \mid \varphi_{f(0)} = f\} \in$ **Mex**, and yet, by a recursion theorem argument from [5], self-referential classes like $\mathcal{S}_0$ are so large they contain a finite variant of each element of $\mathcal{R}$.

[23] The latter features infinitary self-reference arguments employing my ORT Theorem [21, 24].

$f]$. **Bc** or *behaviorally correct* learning features semantic convergence to correct programs; whereas, **Ex**-learning features syntactic convergence to correct programs. **Bc** $\overset{\text{def}}{=} \{\mathcal{S} \subseteq \mathcal{R} \mid (\exists \mathbf{M})[\mathbf{M} \text{ **Bc**-learns each } f \in \mathcal{S}]\}$. Steel [47] showed **Ex**$^* \subseteq$ **Bc**. Bārzdiņš [4] first studied **Bc** and showed that $(\mathbf{Bc} - \mathbf{Ex}) \neq \emptyset$. Harrington and I [47] showed that $(\mathbf{Bc} - \mathbf{Ex}^*) \neq \emptyset$.[24] Anyhow, the *cost* mentioned above of **Bc**-learning is that to realize its full power one has to contend with the final, correct programs being of *un*bounded size.[25]

In this section, we have quite plausibly been taking a predictive scientific explanation to be modeled as a $\varphi$-program for predicting the results of all experiments regarding a phenomenon to be explained.[26] Essentially, in terms of the arithmetical hierarchy [148], $\varphi$-programs are intercompilable with $\Sigma_1^0$-definitions of the corresponding (partial) functions. In [37, 10, 48, 49] the learning or inductive inference of $\Sigma_2^0$-definitions is also considered. Of course, from such definitions one may not be able to extract predictions about the outcomes of associated experiments, *but* in *some* cases, some higher order information may be extractable. Even if, from such a definition, one cannot calculate values for an $f$ so defined, one may be able to extract data *refutable* global or shape information about the curve of $f$, for example, that $f$ is monotone increasing.[27]

---

[24] Our proof was an infinitary recursion theorem argument based on my ORT [21, 24].

[25] Anomalous variants of **Bc**-learning, for example, **Bc**$^a$ for allowing up to $a$ anomalies in final programs, and corresponding hierarchies of learning/inferring power are studied in [47]. Harrington showed [47] that some machine witnesses $\mathcal{R} \in$ **Bc**$^*$, and [27] shows that such machines on infinitely many computable functions have their anomalies occurring in undesirable positions.

[26] Fulk [70] argues that the set of distinguishable experiments *one can actually do and record* on a phenomenon is countable: lab manuals can and do contain only finite notations from a finite alphabet and/or bounded-size, finite-precision images.

[27] The difference is somewhat analogous to the difference between predicting the location of planet at any time and predicting the shape of the planet's orbit [37, 10].

The *limiting-computable partial functions* are those computable by a total mind-changing algorithm, i.e., those that are the limit of a computable function [**157**]. With care one can intercompile between $\Sigma_2^0$-definitions and some form of programs for the limiting-computable partial functions.[28] For this reason we write **LimEx** for the inference criterion just like **Ex** *except* that the "programs" output and converged to are $\Sigma_2^0$-definitions (or a suitable form of limiting programs). N.B. We are, of course, interested in **LimEx** for classes of *computable* functions only. We take $\varphi^2$ to be an acceptable programming system (numbering) for the $\Sigma_2^0$ partial functions, where $\varphi_p^2$ is the partial function defined by suitable form of limiting program $p$.

For expressions E admitting translation into the language of first order arithmetic [**113**] we write $\ll E \gg$ for a fixed, natural such translation.

$$\eta =^\infty \theta \overset{\text{def}}{\Leftrightarrow} \| \{x \mid \eta(x) = \theta(x)\} \| \text{ is infinite.}$$

**Theorem 3.5** (Case and Suraj [**48, 49**]). *Suppose* **T** *is a computably axiomatizable first order theory which extends Peano Arithmetic* (**PA**) *and in which one cannot prove sentences that are false in the standard model* ([**113**]). *Then there is a class of monotone computable functions* $C$ *such that*

(1) $C \not\subseteq \cup_{k \in N} \mathbf{Bc}^k$,
(2) $(\forall f \in C)(\forall p \mid \varphi_p =^\infty f)[\mathbf{T} \not\vdash \ll \varphi_p \text{ is monotone} \gg],$[29] *and*
(3) *there exists a machine* **M** *which* **LimEx**-*learns every function in* $C$ *and, for every* $f \in C$, *for every* $e$ *such that* $\mathbf{M}(f){\downarrow} = e$,
    (a) $\mathbf{PA} \vdash \ll \varphi_e^2 \text{ is monotone} \gg$,
    (b) $(\forall x, y)[\mathbf{T} \not\vdash \ll \varphi_e^2(x) = y \gg]$[30], *and*
    (c) $\mathbf{PA} \vdash \ll \varphi_e^2 \text{ is computable} \gg$.

---

[28] The trick is to express the limiting computable partial functions as the *uniform* limit of a single, suitable computable function [**143, 49**].
[29] Therefore, $(\forall f \in C)(\forall p \mid \varphi_p =^\infty f$ and $\varphi_p$ is monotone $)[\mathbf{T} \not\vdash \ll \varphi_p \text{ is monotone} \gg]$. This is, perhaps, surprisingly strong.
[30] Therefore, $(\forall x, y \mid \varphi_e^2(x) = y)[\mathbf{T} \not\vdash \ll \varphi_e^2(x) = y \gg]$.

This theorem (Theorem 3.5), then, provides some strong trade-offs between inferring $\Sigma_1^0$- vs. $\Sigma_2^0$-definitions. For $\mathcal{C}$, one can employ the latter, but *not* the former for successful inference *and* to prove monotonicity. But, by the important Clause 3b in Theorem 3.5 just above, one cannot predict, from the latter and $\mathbf{T}$, for the sake of Popper's Refutability Principle for science, any data points at all in the graphs of the $\varphi_e^2$'s! The output $\Sigma_2^0$-definitions are not, in principle, refutable by incorrect data point predictions, but *in principle*, they admit of being refuted by non-monotonicity (in the input data itself). Hence this result exhibits, then, some new subtleties re the application of Popper's Refutability Principle: one cannot inductively infer $\mathcal{C} \subseteq \mathcal{R}$ without being forced to accept a weakened refutability principle.

I would like to see more CT learning theory theorems like those above with some insight and/or shock value for philosophy of science.

**3.2.2. Cognitive science and language learning.** In this section, we look at learning grammars for (formal) languages from positive information about them. The original paradigm was Gold's [**76**]. Thanks to code numbering and for mathematical convenience we can and will take our languages to be r.e. subsets of $N$. Grammars will be type 0 [**81**], and, hence, we can take a grammar $g$ for an r.e. language $L$ to be an r.e. index for $L$, i.e., such that $W_g = L$.

We say $T$ is a *text* for $L \overset{\text{def}}{\Leftrightarrow} \{T(0), T(1), \ldots\} = L$.[31] We say, in this case, $T$ is *for* $L$. In this section, $\mathbf{M}$s will computably map finite initial segments of texts into grammars/r.e. indices, and, without loss of generality for what we want to do, we take $\mathbf{M}$s to be total. Next are defined some criteria of successful language learning.

---

[31] In more formal expositions, we allow $\rho T$ to contain also #s, where a # models a *pause* and is not part of the language $L$. Then the text with successive values consisting only of #s is the only text for the empty language. Herein we need not be so careful about handling the empty language.

**Definition 3.4** ([**42, 132, 25**]). Suppose $b \in (N^+ \cup \{*\})$, where $N^+ = \{1, 2, \ldots\}$ and $x \leqslant *$ means $x < \infty$.

(1) $\mathcal{L} \in \mathbf{TxtEx} \Leftrightarrow (\exists M)(\forall L \in \mathcal{L})(\forall T \text{ for } L)[\mathbf{M} \text{ on } T \text{ outputs } g_0, g_1, g_2, \ldots \Rightarrow (\exists t)[g_t = g_{t+1} = \cdots \wedge W_{g_t} = L]].$

(2) $\mathcal{L} \in \mathbf{TxtBc} \Leftrightarrow (\exists M)(\forall L \in \mathcal{L})(\forall T \text{ for } L)[\mathbf{M} \text{ on } T \text{ outputs } g_0, g_1, g_2, \ldots \Rightarrow (\exists t) [g_t, g_{t+1}, \ldots \text{ each generates/enumerates } L]].$

(3) $\mathcal{L} \in \mathbf{TxtFex}_b \Leftrightarrow (\exists M)(\forall L \in \mathcal{L})(\forall T \text{ for } L)[\mathbf{M} \text{ on } T \text{ outputs } g_0, g_1, g_2, \ldots \Rightarrow (\exists t) [g_t, g_{t+1}, \ldots \text{ each generates/ enumerates } L \wedge \|\{g_t, g_{t+1}, \ldots\}\| \leqslant b]].$

The class $\mathcal{F}$ of all finite languages $\in \mathbf{TxtEx}$, but the class of all regular languages (from automata theory [**81**]) is not [**76**]. $\mathcal{K} = \{K \cup \{x\} \mid x \in N\} \in (\mathbf{TxtBc} - \mathbf{TxtEx})$, where $K$ is the diagonal halting problem [**148**]. $\mathbf{TxtFex}_b$ is like $\mathbf{TxtBc}$ except the set of final, correct programs has cardinality $\leqslant b$. $\mathbf{TxtFex}_1 = \mathbf{TxtEx}$ & $\mathcal{K} \notin \mathbf{TxtFex}_b$.

We have $\mathbf{TxtFex}_1 \subset \mathbf{TxtFex}_2 \subset \ldots \subset \mathbf{TxtFex}_* \subset \mathbf{TxtBc}$ (see [**25**]).[32][33]

Some sample publications in computational learning theory re formal language learning are [**76, 3, 22, 131, 70, 71, 96, 8, 86, 25, 9, 29**].

The language learning model of the present section, although obviously limited as a model for human language learning, has, nonetheless, been influential in cognitive science and in contemporary theories of natural languages, for example, [**136, 181, 128, 182, 130, 11, 74, 94**].

Regarding this model, I gradually acquired the belief that, in spite of its limitations, there was the possibility for theorems with insights into cognitive science. In the rest of this section, I provide

---

[32] [**132**] showed $\mathbf{TxtFex}_1 \subset \mathbf{TxtFex}_* \subset \mathbf{TxtBc}$.
There are also anomaly hierarchies, but we will not go into them here. See [**25**].
[33] This hierarchy result contrasts with what happens with the criteria for learning programs in the limit for $f \in \mathcal{R}$ from Section 3.2.1 above: for **Ex** style learning, converging to finitely many correct programs in the limit offers no more learning power than converging to one. See [**47**].

an example.[34] The motivation comes from empirical observations from child cognitive development.

*U-shaped learning behavior* features the pattern of learning, unlearning, and relearning. It occurs in child development re, for example, verb regularization [**139, 119, 166**] and understanding of various (Piaget-like) conservation principles [**162**], for example, temperature and weight conservation and interaction between object tracking and object permanence. An example regarding irregular verbs in English follows. A child first uses *spoke*, the correct past tense of the *ir*regular verb *to speak*. Then the child overregularizes *in*correctly using *speaked*. Lastly, the child returns to using *spoke*. Our theoretical examples will involve the formal learning, unlearning, and relearning of type 0 grammars for whole formal languages *L*. The main "theoretical" concern of the empirically based cognitive science literature on U-shaped learning is with how to model U-shaped learning. For example, is U-shaped language learning done employing subconscious general rules vs. tables of exceptions [**14**]? That is a nice concern, but not at all my interest. My interest is in the following question. Is U-shaped learning an *un*necessary and harmless accident of human evolution *or* is U-shaped learning advantageous in that some *classes* of tasks *can* be learned in U-shaped way, but *not* otherwise? I.e., are some classes of tasks learnable only by returning to *abandoned* correct, learnable behaviors? Of course, as a question about humans, this is very difficult to answer. So, I sought some learning theory insights about what could *possibly* be true.

---

[34] The proof techniques for learning language grammars from positive data, unlike the results in the just previous section (Section 3.2.1) feature more than considerations of algorithmicity. They also feature finite extension arguments which, of course, can be conceptualized in terms of Baire Category Theory [**122, 148, 83, 129, 131**]. Some of the proofs in [**25**] employ such a mixture but resembling finite injury priority arguments. These are to obtain results for **TxtFex**$_b$ having to do with without-loss-of-generality local and global *in*sensitivity to order of data presentation and whether the texts are restricted to being computable.

Next is the definition of language learning criteria which are restricted by *dis*allowing U-shaped learning behavior. We think of $W_g$ as the [summary of the] *behavior* of $g$.

**Definition 3.5.** Suppose $\mathbf{C} \in \{\mathbf{TxtFex}_b, \mathbf{TxtBc}\}$. Then, $\mathcal{L} \in \mathbf{NonUC} \Leftrightarrow (\exists M)(\forall L \in \mathcal{L})(\forall T$ for $L)[\mathbf{M}$ on $T$ outputs $g_0, g_1, g_2, \ldots \Rightarrow (\forall i, j, k \mid i < j < k)[W_{g_i} = W_{g_k} = L \Rightarrow W_{g_j} = W_{g_i}]]$.

Non-U-shaped learners never abandon *correct behaviors* for learned $L \in \mathcal{L}$ and, then, return to those behaviors.

From [**29**], the transitive closure of the inclusions (denoted by $\longrightarrow$) in Fig. 1 holds *and* no other inclusions hold.

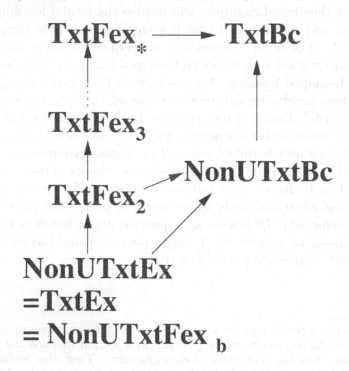

FIGURE 1. Results on U-Shaped Learning

Hence U-shaped learning *is* needed for some class in **TxtBc**; is *not* for **TxtEx** learning, i.e., for learning *one* successful grammar in the limit; *is* needed for some class in **TxtFex**$_2$ even if we allow finitely many grammars in the limit — but *not* if we allow infinitely many grammars in the limit; and *is* needed for some $\mathcal{L} \in \mathbf{TxtFex}_3$ even if we allow infinitely many grammars in the limit.

Now that we know some mathematical *possibilities*, a question for the cognitive scientists is: does the class of tasks humans must learn to be competitive in the genetic marketplace, like this latter $\mathcal{L}$, *necessitate* U-shaped learning?

I would like to see more CT learning theory results like the above which give cognitive science something new to think about.

### 3.2.3. Applied machine learning.

In the context of dealing with the difficulties of actually applying learning in robotics, Drew McDermott [110] says, "Learning makes the most sense when it is thought of as filling in the details in an algorithm that is already nearly right." I suggested to colleagues that we get some corresponding learning theory results regarding learning programs for functions from approximately correct such programs (as well as from data on the functions). Martin Kummer came up with several nice ideas for such approximate programs for a computable 0-1 valued function — including decision programs for *bounded width* trees [148] containing or enveloping the function, and we produced [41]. A sample result from this paper implies that if the approximately correct programs are for enveloping trees of width $n > 0$, then some probabilistic machine (in the sense of [137, 138]) **Ex**-learns every 0-1 valued computable function with probability of success $\frac{1}{n}$. For **Bc** the probability is one.

In the late 90s, I started attending applied machine learning conferences and workshops. Early on I noticed practical interest in so-called *concept drift* and *context sensitive learning*.

A drifting concept to be learned is one which is a moving target. See, for example, [6, 7, 61, 67, 79, 103, 186]. I got some computability theory collaborators together to produce [34] in which we show, for various learnability criteria (including some, suggested by Frank Stephan, for learning Martingale betting strategies), bounds on the speed of the moving target that permit success at all.

Context sensitive learning involves trying to learn $Y$ by first [178, 179, 167, 56, 57, 58, 62, 174, 169] *or* simultaneously [18, 19, 116, 12, 59, 111, 140, 161] trying to learn also $X$ — even in cases where there may be no inherent interest in learning $X$. There is, in many cases, an empirical advantage in doing this for some $X, Y$. It can happen that $Y$ is not learnable by itself, but is learnable if one learns $X$ first or simultaneously. For example, to teach a robot to drive a car, it is useful to train it also to predict the center of the road markings (see, for example, [15, 19]). I realized there was already a CT learning theory paper that I liked very much, [2], which showed mathematically these context sensitivity phenomenon *must* happen for *some* tasks $X, Y$. Later we produced [36] providing a kind of strengthening for the case one learns $X, Y$ simultaneously.[35]

These results regarding context sensitive learning provide mathematical support for the corresponding empirical phenomena suggesting the *possibility* that these empirical phenomena are not just accidental or illusory.

I would like to see more of these kinds of CT learning theory papers.

Next is an interesting four part story.

*Part I of the four part story.* In my visits to the School of Computer Science and Engineering, University of New South Wales, Sydney, Australia, I have learned about the machine learning projects of Claude Sammut there. I became particularly interested in the *behavioral cloning* approach to machine learning of reactive process-control. This is surveyed in [16] and involves using data from the (non-verbal, performance) behavior of master or expert human controllers in order to make machine learning

---

[35] Of course machine learning is an engineering endeavor. However, philosophers of science as well as practitioners in scientific disciplines should, I believe, be considering their relevance to their endeavors.

of complex control feasible/possible. For example, it has been used successfully to teach an autopilot to fly an aircraft simulator [16, 158, 120, 151, 152] and to teach a machine to operate efficiently a (simulated) free-swinging shipyard crane [16, 175].

One of the difficulties Claude made me aware of in the learning-to-fly project was that attempts to make use of the behavioral data from more than one human expert at a time had failed miserably. Different pilots had very different strategies, and it was not clear how to mix them.

*Part II of the four part story.* In a visit to Martin Kummer he put me onto his *theoretical* work on learning, from programs for game trees, etc., winning strategies for infinite reactive process-control games called *closed computable games* [98]. I would not provide here the details but will give the computability-theoretical flavor of these games with two contrasting examples.[36]

**Example 3.1.** Fix $n_0 \in N$. Player I is a digital thermostat, Player II is the temperature (which is subject to a discrete unseen physical disturbance); *winning* for Player I is: past time (= move) $n_0$ keeping the temperature within some pre-assigned integer bounds.

Example 3.1 *is* a closed computable game. Importantly, Player I *can* algorithmically detect if he/she/it has, at any point, lost.

**Example 3.2.** Player I is a digital thermostat, Player II is the temperature (which is subject to a discrete, unseen physical disturbance); *winning* for Player I is: past *some* time (= move) $n$ keeping the temperature within some pre-assigned integer bounds.

Example 3.2 is *not* a closed computable game. Importantly, Player I can *not* algorithmically detect if he/she/it has, at any point, lost.

---

[36] For more on these games, also see [45, 118, 168].

Of course the behavioral cloning games in Part I are not infinite, but there is otherwise some suggestive similarity with the closed computable games.

Kummer's co-author, Matthias Ott, had some ideas already for adding the *behavior* of masters playing winning strategies as additional information for the learning of closed computable games. This looked like behavioral cloning from Part I of the story! We produced [44], and one of the theorems there said there *existed* cases where cloning $n + 1$ disparate masters enable learning to win more games than merely cloning $n$. This was theoretical support, then, for the *possibility* that, in the behavioral cloning experiments, there could be a way to clone behaviorally multiple masters or experts — and with some performance advantage over merely cloning one master.

*Part III of the four part story.* I went to an applied machine learning workshop, and told participants who cared about behavioral cloning about the just above result that there are cases for which cloning more experts is better than cloning fewer. I am not sure if I expected them to say, in effect, Oh, good, I will go home and figure out how to apply that to my behavioral cloning problems. Instead they asked me how to do it for practical problems. Our existence theorem had not provided me just how to do it. I did try after that to get Sammut's group to see what we could do, but I was never around them long enough to get much work done on it.

*Part IV of the four part story.* Some time later I found out, from Mike Bain in Sammut's group, about Dorian Šuc's wonderful doctoral dissertation in Ljubljana, Slovenia, [177]. He had found a way to clone behaviorally more than one human expert simultaneously for the free-swinging shipyard crane problem — by having more than one level of feedback control, *and* he got enhanced performance from cloning the multiple experts! Dorian had not known anything about our suggestive theoretical result, he just solved the problem.

What I would like to see: get more CT learning results which *should* inform machine learning practitioners.

## 3.3. Machine self-reflection

This paragraph is based mostly on [24]. Kleene's (Second) Recursion Theorem can be conceptualized as follows. Given any pre-assigned algorithmic task, there is a $\varphi$-program $e$ which first looks in a mirror[37] to see in detail and exactitude its own code script, flow chart, or wiring diagram, and, then, $e$ uses this image in the mirror as a datum (and its external input as another datum) for input to the pre-assigned algorithmic task — which task it then carries out with these two inputs. Essentially, then, $e$ has a perfect self-model (a copy of itself) and employs it according to the algorithmic pre-assigned task which describes how to use it (together with its external input). No infinite regress is required since $e$'s copy is projected *external to* $e$. Such $e$ are self-reflecting/self-knowing programs.[38] [39]

In the late 70s, I realized that the constructive form of Kleene's Recursion Theorem (I will call it **KRT**) could be conceptualized as a kind of *non*-denotational program control structure [163]. Typical *denotational* control structures are **if–then–else** and **while–loop**. I believed it would be possible to develop a general theory of control structures in the context of CT-style programming systems (numberings). It was. I supervised the doctoral dissertations

---

[37] We can suppose the mirror is a corner mirror so the image in it does not appear left-right reversed.

[38] Examples of using self-knowledge in a simple way were presented in the proof of Theorem 3.1 in Section 3.2.1 above. Examples of using self-knowledge in more complex ways are in [21].

[39] I intend to write the paper version of [26] in which I describe what, I believe, Kleene's Recursion Theorem has to do with the self-reflection component of consciousness. N.B. I will not provide an elucidation of what Dave Chalmers in his very influential book [31] describes as the hard problem of consciousness, for example, the problem of qualia. I will provide some ideas on the problems of why we are not unconscious zombies [31] and how we can be machines and, yet, differ in kind from Searle's famous Chinese Room [155].

[**144, 149**] to help work this out.[40] In the context of programming systems (numberings) for the class of partially computable functions where each system has a universal program inside the system, I showed that the acceptable programming systems [**147, 148**], are characterized as those in which each possible control structure has an implementation [**144, 149**]. One of my principal goals in all this was to try to characterize **KRT** insightfully — in the interest of understanding the utility of self-knowledge. The ancient Greeks thought self-knowledge was important, and, perhaps, one could obtain some mathematical insight into its utility. Characterizations have been elusive, but we have had better luck at insight into what epitomizes the "complement" of **KRT**. Here is one of my favorite theorems of Jim Royer on this latter subject. Again, the programming systems (numberings) considered are for the class of partially computable functions where each has a universal program inside the system.

**Theorem 3.6** (Royer [**149**]). *KRT and if–then–else are complementary in the sense that:*

(1) *For each there is a programming system with an implementation of that one but with no implementation of the other one; and*

(2) *If a programming system has an implementation of both, it is acceptable; i.e., has an implementation of all control structures.*

Hence decision branching and self-reflection are complementary.

I noticed, from the proofs of this theorem (Theorem 3.6) and related ones in [**149**], that one of the crucial elements was the constructivity component of **KRT**, but I wanted to understand the self-knowledge component, period. Let **krt** be the not necessarily constructive Kleene Recursion Theorem. With a new Ph.D. student, Sam Moelius, we have begun to find epitomizers of the

---

[40] For definitions, etc., see [**144, 145, 149**]. For more on this CT approach to control structures, see [**146, 108, 85, 39**].

complement of **krt**. This is work not yet completed. We will see how it goes.

I would like to see more CT work on mathematically under-standing machine self-knowledge.

## 3.4. CT for computational complexity

In this section, we explore a tiny fraction of the available and some-what recent literature. I like very much, though, the early results of abstract complexity theory such as the surprising Blum Speed-Up Theorem [13, 187], its strengthening [114], and [117].[41] Many more recent results in complexity theory involve limiting some CT techniques to severely time or space bounded realms. See, for ex-ample, [143] and its bibliography. Actually, in co-creating [143] I had in mind bringing CT techniques far down into the subre-cursive realm, for example, all the way down to linear time com-putable. Of course, extremely complicated priority arguments or even finite injury priority arguments with no computable bound on the injuries do not seem to fit well this realm.[42] Priority con-structions with bounded finite injury *can* sometimes be used to get complexity theory results, e.g, in [97] at the cost of exponen-tial time. Impressively, [30] applies carefully bounded priorities toward feasible learnability. Employing CT tricks to provide the practitioner with feasible algorithms, while very difficult, would be highly desirable for the future.

[28] presents learnability applications of CT to prove results about the *quality* of the final learned programs. Below is a special case of one of the results. Suppose $k > 0$. Run times are measured

---

[41] [127] surveys much of this work.

Proofs of such results by complicated Kleene Recursion Theorem arguments can be conceptually simplified by employing instead my ORT. See [109, 160] for examples of how I do this.

[42] Possibly, these kinds of arguments could be introduced into this realm by employing Hybrid recursion theorems from [143]. These per-mit, for *example*, self-other reference between low level subrecursive pro-gramming systems and systems for functions partial computable in $K$.

with respect to multi-tape Turing machines, and we suppose $\varphi^{\text{TM}}$ is an acceptable system based on them — with $\Phi_p^{\text{TM}}$ the run time (partial) function of $\varphi^{\text{TM}}$-program $p$ [**13**]. Let $\mathcal{P}^k \overset{\text{def}}{=}$ the set of characteristic functions of sets decidable in $k$-degree polynomial time (in the length of inputs). Pick an inverse $\alpha$ to Ackermann's function computable in linear time — of course $\alpha$ is very slow growing [**43**]. Let $\mathcal{Q}^k \overset{\text{def}}{=}$ the set of characteristic functions of sets decidable in time a $k$-degree polynomial of $n$ times $\log(n)$ times $\alpha(n)$, where $n$ is the length of the input. $\mathcal{P}^k \subset \mathcal{Q}^k$ [**80, 81**], and this is a tightest known separation. Since each of $\mathcal{P}^k$ and $\mathcal{Q}^k$ are r.e. classes of computable functions, by the enumeration technique in [**5**], they are **Ex**-learnable. For example, then, $\mathcal{P}^k$ is so learnable by a machine all of whose output conjectures run in $k$-degree polynomial time.

**Theorem 3.7** (Case, Chen, Jain, Merkle, and Royer [**28**]). *Suppose* **M Ex**-*identifies* $\mathcal{Q}^k$, *where* $k \geqslant 1$. *Then there is an* "*easy*" $f$, *the characteristic function of some finite set, such that*
$$(\forall a)(\overset{\infty}{\forall} x)[\Phi_{\text{M}(f)}^{\text{TM}}(x) > a \cdot (|x| + 1)^k].$$

Hence, to learn $\mathcal{Q}^k$, a little bigger class than $\mathcal{P}^k$, we have severe complexity deficiencies in the final programs on very easy functions $f$.

Theorem 3.7 just above is proved by delayed diagonalization (or slowed simulation) [**104, 143**] with cancellation [**13**] (or zero injury), complexity-bounded self-reference [**143**], and very careful subrecursive programming [**143**].

In [**28**], we have other results of this ilk. For example, *if* the classes polynomial time and non-deterministic polynomial time *do* separate, then **Ex**-learning the latter with output conjectures non-deterministic polynomial time bounded Turing machines will force there to exist some easy functions $f$ (characteristic functions of finite sets) whose final learned programs will have some otherwise unnecessary and undesirable non-determinism. Also obtained is a similar result comparing quantum polyomial time and polynomial time (again, *if* they separate), where, in learning the then larger class, the complexity deficiency in final programs for some easy functions is otherwise unnecessary quantum parallelism. Standard diagonalizations are too rough to be used in these realms where we are not even sure currently if there are separations. We resort instead to lifts of arguments about more delicate $\Sigma_2^0$-*in*separability of certain subrecursive index sets [**23, 143**].

In [**28**], there are additionally results about cases where final programs *are* asymptotically optimal, but they are *informationally deficient*: one cannot prove about them even suboptimal run time bounds.

The lesson for the practioner of such results from [**28**] is: do not try to learn too much (if you do not have to); else, you may get undesirable learned programs.

I would like to see more CT results in complexity theory with even more remarkable advice to the practitioner.

Jim Royer has been working for some time on a program to bridge between European theoretical computer scientists who seek to understand higher types in programming languages, but who generally ignore even issues of algorithmicity and U.S. theoretical computer scientists interested only in feasible algorithms. For example, [**150**] presents an analog of the Kreisel-Lacombe-Shoenfield Theorem [**148**] for *feasible* type-2 functionals [**112, 87, 88, 156, 82**].

I would like to see more of this level attempt to provide some *eventual* advice to the practitioner, for example, to the designer of elegant, new programming languages.

## 3.5. Physics and all the rest

Kreisel has written about the problem of whether the physical world permits calculations beyond the Turing-computable, for example, [**99, 100**]. See [**126**] for nice discussion of the issues. *Hypercomputation* involves allowing infinitely many computation steps in finite time.[43] The problem is whether in our universe such computations are executable. Norman Margolus at MIT whose background includes both physics and computer science explained to me a few years ago that such computations would require an unlimited supply of energy. See also [**52, 53, 54, 55, 64**] for further arguments that this sort of computation is not available in the real world.

Along different lines, we see, though, the impressive and surprising work of Pour-El and Richards [**133, 134, 135**]. In [**134**] they provide a (higher type) *un*computable solution to the wave equation with a (higher type) *computable* boundary condition! [44]

On the other hand: when I first studied Maxwell's Equations as an undergraduate I noticed that they were applied beautifully and elegantly to clouds of electrons. Problem: the clouds are discrete, yet the mathematics is essentially continuous. Of course, it is too hard for practical purposes to model a large cloud of electrons discretely, and the continuous mathematics nicely smooths out the discreteness and provides good enough experimental predictions. My reaction, though, was disillusionment. I naively expected physicists to seek absolute knowledge and at least to apologize for not providing it. Of course, they do not care about such matters. As you may note from some of the things I wrote about above, for example, in Section 3.2.2, I no longer expect absolute knowledge.[45]

So, then, *is* at least *some* of physical reality *absolutely* modeled by continuous mathematics involving real numbers? Perhaps all but physical *space* is discrete? [**78**, pp. 164–165] argues that there must exist as a universal constant in nature a smallest length. It may be that the universe, *including space*, is discrete. Researchers in the cellular automata approach to physics

---

[43] A recursive iteration of the idea would lead to Kreisel's $\aleph_0$-mind computability (characterizing the $\Pi_1^1$-computable partial functions) [**148**].

[44] For a different perspective on this work, see [**180**].

[45] But, anyway, let me at least apologize for not providing it.

(see [**63, 115, 69, 173, 170, 107, 165, 106, 172, 171, 176, 185, 164, 65, 77**])[46] take this idea seriously.

So regarding the work referenced above by Pour-El and Richards, while I admire this work very much, I believe one has to be careful about work on physics *equations* which may be only wonderfully convenient continuous approximations to various discrete realities. The resultant work will not really be about *physics*.

So, I am left with not so many examples of prior applications of CT to physics I would like to see more of in the future. There was at least the quantum computing example in Section 3.4 above.

Anyhow, I would like to see future applications of CT with insights and/or advice to physics (and all the science and engineering disciplines for which I have provided no example prior applications of CT).

## References

1. A. Ambainis, J. Case, S. Jain, and M. Surajm, *Parsimony hierarchies for inductive inference*, J. Symb. Log. **69** (2004), 287–328.

2. D. Angluin, W. Gasarch, and C. Smith, *Training sequences*, Theor. Comput. Sci. **66** (1989), no. 3, 255–272.

3. D. Angluin, *Inductive inference of formal languages from positive data*, Inf. Control **45** (1980), 117–135.

4. J. Bārzdiņš, *Two theorems on the limiting synthesis of functions* (in Russian), Theory of Algorithms and Programs, Riga, Latvian State Univ. **210** (1974), 82–88.

5. L. Blum and M. Blum, *Toward a mathematical theory of inductive inference*, Inf. Control **28** (1975), 125–155.

6. P. Bartlett, S. Ben-David, and S. Kulkarni, *Learning changing concepts by exploiting the structure of change*, In: Proceedings of the Ninth Annual Conference on Computational Learning Theory, ACM Press, 1996, pp. 131–139.

---

[46] Here [**63**] is crucial, and [**115**] lays out the ideas of Ed Fredkin on some of the ways physical space might be discrete.

7. A. Blum and P. Chalasani, *Learning switching concepts*, In: Proceedings of the Fifth Annual Conference on Computational Learning Theory, ACM Press, 1992, pp. 231–242,

8. G. Baliga, J. Case, and S. Jain, *Language learning with some negative information*, J. Comput. Syst. Sci. **51** (1995), 273–285.

9. G. Baliga, J. Case, and S. Jain, *The synthesis of language learners*, Inf. Comput. **152** (1999), no. 1, 16–43.

10. G. Baliga, J. Case, S. Jain, and M. Suraj, *Machine learning of higher order programs*, J. Symb. Log. **59** (1994), no. 2, 486–500.

11. R. Berwick, *The Acquisition of Syntactic Knowledge*, The MIT Press, 1985.

12. K. Bartlmae, S. Gutjahr, and G. Nakhaeizadeh, *Incorporating prior knowledge about financial markets through neural multitask learning*, In: Proceedings of the Fifth International Conference on Neural Networks in the Capital Markets, 1997.

13. M. Blum, *A machine independent theory of the complexity of recursive functions*, J. Assoc. Comput. Mach. **14** (1967), 322–336.

14. M. Bowerman, *Starting to talk worse: Clues to language acquisition from children's late speech errors*, In: U-Shaped Behavioral Growth, S. Strauss and R. Stavy (Eds.), Academic Press, 1982.

15. S. Baluja and D. Pomerleau, *Using the representation in a neural network's hidden layer for task-specific focus of attention*, Technical Report CMU-CS-95-143, School of Computer Science, CMU, May 1995. [To appear in Proceedings of the 1995 IJCAI]

16. M. Bain and C. Sammut, *A framework for behavioural cloning*, In: Machine Intelligence 15, Intelligent Agents, K. Furakawa S. Muggleton, and D. Michie (Eds.), Oxford Univ. Press, 1999, pp. 103–129.

17. A. W. Burks (Ed.), *Essays on Cellular Automata*, Univ. Illinois Press, 1970.

18. R. A. Caruana, *Multitask connectionist learning*, In: Proceedings of the 1993 Connectionist Models Summer School, pp. 372–379.

19. R. A. Caruana, *Algorithms and applications for multitask learning*, In: Proceedings of the 13th International Conference on Machine Learning, 1996, pp. 87–95.

20. J. Case, *A note on the degrees of self-describing Turing machines*, J. Assoc. Comput. Mach. **18** (1971), 329–338.

21. J. Case, *Periodicity in generations of automata*, Math. Syst. Theory **8** (1974), 15–32.

22. J. Case, *Learning machines*, In: Language Learning and Concept Acquisition, W. Demopoulos and A. Marras (Eds.), Ablex Publishing Company, 1986.

23. J. Case, *Effectivizing inseparability*, Z. Math. Logik Grundlagen Math. **37** (1991), no. 2, 97–111. [http://www.cis.udel.edu/~case/papers/mkdelta.pdf corrects missing set complement signs in definitions in the journal version]

24. J. Case, *Infinitary self-reference in learning theory*, J. Exp. Theor. Artif. Intell. **6** (1994), no. 1, 3–16.

25. J. Case, *The power of vacillation in language learning*, SIAM J. Comput. **28** (1999), no. 6, 1941–1969.

26. J. Case, *Machine self-reference and consciousness*, In: Proceedings and Abstracts of the Third Annual Meeting of the Association for the Scientific Study of Consciousness, London, Ontario, 1999. [http://www.cis.udel.edu/~case/slides/krt-consc-slides.pdf]

27. J. Case, K. Chen, and S. Jain, *Costs of general purpose learning*, Theor. Comput. Sci. **259** (2001), no. 1-2, 455–473.

28. J. Case, K. Chen, S. Jain, W. Merkle, and J. Royer, *Generality's price: Inescapable deficiencies in machine-learned programs*, Ann. Pure Appl. Logic **139** (2006), no. 1-3, 303–326.

29. L. Carlucci, J. Case, S. Jain, and F. Stephan, *Non U-shaped vacillatory and team learning*, In: Algorithmic Learning Theory: 16th International Conference, ALT 2005, Singapore, October 8-11, 2005. Proceedings, S. Jain, H. U. Simon, and E. Tomita (Eds.), Lect. Notes Comput. Sci. **3734**, Springer, 2005,

30. Z. Chen and S. Homer, *The bounded injury priority method and the learnability of unions of rectangles*, Ann. Pure Appl. Logic **77** (1996), no. 2, 143–168.

31. D. Chalmers, *The Conscious Mind: In Search of a Fundamental Theory*, Oxford, Oxford University Press, 1996.

32. K. Chen, *Tradeoffs in Machine Inductive Inference*, PhD Thesis, Computer Science Department, SUNY at Buffalo, 1981.

33. K. Chen, *Tradeoffs in the inductive inference of nearly minimal size programs*, Inf. Control **52** (1982), 68–86.

34. J. Case, S. Jain, S. Kaufmann, A. Sharma, and F. Stephan, *Predictive learning models for concept drift*, Theor. Comput. Sci. **268** (2001), no. 2, 323–349.

35. J. Case, S. Jain, and S. Ngo Manguelle, *Refinements of inductive inference by Popperian and reliable machines*, Kybernetika **30** (1994), no. 1, 23–52.

36. J. Case, S. Jain, M. Ott, A. Sharma, and F. Stephan, *Robust learning aided by context*, J. Comput. Syst. Sci. **60** (2000), 234–257.

37. J. Case, S. Jain, and A. Sharma, *On learning limiting programs*, Int. J. Found. Comput. Sci. **3** (1992), no. 1, 93–115.

38. J. Case, S. Jain, and A. Sharma, *Machine induction without revolutionary changes in hypothesis size*, Inf. Comput. **128** (1996), no. 2, 73–86.

39. J. Case, S. Jain, and M. Suraj, *Control structures in hypothesis spaces: The influence on learning*, Theor. Comput. Sci. **270** (2002), no. 1-2, 287–308.

40. J. Case, S. Jain, F. Stephan, and R. Wiehagen, *Robust learning – rich and poor*, J. Comput. Syst. Sci. **69** (2004), 123–165.

41. J. Case, S. Kaufmann, E. Kinber, and M. Kummer, *Learning recursive functions from approximations*, J. Comput. Syst. Sci. **55** (1997), 183–196.

42. J. Case and C. Lynes, *Machine inductive inference and language identification*, In: Automata, Languages and Programming: Ninth Colloquium Aarhus, Denmark, July 12-16, 1982, M. Nielsen and E. M. Schmidt (Eds.), Lect. Notes Comput. Sci. **140** Springer, 1982, pp. 107–115.

43. T. Cormen, C. Leiserson, R. Rivest, and C. Stein, *Introduction to Algorithms*, The MIT Press, 2001.

44. J. Case, M. Ott, A. Sharma, and F. Stephan, *Learning to win process-control games watching game-masters*, Inf. Comput. **174** (2002), no. 1, 1–19.

45. D. Cenzer and J. Remmel, *Recursively presented games and strategies*, Math. Soc. Sci. **24** (1992), no. 2-3, 117–139.

46. J. Case and C. Smith, *Anomaly hierarchies of mechanized inductive inference*, In: Conference Record of the Tenth Annual ACM Symposium on Theory of Computing, San Diego, California, 1-3 May 1978, pp. 314–319.

47. J. Case and C. Smith, *Comparison of identification criteria for machine inductive inference*, Theor. Comput. Sci. **25** (1983), 193–220.

48. J. Case and M. Suraj, *Inductive inference of $\Sigma_1^0$-vs. $\Sigma_2^0$-definitions for computable functions*, In: Proceedings of the International Conference on Mathematical Logic, Novosibirsk, Russia, 1999.

49. J. Case and M. Suraj, *Weakened refutability for machine learning of higher order definitions* 2006. [Working paper for eventual journal submission]

50. M. Davis, *Is mathematical insight algorithmic?* Behav. Brain. Sci. **3** (1990), 659–660.

51. M. Davis, *How subtle is Gödel's theorem? More on Roger Penrose* Behav. Brain. Sci. **16** (1993), 611–612.

52. M. Davis, *The myth of hypercomputation* In: Alan Turing: Life and Legacy of a Great Thinker, C. Teuscher (Ed.), Springer, 2004, pp. 195–212.

53. M. Davis, *Computability, computation and the real world*, In: Imagination and Rigor: Essays on Eduardo R. Caieniello's Scientific Heritage, S. Termini (Ed.), Springer, 2005, pp. 63–70.

54. M. Davis, *Why there is no such subject as hypercomputation*, Appl. Math. Comput., 2006. [To appear]

55. M. Davis, *The Church-Turing thesis: Consensus and opposition*, In: Proceedings cCiE 2006, Springer Notes on Computer Science, Swansee, July 2006.

56. H. de Garis, *Genetic programming: Building nanobrains with genetically programmed neural network modules*, In: IJCNN: International Joint Conference on Neural Networks, Vol. 3, IEEE Service Center, Piscataway, New Jersey, June 17–21, 1990, pp. 511–516.

57. H. de Garis, *Genetic programming: Modular neural evolution for Darwin machines*, In: International Joint Conference on Neural Networks, Vol. 1, M. Caudill (Ed.), Lawrence Erlbaum Associates, Publishers, Hillsdale, New Jersey, January 1990. pp. 194–197.

58. H. de Garis, *Genetic programming: Building artificial nervous systems with genetically programmed neural network modules*, In: Neural and Intelligenct Systems Integeration: Fifth and Sixth Generation Integerated Reasoning Information Systems, B. Souček and The IRIS Group (Eds.), John Wiley and Sons, 1991, Chapt. 8, pp. 207–234.

59. T. G. Dietterich, H. Hild, and G. Bakiri, *A comparison of ID3 and backpropogation for English text-to-speech mapping*, Mach. Learn. **18** (1995), no. 1, 51–80.

60. K. deLeeuw, E. Moore, C. Shannon, and N. Shapiro, *Computability by probabilistic machines*, Automata Studies, Ann. Math. Studies **34** (1956), 183–212.

61. M. Devaney and A. Ram, *Dynamically adjusting concepts to accommodate changing contexts*, In: Proceedings of the ICML-96 Pre-Conference Workshop on Learning in Context-Sensitive Domains, Bari, Italy, M. Kubat and G. Widmer (Eds.), 1994. [Journal submission]

62. S. Fahlman, *The recurrent cascade-correlation architecture*, In: Advances in Neural Information Processing Systems 3, R. Lippmann, J. Moody, and D. Touretzky (Eds.), Morgan Kaufmann, 1991, pp. 190–196.

63. R. Feynman, *Simulating physics with computers*, Int. J. Theor. Phys. **21** (1982), no. 6/7.

64. R. Feynman, *Feynman Lectures on Computation*, A. Hey and R. Allen (Eds.), Perseus Books, 2000.

65. U. Frisch, B. Hasslacher, and Y. Pomeau, *Lattice-gas automata for the Navier Stokes equation*, Phys. Rev. Letters **56** (1986), no. 14, 1505–1508.

66. M. Fulk and S. Jain, *Approximate inference and scientific method*, Inf. Comput. **114** (1994), no. 2, 179–191.

67. Y. Freund and Y. Mansour, *Learning under persistent drift*, In: Proceedings of the Third European Conference on Computational Learning Theory (EuroCOLT'97), S. Ben-David (Ed.), Lect. Notes Artif. Intell. **1208**, Springer, 1997, pp. 94–108.

68. R. Freivalds, *Minimal Gödel numbers and their identification in the limit*, In: Mathematical Foundations of Computer Science 1975 4th Symposium, Marianske Lazne, September 1-5, 1975, J. Becvar (Ed.), Lect. Notes Comput. Sci. **32**, Springer, 1975, pp. 219–225.

69. E, Fredkin and T. Toffoli, *Conservative logic*, Int. J. Theor. Phys. **21** (1982), no. 3/4.

70. M. Fulk, *A Study of Inductive Inference Machines*, PhD Thesis, SUNY at Buffalo, 1985.

71. M. Fulk, *Prudence and other conditions on formal language learning*, Inf. Comput. **85** (1990), no. 1, 1–11.

72. J. Gill, *Probabilistic Turing Machines and Complexity of Computation*, PhD Thesis, University of California, Berkeley, 1972.

73. J. Gill, *Computational complexity of probabilistic Turing machines*, SIAM J. Comput. **6** (1977), 675–695.

74. L. Gleitman, *Biological dispositions to learn language*, In: Language Learning and Concept Acquisition, W. Demopoulos and A. Marras (Eds.), Ablex Publ. Co., 1986.

75. C. Glymour, *Inductive inference in the limit*, Erkenntnis, **22** (1985), 23–31.

76. E. Gold, *Language identification in the limit*, Inf. Control **10** (1967), 447–474.

77. B. Hasslacher, *Discrete fluids*, Los Alamos Sci. **15)** (1987), 175–217.

78. W. Heisenberg, *Physics and Philosophy*, Harper and Brothers Publishers, 1958.

79. D. Helmbold and P. Long, *Tracking drifting concepts by minimizing disagreements*, Mach. Learn. **14** (1994), no. 1, 27–45.

80. J. Hartmanis and R. Stearns, *On the computational complexity of algorithms*, Trans. Am. Math. Soc. **117** (1965), 285–306.

81. J. Hopcroft and J. Ullman, *Introduction to Automata Theory Languages and Computation*, Addison-Wesley, 1979.

82. R. Irwin, B. Kapron, and J. Royer, *On characterizations of the basic feasible functional (Part I)*, J. Funct. Program. **11** (2001), no. 1, 117–153.

83. T. Jech, *Set Theory*, Academic Press, 1978.

84. N. Jessop, *Biosphere: A Study of Life*, Prentice-Hall, 1989.

85. S. Jain and J. Nessel, *Some independence results for control structures in complete numberings*, J. Symb. Log. **66** (2001), no. 1, 357–382.

86. S. Jain, D. Osherson, J. Royer, and A. Sharma, *Systems that Learn: An Introduction to Learning Theory*, The MIT Press, 1999.

87. B. Kapron and S. Cook, *A new characterization of Mehlhorn's polynomial time functionals*, In: Proceedings of the 32nd Annual Symposium on Foundations of Computer Science, San Juan, Puerto Rico, 1-4 October 1991. IEEE Computer Society 1991, pp. 342–347.

88. B. Kapron and S. Cook, *A new characterization of type-2 feasibility*, SIAM J. Comput. **25** (1996), no. 1, 117–132.

89. K. Kelly, *The Logic of Reliable Inquiry*, Oxford Univ. Press, 1996.

90. K. Kelly, *The logic of success*, Br. J. Philos. Sci. **51** (2001), 639–666.

91. K. Kelly and C. Glymour, *Convergence to the truth and nothing but the truth*, Philos. Sci. **56** (1989), 185–220.

92. K. Kelly and C. Glymour, *Theory discovery from data with mixed quantifiers*, J. Philos. Logic **19** (1990), no. 1, 1–33.

93. E. Kinber, *On a theory of inductive inference*, In: Fundamentals of Computation Theory: Proceedings of the 1977 International FCT-Conference, Poznan-Kornik, Poland September 19-23, 1977, M. Karpinski (Ed.), Lect. Notes Comput. Sci. **56**, Springer, 1977, pp. 435–440.

94. D. Kirsh, *PDP learnability and innate knowledge of language*, In: Connectionis: Theory and Practice, S. Davis (Ed.), Oxford Univ. Press, 1992, pp. 297–322.

95. S. Kleene, *Origins of recursive function theory*, Ann. Hist. Comput. **3** (1981), no. 1, 52–67.

96. S. Kapur, B. Lust, W. Harbert, and G. Martohardjono, *Universal grammar and learnability theory: The case of binding domains and the 'subset principle'*, In: Knowledge and Language, Vol. I, E. Reuland and W. Abraham (Eds.), Kluwer, 1993, pp. 185–216.

97. S. Kurtz, S. Mahaney, and J. Royer, *The structure of complete degrees*, In: Complexity Theory Retrospective, A. Selman (Ed.), Springer, 1990, pp. 108–146.

98. M. Kummer and M. Ott, *Learning branches and learning to win closed games*, In: Proceedings of the Ninth Annual Conference on Computational Learning Theory, ACM Press, 1996, pp. 280–291.

99. G. Kreisel. *Mathematical logic*, In: Lectures in Modern Mathematics III, T. L. Saaty (Ed.), J. Wiley and Sons, 1965, pp. 95–195.

100. G. Kreisel, *A notion of mechanistic theory*, Int. J. Theor. Phys. **29** (1974), 11–26.

101. K. Kelly and O. Schulte, *The computable testability of theories with uncomputable predictions* Erkenntnis, **43** (1995), 29–66.

102. K. Kelly, O. Schulte, and C. Juhl, *Learning theory and philosophy of science*, Philos. Sci. **64** (1997), 245–267.

103. M. Kubat, *A machine learning based approach to load balancing in computer networks*, Cybernet. Syst. **23** (1992), 389–400.

104. R. Ladner, *On the structure of polynomial time reducibility*, J. Assoc. Comput. Mach. **22** (1975), 155–171.

105. S. Lange and P. Watson, *Machine discovery in the presence of incomplete or ambiguous data*, In: Algorithmic Learning Theory, K. Jantke and S. Arikawa (Eds.), Lect. Notes Artif. Intell. **872** , Springer, 1994, pp. 438–452.

106. Thinking Machines. Introduction to data level parallelism. Technical Report 86.14, Thinking Machines, April 1986.

107. N. Margolus, *Physics–like models of computation*, Physica 10D, (1984), 81–95.

108. Y. Marcoux, *Composition is almost (but not quite) as good as s-1-1*, Theor. Comput. Sci. **120** (1993), no. 2, 169–195.

109. D. Moore and J. Case, *The complexity of total order structures*, J. Comput. Syst. Sci. **17** (1978), 253–269.

110. D. McDermott, *Robot planning*, AI Magazine, **13** (1992), no. 2, 55–79.

111. T. Mitchell, R. Caruana, D. Freitag, J. McDermott, and D. Zabowski, *Experience with a learning, personal assistant*, Commun. ACM **37** (1994), no. 7, 81–91.

112. K. Mehlhorn, *Polynomial and abstract subrecursive classes*, J. Comput. Syst. Sci. **12** (1976), 147–178.

113. E. Mendelson, *Introduction to Mathematical Logic*. Chapman and Hall, London, 1997.

114. A. Meyer and P. Fischer, *Computational speed-up by effective operators*, J. Symb. Log. **37** (1972), 48–68.

115. M. Minsky, *Cellular vacuum*, Int. J. Theor. Phys. **21** (1982), no. 6/8, 537–551.

116. S. Matwin and M. Kubat, *The role of context in concept learning*, In: Proceedings of the ICML-96 Pre-Conference Workshop on Learning in Context-Sensitive Domains, Bari, Italy, 1996, M. Kubat and G. Widmer (Eds.), pp. 1–5.

117. E. McCreight and A. Meyer, *Classes of computable functions defined by bounds on computation*, In: Proceedings of the First Annual ACM Symposium on Theory of Computing, 1969, pp. 79–88.

118. O. Maler, A. Pnueli, and J. Sifakis, *On the synthesis of discrete controllers for timed systems*, In: STACS 95: 12th Annual Symposium on Theoretical Aspects of Computer Science Munich, Germany, March 2-4, 1995 Proceedings, E. W. Mayr and C. Puech (Eds.), Lect. Notes Comput. Sci. **900**, Springer, 1995, pp. 229–242.

119. G. Marcus, S. Pinker, M. Ullman, M. Hollander, T. J. Rosen, and F. Xu, *Overregularization in Language Acquisition*, Univ. Chicago Press, 1992. [Includes commentary by H. Clahsen]

120. D. Michie and C. Sammut, *Machine learning from real-time input-output behavior*, In: Proceedings of the International Conference on Design to Manufacture in Modern Industry, 1993, pp. 363–369.

121. J. Myhill, *Some philosophical implications of mathematical logic: I. three classes of ideas*, Rev. Metaphysics **6** (1952), no. 2.

122. J. Myhill, *A note on the degrees of partial functions*, Proc. Am. Math. Soc. **12** (1961), 519–521.

123. J. Myhill, *Abstract theory of self-reproduction*, In: Views on General Systems Theory, M. D. Mesarović (Ed.), J. Wiley and Sons, 1964, pp. 106–118.

124. J. Von Neumann, *Theory of Self–Reproducing Automata*, Univ. Illinois Press, 1966. [Edited and completed by A. W. Burks]

125. *Report* of the assessment panel for the international assessment of the U.S. math sciences, Technical Report NSF9895, National Science Foundation, March 1998. [http://www.nsf.gov/publications/pub_summ.jsp?ods_key=nsf9895]

126. P. Odifreddi, *Classical Recursion Theory*, North-Holland, 1989.

127. P. Odifreddi, *Classical Recursion Theory. Vol. II*, Elsivier, 1999.

128. D. Osherson, M. Stob, and S. Weinstein, *Ideal learning machines*, Cognitive Sci. **6** (1982), 277–290.

129. D. Osherson, M. Stob, and S. Weinstein, *Note on a central lemma of learning theory*, J. Math. Psychol. **27** (1983), 86–92.

130. D. Osherson, M. Stob, and S. Weinstein, *Learning theory and natural language*, Cognition **17** (1984), no. 1, 1–28.

131. D. Osherson, M. Stob, and S. Weinstein, *Systems that Learn: An Introduction to Learning Theory for Cognitive and Computer Scientists*, The MIT Press, 1986.

132. D. Osherson and S. Weinstein, *Criteria of language learning*, Inf. Control **52** (1982), 123–138.

133. M. Pour-El and M. B. Richards, *A computable ordinary differential equation which possesses no computable solution* Ann. Math. Logic **17** (1979), 61–90.

134. M. Pour-El and M. B. Richards, *The wave equation with computable initial data such that its unique solution is not computable*, Adv. Math. **39** (1981), 215–239.

135. M. Pour-El and M. B. Richards, *Computability in Analysis and Physics*, Springer, 1989.

136. S. Pinker, *Formal models of language learning*, Cognition **7** (1979), no. 3, 217–283.

137. L. Pitt, *A Characterization of Probabilistic Inference*, PhD Thesis, Yale University, 1984.

138. L. Pitt, *Probabilistic inductive inference*, J. Assoc. Comput. Mach. **36** (1989), 383–433.

139. K. Plunkett and V. Marchman, *U-shaped learning and frequency effects in a multi-layered perceptron: Implications for child language acquisition*, Cognition **38** (1991), no. 1, 43–102.

140. L. Pratt, J. Mostow, and C. Kamm, *Direct transfer of learned information among neural networks*, In: Proceedings of the 9th National Conference on Artificial Intelligence (AAAI-91), 1991.

141. H. Putnam, *Probability and confirmation*, In: Voice of America, Forum on Philosophy of Science, Vol. 10, 1963. [Reprinted as [**142**]]

142. H. Putnam, *Probability and confirmation*, In: *Mathematics, Matter, and Method*, Cambridge Univ. Press, 1975.

143. J. Royer and J. Case, *Subrecursive Programming Systems: Complexity and Succinctness*, Birkhäuser, 1994.

144. G. Riccardi, *The Independence of Control Structures in Abstract Programming Systems*, PhD Thesis, SUNY Buffalo, 1980.

145. G. Riccardi, *The independence of control structures in abstract programming systems*, J. Comput. Syst. Sci. **22** (1981), 107–143.

146. G. Riccardi, *The independence of control structures in programmable numberings of the partial recursive functions*, Z. Math. Logik Grundlagen Math. **48** (1982), 285–296.

147. H. Rogers, *Gödel numberings of partial recursive functions*, J. Symb. Log. **23** (1958), 331–341.

148. H. Rogers, *Theory of Recursive Functions and Effective Computability*, McGraw Hill, 1967. [Reprinted: The MIT Press, 1987]

149. J. Royer, *A Connotational Theory of Program Structure*, Lect. Notes Comput. Sci. **273**, Springer, 1987.

150. J. Royer, *Semantics versus syntax versus computations: Machine models for type-2 polynomial-time bounded functionals*, J. Comput. Syst. Sci. **54** (1997), 424–436.

151. C. Sammut, *Acquiring expert knowledge by learning from recorded behaviors*, In: Japanese Knowledge Acquisition Workshop, 1992.

152. C. Sammut, *Automatic construction of reactive control systems using symbolic machine learning*, Knowledge Engineering Rev. **11** (1996), no. 1, 27–42.

153. O. Schulte, *Means-ends epistemology*, Br. J. Philos. Sci. **50** (1999), 1–31.

154. O. Schulte, *Inferring conservation principles in particle physics: A case study in the problem of induction*, Br. J. Philos. Sci. **51** (2000), 771–806.

155. J. Searle, *Minds, brains, and programs*, Behav. Brain. Sci. **3** (91980), 417–424.

156. A. Seth, *Complexity Theory of Higher Type Functionals*, PhD Thesis, University of Bombay, 1994.

157. N. Shapiro, *Review of "Limiting recursion" by E.M. Gold and "Trial and error predicates and the solution to a problem of Mostowski" by H.Putnam*, J. Symb. Log. **36** (1971), 342.

158. C. Sammut, S. Hurst, D. Kedzier, and D. Michie. *Learning to fly*, In: Proceedings of the Ninth International Conference on Machine Learning, D. Sleeman and P. Edwards (Eds.), Morgan Kaufmann, 1992, pp. 385–393.

159. T. Slaman, *Long range goals*, COMP-THY Archives, #13, April 1998. [http://listserv.nd.edu/archives/comp-thy.html]

160. C. Smith, *A Recursive Introduction to the Theory of Computation*, Springer, 1994.

161. T. J. Sejnowski and Ch. Rosenberg, *NETtalk: A parallel network that learns to read aloud*, Technical Report JHU-EECS-86-01, Johns Hopkins University, 1986.

162. S. Strauss and R. Stavy (Eds.), *U-Shaped Behavioral Growth*, Academic Press, 1982.

163. J. Stoy, *Denotational Semantics: The Scott-Strachey Approach to Programming Language Theory*, The MIT Press, 1977.

164. K. Svozil, *Are quantum fields cellular automata?* Physics Letters A, **119** (1986), no. 4, 153–156.

165. J. B. Salem and S. Wolfram, *Thermodynamics and hydrodynamics with cellular automata*, In: Theory and Applications of Cellular Automata, S. Wolfram (Ed.), World Scientific, 1986.

166. N. A. Taatgen and J. R. Anderson, *Why do children learn to say "Broke"? A model of learning the past tense without feedback*, Cognition, **86** (2002), no. 2, 123–155.

167. F. Tsung and G. Cottrell, *A sequential adder using recurrent networks*, In: IJCNN-89-WASHINGTON D.C.: International Joint Conference on Neural Networks. Vol. 2, IEEE Service Center, Piscataway, New Jersey, June 18–22, 1989, pp. 133–139.

168. W. Thomas, *On the synthesis of strategies in infinite games*, In: STACS 95: 12th Annual Symposium on Theoretical Aspects of Computer Science Munich, Germany, March 2-4, 1995 Proceedings, E. W. Mayr and C. Puech (Eds.), Lect. Notes Comput. Sci. **900**, Springer, 1995, pp. 1–13.

169. S. Thrun, *Is learning the n-th thing any easier than learning the first*, In: Advances in Neural Information Processing Systems, 8, Morgan Kaufmann, 1996.

170. T. Toffoli and N. Margolus, *Cellular Automata Machines*, The MIT Press, 1987.

171. T. Toffoli, *Cellular automata machines*, Technical Report 208, Comp. Comm. Sci. Dept., University of Michigan, 1977.

172. T. Toffoli, *Computation and construction universality of reversible cellular automata*, J. Comput. Syst. Sci. **15** (1997), 213–231.

173. T. Toffoli, *CAM: A high–performance cellular–automaton machine*, Physica 10D, (1984), 195–204.

174. S. Thrun and J. Sullivan, *Discovering structure in multiple learning tasks: The TC algorithm*, In: Proceedings of the Thirteenth International Conference on Machine Learning (ICML-96), Morgan Kaufmann, 1996, pp. 489–497.

175. T. Urbančič and I. Bratko, *Reconstructing human skill with machine learning*, In: Proceedings of the Eleventh European Conference on Artificial Intelligence, A. Cohn (Ed.), John Wiley and Sons, 1994.

176. G. Y. Vichniac, *Simulating physics with cellular automata*, Physica 10D, (1984), 96–116.

177. D. Šuc, *Machine reconstruction of human control strategies*, In: Frontiers in Artificial Intelligence and Applications. Vol. 9, IOS Press, 2003.

178. A. Waibel *Connectionist glue: Modular design of neural speech systems*, In: Proceedings of the 1988 Connectionist Models Summer School, D. Touretzky, G. Hinton, and T. Sejnowski (Eds.), Morgan Kaufmann, 1989. pp. 417–425.

179. A. Waibel, *Consonant recognition by modular construction of large phonemic time-delay neural networks*, In: Advances in Neural Information Processing Systems I, D. S. Touretzky (Ed.), Morgan Kaufmann, 1989, pp. 215–223.

180. K. Weihrauch and N. Zhong, *Is wave propagation computable or can wave computers beat the Turing machine*, Proc. London Math. Soc. **85** (2002), 312–332.

181. K. Wexler and P. Culicover, *Formal Principles of Language Acquisition*, The MIT Press, 1980.

182. K. Wexler, *On extensional learnability*, Cognition, **11** (1982), no. 1, 89–95.

183. R. Wiehagen, *Limes-Erkennung rekursiver Funktionen durch spezielle Strategien*, Electron. Inform.-verarb. Kybernetik **12** (1976), 93–99.

184. R. Wiehagen, *Zur Theorie der Algorithmischen Erkennung*, PhD Thesis, Humboldt University of Berlin, 1978.

185. S. Wolfram, *Statistical mechanics of cellular automata*, Rev. Modern Phys. **55** (1983), no. 33, 601–644.

186. S. Wrobel, *Concept Formation and Knowledge Revision*, Kluwer, 1994.

187. P. Young, *Easy constructions in complexity theory: Gap and speedup theorems*, Proc. Am. Math. Soc. **37** (1973), 555–563.

# Computability and Computable Models

## Sergei S. Goncharov [†]

*Sobolev Institute of Mathematics SB RAS*
*Novosibirsk State University*

*Novosibirsk, Russia*

The intuitive notion of computability was formalized in the XXth century, which strongly affected the development of mathematics and applications, new computational technologies, various aspects of the theory of knowledge, etc. A rigorous mathematical definition of computability and algorithm generated new approaches to understanding a solution to a problem and new mathematical disciplines such as computer science, algorithmical complexity, linear programming, computational modeling and simulation databases and search algorithms, automatical cognition, program languages and semantics, net security, coding theory, cryptography in open systems, hybrid control systems, information systems, etc.

[†] The author was supported by the Russian Foundation for Basic Research (grant no. 05-01-00819) and the President grant of Scientific Schools (grant no. 4413.2006.01).

**Mathematical Problems from Applied Logic. Logics for the XXIst Century. II.** Edited by Dov M. Gabbay *et al.* / International Mathematical Series, 5, Springer, 2007

Computability theory is traditionally understood as a branch of mathematical logic. However, owing to the ubiquitous use of computers and other electronic devices, many aspects usually studied within the framework of computability theory have become actual in numerous various areas even very far from mathematics.

In view of the wide range of applications, the two following directions of the further development of computability theory are of great interest.

- Investigate and determine bounds for the applicability of given computable model and algorithm to an real object existing in reality and processes flowing there.

- Create computability theory over abstract structures which could provides a unique approach to both computational processes in continuous models in reality and their discrete analogs.

In this paper, we discuss the first direction. We review recent important results and formulate more than 30 actual problems and open questions dictated by applications of the theory of computable models.

## 1. Preliminaries

The theory of constructive and computable models dates back to the works of Fröhlich and Shepherdson [44], Mal'tsev [101], Rabin [137], and Vaught [152] in the 1950's. This theory studies algorithmic properties of abstract models by constructing representations of the models on natural numbers and clarify relationships between properties and structural properties of the models. A systematic study of constructive and computable algebraic systems was initiated by Mal'tsev [100]. Historically, there were two approaches to the study of computable models.

The first approach is based on the notion of a numbering of the basic set of a model by natural numbers. Model properties are expressed in the binary form of natural numbers and thereby can be handled with computer technologies. In general, instead of numbers, names in some finite alphabet are ascribed to elements of a model. Since an element can possess several names, recognition algorithms are required. Such algorithms must recognize the names of given elements and determine whether certain properties are realized on elements with given names.

The second approach deals with models whose basic sets consist of natural numbers [**101**]. This approach leads to the notion of a recursive (computable) model.

Due to R. Soare and his critical revision of the terminology used in computability theory, the term "computable model" becomes common last years. Indeed, this choice reflects our intuitive impression of computability.

Both approaches were developed simultaneously and are closely connected. In fact, they are equivalent from the mathematical point of view.

In this section, we recall basic facts in model theory, numberings, computability theory which are necessary for discussing current problems in the theory of computable models. The material of this section mainly follows [**40**].

Throughout the paper, we use the standard set-theoretic notation: $P(M)$ is the set of all subsets of a set $M$, $\mathrm{id}_M$ is the identity mapping of a set $M$, and $\omega = \{0, 1, \ldots\}$. If $f$ is a (partial) mapping, then $\mathrm{Rang}\, f$ ($\mathrm{Dom}\, f$) denotes the range (domain) of $f$. The symbols "$\Rightarrow$" and "$\Leftrightarrow$" mean the expressions "if ..., then ..." and "... if and only if ...". The expression $a \leftrightharpoons b$ means $b$ is denoted by $a$. For category theory we refer to [**12**] and [**24**].

## 1.1. Algebraic structures, models, and theories

The classical theory of models and algebraic systems, founded by Mal'tsev and Tarski, was one of the main directions in mathematical logic where the key results were obtained during the second half of the XXth century.

A *signature* $\sigma$ of the language of the first-order predicate calculus is the pair consisting of the triple of disjoint sets $\sigma^P$, $\sigma^F$, $\sigma^C$ and a mapping $\mu \colon \sigma^P \cup \sigma^F \to \omega^+$, where $\omega^+ \leftrightharpoons \{1, 2, \ldots\}$. If $P \in \sigma^P$ and $\mu(P) = n$, then $P$ is called an *n-ary predicate symbol*. If $f \in \sigma^F$ and $\mu(f) = m$, then $f$ is called an *m-ary functional symbol*. Elements of $\sigma^C$ are called *constant symbols*. We often write $\sigma$ in the form $\sigma = \langle P_1^{n_1}, \ldots, P_k^{n_k}; f_1^{m_1}, \ldots, f_s^{m_s}; c_1, \ldots, c_t \rangle$, where the superscripts are the values of $\mu$ for the corresponding symbols. The expression $P \in \sigma$ ($f \in \sigma$ or $c \in \sigma$) means that $P$ ($f$ or $c$) is a predicate (functional or constant) symbol of the signature $\sigma$.

The set of all formulas of the language of the first-order predicate calculus of a signature $\sigma$ is denoted by $L_\sigma$ (cf. definitions in [16] and [24]). We write $\Phi(x_1, \ldots, x_n)$ if every free variable of a formula $\Phi$ belongs to the set $\{x_1, \ldots, x_n\}$.

An *algebraic structure* (or *model*) $\mathfrak{A}$ of a signature $\sigma$ is the pair consisting of a nonempty set $|\mathfrak{A}|$, called the *basic set* of $\mathfrak{A}$, and a family of (*basis*) predicates $P^{\mathfrak{A}} \subseteq |\mathfrak{A}|^{\mu(P)}$ ($P \in \sigma^P$), operations $f^{\mathfrak{A}} \colon |\mathfrak{A}|^{\mu(f)} \to |\mathfrak{A}|$ ($f \in \sigma^F$), and constants $c^{\mathfrak{A}} \in |\mathfrak{A}|$ ($c \in \sigma^C$).

For a formula $\Phi(x_1, \ldots, x_n)$ of a signature $\sigma$ and an algebraic structure $\mathfrak{A}$ of the same signature $\sigma$ we introduce the notion of the *truth* of $\Phi$ in $\mathfrak{A}$ for $x_i \to a_i \in |\mathfrak{A}|$, $i = 1, \ldots, n$. We write $\mathfrak{A} \vDash \Phi(a_1, \ldots, a_n)$ if $\Phi$ is true in $\mathfrak{A}$ on $a_1, \ldots, a_n$. If $T$ is a system of *sentences* (i.e., formulas without free variables), then $\mathfrak{A} \vDash T$ means $\mathfrak{A} \vDash \Phi$ for all $\Phi \in T$.

A set $T$ of sentences of a signature $\sigma$ is called a *theory* if for any sentence $\Phi$ and model $\mathfrak{A}$ of the signature $\sigma$ from $\mathfrak{A} \vDash T \Rightarrow \mathfrak{A} \vDash \Phi$ it follows that $\Phi \in T$. Using the notion of the deducibility $\vdash$ in the first-order predicate calculus (cf. [24]), we can define a theory $T$ as follows: $T \vdash \Phi \Rightarrow \Phi \in T$. It is clear that for every class $K$ of algebraic structures of a signature $\sigma$ the set of all sentences $\Phi$ such that $\mathfrak{A} \in K \Rightarrow \mathfrak{A} \vDash \Phi$ is a theory, called the *elementary theory* of $K$ and denoted by $\mathrm{Th}(K)$.

A subset $A$ of a theory $T$ is called a *system of axioms* of $T$ and is denoted by $T = [A]$ if $\mathfrak{A} \vDash A$ implies $\mathfrak{A} \vDash T$ for any algebraic structure $\mathfrak{A}$ or, which is the same, $T = \{\Phi \mid \Phi$ is a sentence of the signature $\sigma$ and $A \vdash \Phi\}$.

A theory $T$ is *consistent* if it differs from the set of all sentences. A theory $T$ is *complete* if it is consistent and $\Phi \in T$ or $\neg\Phi \in T$ for any sentence $\Phi$.

Consider an algebraic structure $\mathfrak{A}$ of a signature $\sigma$. Let a nonempty subset $B \subseteq |\mathfrak{A}|$ be closed with respect to the basic operations and constants, i.e., $f^{\mathfrak{A}}(a_1, \ldots, a_m) \in B$ for any $a_1, \ldots, a_m \in B$, $f^m \in \sigma^F$, and $c^{\mathfrak{A}} \in B$ for any $c \in \sigma^C$. On $B$, we introduce an algebraic structure of the signature $\sigma$ and denote it by $\mathfrak{A} \restriction B$. If $\mathfrak{A}_0$ and $\mathfrak{A}_1$ are algebraic structures of the signature $\sigma$, $|\mathfrak{A}_0| \subseteq |\mathfrak{A}_1|$, and $\mathfrak{A}_0 = \mathfrak{A}_1 \restriction |\mathfrak{A}_0|$, then $\mathfrak{A}_0$ is called a *substructure* of $\mathfrak{A}_1$ and is denoted by $\mathfrak{A}_0 \leqslant \mathfrak{A}_1$.

A substructure $\mathfrak{A}_0$ of an algebraic structure $\mathfrak{A}_1$ of a signature $\sigma$ is said to be *elementary* and is denoted by $\mathfrak{A}_0 \preccurlyeq \mathfrak{A}_1$ if $\mathfrak{A}_0 \models \Phi(a_1, \ldots, a_n) \Leftrightarrow \mathfrak{A}_1 \models \Phi(a_1, \ldots, a_n)$ for any formula $\Phi(x_1, \ldots, x_n)$ of the signature $\sigma$ and $a_1, \ldots, a_n \in |\mathfrak{A}_0|$.

Let $\mathfrak{A}$ and $\mathfrak{B}$ be algebraic structures of a signature $\sigma$. A mapping $\varphi \colon |\mathfrak{A}| \to |\mathfrak{B}|$ is called a *homomorphism* from $\mathfrak{A}$ into $\mathfrak{B}$ and is denoted by $\varphi \colon \mathfrak{A} \to \mathfrak{B}$ if

- $\langle a_1, \ldots, a_n \rangle \in P^{\mathfrak{A}} \Rightarrow \langle \varphi a_1, \ldots, \varphi a_n \rangle \in P^{\mathfrak{B}}$ for any predicate symbol $P^n \in \sigma$ and $a_1, \ldots, a_n \in |\mathfrak{A}|$,
- $\varphi f^{\mathfrak{A}}(a_1, \ldots, a_m) = f^{\mathfrak{B}}(\varphi a_1, \ldots, \varphi a_m)$ for any functional symbol $f^m \in \sigma$ and $a_1, \ldots, a_m \in |\mathfrak{A}|$,
- $\varphi(c^{\mathfrak{A}}) = c^{\mathfrak{B}}$ for any constant symbol $c \in \sigma$.

An equivalence relation $\eta$ on the basic set $|\mathfrak{A}|$ of an algebraic structure $\mathfrak{A}$ of a signature $\sigma$ is called a *congruence* on $\mathfrak{A}$ if $\langle f^{\mathfrak{A}}(a_1, \ldots, a_m), f^{\mathfrak{A}}(b_1, \ldots, b_m) \rangle \in \eta$ for every functional symbol $f^m \in \sigma$ and $\langle a_1, b_1 \rangle, \ldots, \langle a_m, b_m \rangle \in \eta$. A congruence $\eta$ on $\mathfrak{A}$ is said to be *strict* if $\langle a_1, \ldots, a_n \rangle \in P^{\mathfrak{A}} \Leftrightarrow \langle b_1, \ldots, b_n \rangle \in P^{\mathfrak{A}}$ for any predicate symbol $P^n \in \sigma$ and $\langle a_1, b_1 \rangle, \ldots, \langle a_n, b_n \rangle \in \eta$.

If $\eta$ is a congruence on an algebraic structure $\mathfrak{A}$ of a signature $\sigma$, then, on the set $A^* \rightleftharpoons |\mathfrak{A}|/\eta$, we can introduce an algebraic structure $\mathfrak{A}^*$, called a *quotient structure* and denoted by $\mathfrak{A}/\eta$, of the signature $\sigma$ as follows:

- if $P^n \in \sigma$, then $P^{\mathfrak{A}^*} \leftrightharpoons \{\langle [a_1]_\eta, \ldots, [a_n]_\eta \rangle \mid \text{there exist } b_i \in [a_i]_\eta,\ i = 1, \ldots, n,\ \text{such that } \langle b_1, \ldots, b_n \rangle \in P^{\mathfrak{A}} \}$,

- if $f^m \in \sigma$ and $[a_1]_\eta, \ldots, [a_m]_\eta \in A^*$, then $f^{\mathfrak{A}^*}([a_1]_\eta, \ldots, [a_m]_\eta) \leftrightharpoons [f^{\mathfrak{A}}(a_1, \ldots, a_m)]_\eta$,

- if $c$ is a constant symbol of $\sigma$, then $c^{\mathfrak{A}^*} \leftrightharpoons [c^{\mathfrak{A}}]_\eta$.

Here, $[a]_\eta$ denotes the set of all elements that are $\eta$-equivalent to $a$. The mapping $a \mapsto [a]_\eta$, $a \in |\mathfrak{A}|$, is a homomorphism. If $\varphi \colon \mathfrak{A} \to \mathfrak{B}$ is a homomorphism from $\mathfrak{A}$ into $\mathfrak{B}$, then $\eta_\varphi \leftrightharpoons \{\langle a, b \rangle \mid a, b \in |\mathfrak{A}|,\ \varphi a = \varphi b\}$ is a congruence relation on $\mathfrak{A}$.

If a homomorphism $\varphi \colon \mathfrak{A} \to \mathfrak{B}$ is a one-to-one mapping from $|\mathfrak{A}|$ onto $|\mathfrak{B}|$ and the inverse mapping $\varphi^{-1}$ is a homomorphism from $\mathfrak{B}$ into $\mathfrak{A}$, then $\varphi$ is called an *isomorphism* (from $\mathfrak{A}$ into $\mathfrak{B}$). Two algebraic structures $\mathfrak{A}$ and $\mathfrak{B}$ are said to be *isomorphic* ($\mathfrak{A} \simeq \mathfrak{B}$) if there exists an isomorphism $\varphi \colon \mathfrak{A} \to \mathfrak{B}$. If $\mathfrak{A} \leqslant \mathfrak{B}_0$, $\mathfrak{A} \leqslant \mathfrak{B}_1$, and $\varphi \colon \mathfrak{B}_0 \to \mathfrak{B}_1$ is an isomorphism such that $\varphi \restriction |\mathfrak{A}| = \mathrm{id}_{|\mathfrak{A}|}$, then $\varphi$ is called an $\mathfrak{A}$-*isomorphism*.

For signatures $\sigma$ and $\sigma'$ we write $\sigma \subseteq \sigma'$ if every functional (predicate, constant) symbol of $\sigma$ is a functional (predicate, constant) symbol of $\sigma'$ with the same arity. If $\sigma \subseteq \sigma'$ and $\mathfrak{A}'$ is an algebraic structure of the signature $\sigma'$, then we can construct an algebraic structure of the signature $\sigma$ by "forgetting" the values of symbols of $\sigma' \setminus \sigma$. This structure, denoted by $\mathfrak{A}' \restriction \sigma$, is called the $\sigma$-*restriction* of $\mathfrak{A}'$, and $\mathfrak{A}'$ is called the $\sigma'$-*enrichment* of $\mathfrak{A}' \restriction \sigma$. We write $\mathfrak{A} \leqslant \mathfrak{A}'$ if $\mathfrak{A} \leqslant \mathfrak{A}' \restriction \sigma$.

Let $\mathfrak{A}$ be an algebraic structure of a signature $\sigma$. We extend $\sigma$ by adding constant symbols $\langle c_a \mid a \in |\mathfrak{A}| \rangle$. We set $\sigma^* \leftrightharpoons \sigma \cup \langle c_a \mid a \in |\mathfrak{A}| \rangle$. Setting $c_a^{\mathfrak{A}^*} \leftrightharpoons a$, we obtain the natural $\sigma^*$-enrichment $\mathfrak{A}^*$ of $\mathfrak{A}$. A *diagram* $D(\mathfrak{A})$ of $\mathfrak{A}$ is a set of sentences of the signature $\sigma^*$ such that every sentence in $D(\mathfrak{A})$ is an atomic formula or the negation of an atomic formula and is true in $\mathfrak{A}^*$. By a *complete diagram* $FD(\mathfrak{A})$ we mean the set of sentences of the signature $\sigma^*$ that are true in $\mathfrak{A}^*$.

Two algebraic structures $\mathfrak{A}_0$ and $\mathfrak{A}_1$ of a signature $\sigma$ are *elementarily equivalent* ($\mathfrak{A}_0 \equiv \mathfrak{A}_1$) if $\mathrm{Th}(\mathfrak{A}_0)$ ( $\leftrightharpoons \mathrm{Th}(\{\mathfrak{A}_0\})$) $= \mathrm{Th}(\mathfrak{A}_1)$ ( $\leftrightharpoons \mathrm{Th}(\{\mathfrak{A}_1\})$) or, in other words, $\mathfrak{A}_0 \vDash \Phi$ if and only if $\mathfrak{A}_1 \vDash \Phi$ for any sentence $\Phi$ of the signature $\sigma$.

If $\sigma' \subseteq \sigma$ and $\mathfrak{A}_0 \equiv \mathfrak{A}_1$, then $\mathfrak{A}_0 \restriction \sigma' \equiv \mathfrak{A}_1 \restriction \sigma'$.

We describe a canonical approach to the study of models without functional symbols. For every $m$-ary functional symbol $f \in \sigma$ we introduce a new $(m+1)$-ary predicate symbol $P_f$. Let $\sigma^*$ be obtained from $\sigma$ by replacing every functional symbol $f$ with a predicate symbol $P_f$ ($\sigma^{*P} \leftrightharpoons \sigma^P \cup \langle P_f \mid f \in \sigma^F \rangle$, $\sigma^{*F} \leftrightharpoons \varnothing$, $\sigma^{*C} \leftrightharpoons \sigma^C$, $\mu^* \restriction \sigma^P \leftrightharpoons \mu \restriction \sigma^P$, $\mu^*(P_f) \leftrightharpoons \mu(f) + 1$). Any algebraic structure $\mathfrak{A}$ of the signature $\sigma$ can be "transformed" to a model $\mathfrak{A}^*$ of the signature $\sigma^*$ by setting

$$P_f^{\mathfrak{A}^*} \leftrightharpoons \left\{ \langle a_1, \ldots, a_m, b \rangle \mid a_1, \ldots, a_m, b \in |\mathfrak{A}|,\ f^{\mathfrak{A}}(a_1, \ldots, a_m) = b \right\}$$

for $f^m \in \sigma^F$. It is obvious that $\mathfrak{A}_0 \equiv \mathfrak{A}_1$ if and only if $\mathfrak{A}_0^* \equiv \mathfrak{A}_1^*$.

Therefore, in order to obtain a criterion for the elementary equivalence of two algebraic structures, it suffices to find such a criterion in the case of a finite signature.

We recall some model-theoretic methods of proving the completeness of theories.

**Proposition 1.1.** *A consistent theory $T$ is complete if and only if there exists a model $\mathfrak{M}$ such that $T = \mathrm{Th}(\mathfrak{M})$.*

**Corollary 1.2.** *A consistent theory $T$ is complete if and only if any models $\mathfrak{M}_0$ and $\mathfrak{M}_1$ of $T$ are elementarily equivalent, i.e., $\mathrm{Th}(\mathfrak{M}_0) = \mathrm{Th}(\mathfrak{M}_1)$.*

A theory $T$ is *categorical in power $\alpha$* if any two models of $T$ of power $\alpha$ are isomorphic.

The following assertion is often used in the proof of completeness and decidability.

**Proposition 1.3.** *If a theory $T$ has no finite models and is categorical in some infinite power, then $T$ is complete.*

A theory $T$ is said to be *model-complete* if for every model $\mathfrak{M}$ of $T$ the theory of the signature $\sigma^* = \sigma \cup \langle c_a \mid a \in |\mathfrak{M}| \rangle$ defined by the system of axioms $T \cup D(\mathfrak{M})$ is complete.

We indicate properties equivalent to the model completeness.

**Theorem 1.4.** *Let $T$ be a theory. The following assertions are equivalent.*

(1) *The theory $T$ is model complete.*

(2) *Let $\mathcal{M}$ and $\mathcal{N}$ be models of $T$. If $\mathcal{M}$ is a submodel of $\mathcal{N}$, then $\mathcal{M}$ is an elementary submodel of $\mathcal{N}$.*

(3) *Let $\mathcal{M}$ and $\mathcal{N}$ be models of $T$ with fixed infinite cardinality $\varkappa$. If $\mathcal{M}$ is a submodel of $\mathcal{N}$, then $\mathcal{M}$ is an elementary submodel of $\mathcal{N}$.*

(4) *For any formula $\varphi(\overline{x})$ there exists $\exists$-formula $\psi(\overline{x})$ such that $T \vdash \varphi(\overline{x}) \Leftrightarrow \psi(\overline{x})$.*

Note that a complete theory is not necessarily model-complete and, conversely, a model-complete theory is not necessarily complete. However, there is a canonical method of obtaining a model-complete theory from an arbitrary theory. To demonstrate it, we need the definition of a first-order definable enrichment of a theory of a signature $\sigma$ for a family of formulas. Let $\Phi(x_1, \ldots, x_n)$ be a formula of the signature $\sigma$, and let $\sigma'$ be obtained from $\sigma$ by adding an $n$-ary predicate symbol $P_\Phi$. By a *first-order definable enrichment* of a theory $T$ of a signature $\sigma$ for the formula $\Phi(x_1, \ldots, x_n)$ we mean the theory $T'$ of the signature $\sigma'$ defined by the following system of axioms:

$$T \cup \big\{ \forall x_1 \ldots x_N \big( P_\Phi(x_1, \ldots, x_N) \longleftrightarrow \Phi(x_1, \ldots, x_n) \big) \big\}.$$

A first-order definable enrichment of a theory for a family of formulas $\Phi$ is defined in a similar way. A first-order definable enrichment $T'$ of $T$ is *complete* if it is obtained by adding new predicate symbols for all formulas of the signature $\sigma$. If $T'$ is a first-order definable enrichment of $T$, then $T$ and $T'$ have the same models in the following sense: a model $\mathfrak{M}$ of $T$ admits a unique $\sigma'$-enrichment to a model of $T'$.

**Theorem 1.5.** *The complete first-order definable enrichment of a theory $T$ is a model-complete theory.*

Models $\mathfrak{M}_0$ and $\mathfrak{M}_1$ of a signature $\sigma$ are *universally equivalent* if $\mathfrak{M}_0 \vDash \Phi \Leftrightarrow \mathfrak{M}_1 \vDash \Phi$ for any universal sentence $\Phi$ of the signature $\sigma$.

**Proposition 1.6.** *If any two models of a model-complete theory $T$ are universally equivalent, then $T$ is complete.*

Owing to these assertions, it is reasonable to introduce the following definition. Let $T$ be a consistent theory of a signature $\sigma$. A theory $T^* \supseteq T$ of the signature $\sigma$ is called the *model completion* of $T$ if $T^*$ is a model-complete theory relative to $T$.

The existence of the model completion of an arbitrary theory is not a trivial question. The following condition is sufficient for existing the model completion of a universally axiomatizable theory $T$ of a finite signature. If $\mathfrak{M}$, $\mathfrak{M}_0$, and $\mathfrak{M}_1$ are models of $T$ and $\varphi_0 \colon \mathfrak{M} \to \mathfrak{M}_0$, $\varphi_1 \colon \mathfrak{M} \to \mathfrak{M}_1$ are isomorphic embeddings, then there exists a model $\mathfrak{M}^*$ of $T$ and isomorphic embeddings $\psi_0 \colon \mathfrak{M}_0 \to \mathfrak{M}^*$, $\psi_1 \colon \mathfrak{M}_1 \to \mathfrak{M}^*$ such that $\psi_0 \varphi_0 = \psi_1 \varphi_1$. In this case, the model completion $T^*$ of $T$ exists and is a complete countably categorical theory admitting the quantifier elimination.

Consider types of models. Let $T$ be a (consistent) theory of a signature $\sigma$. Denote by $Fr_n$ the set of all formulas of the signature $\sigma$ with free variables in $\{x_0, \ldots, x_{n-1}\}$, $n \in \omega$. Let $F_n(T)$ be the quotient set of $Fr_n$ by the equivalence relation $\eta_T$ defined as follows: $\langle \varphi, \psi \rangle \in \eta_T \leftrightharpoons \forall x_0 \ldots \forall x_{n-1} \left[ (\varphi \to \psi) \, \& \, (\psi \to \varphi) \right] \in T$ for $\varphi, \psi \in Fr_n$. For $\varphi \in Fr_n$ denote by $[\varphi]$ the element of $F_n(T)$ containing $\varphi$. The following natural embeddings hold:

$$S_0(T) \subseteq S_1(T) \subseteq \ldots, \quad F_0(T) \subseteq F_1(T) \subseteq \ldots$$

REMARK. The set $F_n(T)$ can be regarded as a Boolean algebra (cf. [**40**]) if $[\varphi] \sqcup [\psi] \leftrightharpoons [\varphi \vee \psi]$, $[\varphi] \sqcap [\psi] \leftrightharpoons [\varphi \, \& \, \psi]$, $c[\varphi] \leftrightharpoons [\neg \varphi]$, $0 \leftrightharpoons [\forall x \, (x \neq x)]$, and $1 \leftrightharpoons [\exists x \, (x = x)]$.

By an $n$-*type* of a theory $T$ we mean any maximal $T$-inconsistent subset $S \subseteq Fr_n$ (i.e., the sentence $\exists x_0 \ldots \exists x_{n-1} \left( \overset{k}{\underset{i=1}{\&}} \varphi_i \right)$ belongs to $T$ for any $\varphi_1, \ldots, \varphi_k \in S$). If $\mathfrak{M} \vDash T$ and $a_0, \ldots, a_{n-1} \in |\mathfrak{M}|$, then $S \leftrightharpoons \{\varphi \mid \varphi \in Fr_n, \mathfrak{M} \vDash \varphi(a_0, \ldots, a_{n-1})\}$ is an $n$-type, called the *type* of the $n$-tuple $\langle a_0, \ldots, a_{n-1} \rangle$ of elements of $\mathfrak{M}$. If a type $S$ of $T$ is the type of some $n$-tuple of elements of $\mathfrak{M}$, we say that $S$ *is realized* in $\mathfrak{M}$. Every $n$-type of a theory $T$ is realized in some model of $T$.

An $n$-type $S$ is *principal* if there exists a formula $\varphi \in S$, called the *complete formula of the type $S$*, such that $S$ is a unique $n$-type containing $\varphi$. Let $\mathfrak{M}$ be a model of a theory $T$, and let $S$ be an $n$-type of $T$. We say that $\mathfrak{M}$ *omits* the type $S$ if $S$ is not the type of any $n$-tuple of elements $a_0, \ldots, a_{n-1}$ of $|\mathfrak{M}|$. Any principal type is realized in any model, but this is not true for nonprincipal types in view of the omitting type theorem. As is known (cf. [16]), if $\sigma$ is an at most countable signature and $S_0, S_1, \ldots$ is a countable family of nonprincipal types of a theory $T$, then there exists a countable model $\mathfrak{M}$ of $T$ omitting all the types $S_0, S_1, \ldots$.

If $S$ is an $n$-type and $k < n$, then $S \cap S_k(T)$ is a $k$-type. Suppose that $k < n$, $S$ is a $k$-type, $S'$ is an $n$-type, and $S \subseteq S'$.

The type $S'$ is *principal over the type $S$* if there exists a formula $\varphi \in S'$ such that $S'$ is a unique $n$-type containing $S \cup \{\varphi\}$.

There is the natural one-to-one correspondence between $n$-types of a theory $T$ and ultrafilters of Boolean algebras $F_n(T)$: if $U \subseteq F_n(T)$ is an ultrafilter, then $\pi^{-1}(U)$ is an $n$-type, where $\pi \colon Fr_n \to F_n(T)$ is the natural projection.

Now, we can characterize countably categorical theories.

**Theorem 1.7** ([16]). *Let $T$ be a complete theory of an at least countable signature. The following assertions are equivalent.*

(a) *$T$ is categorical in countable power.*

(b) *For every $n \in \omega$ the theory $T$ has finitely many $n$-types.*

(c) *For every $n \in \omega$ the set $F_n(T)$ is finite.*

A model $\mathfrak{M}$ is *homogeneous* if for any $a_0, \ldots, a_{n-1}, a_n,$ $b_0, \ldots, b_{n-1} \in |\mathfrak{M}|$ such that the types of the $n$-tuple $\langle a_0, \ldots, a_{n-1} \rangle$ and the $n$-tuple $\langle b_0, \ldots, b_{n-1} \rangle$ coincide there exists an element $b \in |\mathfrak{M}|$ such that the types of $\langle a_0, \ldots, a_{n-1}, a_n \rangle$ and $\langle b_0, \ldots, b_{n-1}, b \rangle$ coincide.

One of the most pleasant properties of homogeneous countable models is presented by the following assertion (cf. the proof in [**16**]).

**Proposition 1.8.** *Let $\mathfrak{M}_0$ and $\mathfrak{M}_1$ be homogeneous countable models of the same signature. The following assertions are equivalent.*

(a) *The models $\mathfrak{M}_0$ and $\mathfrak{M}_1$ are isomorphic.*

(b) *The same types are realized in $\mathfrak{M}_0$ and $\mathfrak{M}_1$.*

Any prime model is homogeneous. Recall that a model $\mathfrak{M}$ of a theory $T$ is *prime* if every model $\mathfrak{M}'$ of $T$ has an elementary submodel $\mathfrak{M}_0$ isomorphic to $\mathfrak{M}$. Only finite principal types (if they exist) are realized in a prime model. Therefore, a prime model is unique up to an isomorphism.

We formulate an important sufficient existence condition for prime models. A theory $T$ of a signature $\sigma$ is called a *Henkin theory* if for any sentence of the form $\exists x\, \Phi(x)$ in $T$ there exists a constant $c$ of $\sigma$ such that $\Phi(c) \in T$.

**Proposition 1.9.** *Let $T$ be a complete Henkin theory, and let $\mathfrak{M}$ be a model of $T$. The submodel $\mathfrak{M}_0$ of $\mathfrak{M}$ determined by the set of the values of constants of $\sigma$ is a prime model of $T$.*

To formulate an existence criterion for prime models, we need the following definition. A family $\mathfrak{S}$ of types of a theory $T$ of a signature $\sigma$ is *dense* if the following conditions hold.

- Let $p \in \mathfrak{S}$ be an $n$-type, and let $k \leqslant n$. Then $q \leftrightharpoons p \cap S_k(T) \in \mathfrak{S}$. If $\tau \colon \{x_0, \ldots, x_{n-1}\} \to \{x_0, \ldots, x_{n-1}\}$ is a permutation, then $[p]_{\tau x_0, \ldots, \tau x_{n-1}}^{x_0, \ldots, x_{n-1}} \in \mathfrak{S}$.

- If $p \in \mathfrak{S}$ is an $n$-type and $\varphi(x_0, \ldots, x_n)$ is a formula of the signature $\sigma$ such that $\exists x_n \, \varphi \in p$, then there exists an $(n+1)$-type $q \in \mathfrak{S}$ such that $p \cup \{\varphi\} \subseteq q$.

For a model $\mathfrak{M}$ of a theory $T$ we introduce the family $\mathfrak{S}(\mathfrak{M})$ of types realized in $\mathfrak{M}$. The family $\mathfrak{S}(\mathfrak{M})$ is dense.

Using the Henkin construction, we can prove the following assertion.

**Proposition 1.10.** *If $\mathfrak{S}$ is a dense countable family of types of a complete theory $T$, then there exists an at most countable model $\mathfrak{M}$ of $T$ such that every type realized in $\mathfrak{M}$ belongs to $\mathfrak{S}$.*

**Corollary 1.11.** *A complete theory $T$ of an at most countable signature $\sigma$ has a prime model if and only if the family $\mathfrak{S}_0$ of all principal types of $T$ is dense.*

REMARK. The assumption of Corollary 1.11 is equivalent to the condition that every Boolean algebra $F_n(T)$, $n \in \omega$, is atomic.

A saturated model is homogeneous. Denote by $\varkappa$ a cardinal. A model $\mathfrak{M}$ of a signature $\sigma$ is said to be $\varkappa$-*saturated* if for any subset $X \subseteq |\mathfrak{M}|$ of power less than $\varkappa$, any 1-type of $\mathrm{Th}(\mathfrak{M}, X)$ is realized in the model $\langle \mathfrak{M}, X \rangle$ obtained by the natural enrichment of $\mathfrak{M}$ to a model of the signature $\sigma_X \leftrightharpoons \sigma \cup \langle c_a \mid a \in X \rangle$ $(c_a^{\langle \mathfrak{M}, X \rangle} \leftrightharpoons a)$. A countable $\omega$-saturated model $\mathfrak{M}$ of $T$ is called a *countably saturated model* of $T$.

The following criterion was established in [**16**].

**Criterion 1.12** (existence of a countably saturated model). *A theory $T$ has a countably saturated model if and only if the Boolean algebra $F_n(T)$, $n \in \omega$, is superatomic.*

The following assertion describes the family of all types of homogeneous countable models of a complete theory $T$.

**Proposition 1.13.** *Let $T$ be a complete theory. Then $\mathfrak{S}$ is a family of all types of $T$ that are realized in some homogeneous countable model of $T$ if and only if $\mathfrak{S}$ is a countable dense family*

*of types possessing the following property: if* $p, q \in \mathfrak{S}$ *are* $(n+1)$-
*types such that* $p \cap Fr_n = q \cap Fr_n$, *then there exists an* $(n+2)$-*type*
$s \in \mathfrak{S}$ *such that* $p \cup [q]_{x_{n+1}}^{x_n} \subseteq s$.

## 1.2. Numberings

The theory of constructive models studies algorithmic properties
of algebraic structures. For this purpose, effective representations
of constructive models are considered and the study is based on
computability theory and the theory of algorithms and computable
functions. We refer to the monographs [103, 140, 146] for basic
methods and details of algorithm theory. In this section, we formulate only the main results which will be used in the following
sections. We follow [140] in presentations of computable functions.

By a *numbering* of a nonempty set $S$ we mean any mapping $\nu$
from $\mathbb{N}$ onto $S$. Let $S_0$ and $S_1$ be nonempty sets such that $S_0 \subseteq S_1$,
and let $\nu_0$ and $\nu_1$ be numberings of $S_0$ and $S_1$ respectively. We say
that the numbering $\nu_0$ is *reduced* to the numbering $\nu_1$ (and write
$\nu_0 \leqslant \nu_1$) if there exists a computable function $f$ from $\mathbb{N}$ into $\mathbb{N}$
such that $\nu_0(n) = \nu_1 f(n)$ for any $n \in \mathbb{N}$. Numberings $\nu_0$ and $\nu_1$
are *computably equivalent* or *equivalent* ($\nu_0 \equiv \nu_1$) if $\nu_0 \leqslant \nu_1$ and
$\nu_1 \leqslant \nu_0$. In this case, $S_0$ and $S_1$ coincide. Numberings $\nu_0$ and $\nu_1$
are *computably isomorphic* ($\nu_0 \underset{rec}{\simeq} \nu_1$) if there exists a computable
permutation $f$ of $\mathbb{N}$ such that $\nu_0(n) = \nu_1 f(n)$ for every $n$. Note
that $\nu_0 \underset{rec}{\simeq} \nu_1$ implies $\nu_0 \equiv \nu_1$. The converse assertion does not
hold in general.

On the class $\mathrm{Num}(S)$ of all numberings of a set $S$, the relation
$\equiv$ is an equivalence relation and the reducibility $\leqslant$ induces a partial
order on the equivalence classes by $\equiv$. Let

$$\mathbf{Num}(S) = \langle \mathrm{Num}(S)/\equiv, \leqslant \rangle.$$

If $\nu_0$ and $\nu_1$ are numberings of $S_0$ and $S_1$ respectively, then the
numbering $\nu_0 \oplus \nu_1$ of the union $S_0 \cup S_1$ is defined as follows: $\nu_0 \oplus$

$\nu_1(2n) = \nu_0(n)$ and $\nu_0 \oplus \nu_1(2n + 1) = \nu_1(n)$. If $\nu$ and $\mu$ are numberings of $S$, then $\nu \oplus \mu$ is a numbering of $S$ and determines the least upper bound of the pair $\nu/\equiv$, $\mu/\equiv$ in $\mathbf{Num}(S)$. Thus, $\mathbf{Num}(S)$ can be regarded as an upper semilattice.

For a language $L$ and an interpretation $\underline{\mathrm{int}}_L$ of $L$ on a set $S$ ($\underline{\mathrm{int}}_L : L \to S$) we say that a numbering $\nu_0$ of a subset $S_0 \subseteq S$ is *computable* relative to $\underline{\mathrm{int}}_L$ if there exists a computable function $f$ from $\mathbb{N}$ into $L$ such that $\nu_0(n) = \underline{\mathrm{int}}_L(f(n))$ for any $n \in \mathbb{N}$. If $\nu_0 \leqslant \nu_1$, where $\nu_1$ is a computable numbering relative to $\underline{\mathrm{int}}_L$, then $\nu_0$ is a computable numbering relative to the same interpretation $\underline{\mathrm{int}}_L$. If $\nu_0$ and $\nu_1$ are computable numberings relative to $\underline{\mathrm{int}}_L$, then the sum $\nu_0 \oplus \nu_1$ is also computable relative to the interpretation $\underline{\mathrm{int}}_L$. Thus, if an equivalence class contains some computable numbering relative to $\underline{\mathrm{int}}_L$, then any numbering of this class is computable relative to the same interpretation $\underline{\mathrm{int}}_L$.

Denote by $\mathbf{R}(S, \underline{\mathrm{int}}_L)$ a submodel of $\mathbf{Num}(S)$ consisting of classes containing numberings computable relative to $\underline{\mathrm{int}}_L$. The upper semilattice $\mathbf{R}(S, \underline{\mathrm{int}}_L)$ is called the *Rogers semilattice* of the class of numberings computable relative to $\underline{\mathrm{int}}_L$. If $\nu$ is a numbering of $S$ and $\Xi$ is some class of subsets of $N^{<\infty}$, then $P \subseteq S^k$ is referred to as an $\Xi$-*set* provided that there exists a set $A \in \Xi$ such that $P \leftrightharpoons \{\langle \theta n_1, \ldots, \theta n)k \rangle \mid \langle n_1, \ldots, n_k \rangle \in A\}$.

Consider a family $S$ of partial computable functions. For $L$ we take the language of Turing machines and for $\underline{\mathrm{int}}_L$ we take the function $\underline{\mathrm{int}}_{\mathrm{p.r.}}(M)$ computable by a Turing machine $M$. Thus, we arrive at a standard computable numbering of partial computable functions. In this case, we say that the numbering is *computable*. Note that $\nu$ is computable relative to $\underline{\mathrm{int}}_{\mathrm{p.r.}}$ if and only if there exists a partial computable function $g(n, x)$ such that $\nu(n)$ and $\lambda x g(n, x)$ coincide for any $n \in \mathbb{N}$.

Consider a family $S$ of computably enumerable sets. For $L$ we take the language of Turing machines and for $\underline{\mathrm{int}}_L$ we take the function $\underline{\mathrm{int}}_{\mathrm{r.e.}}(M) \leftrightharpoons \mathrm{Dom}(\underline{\mathrm{int}}_{\mathrm{p.r.}}(M))$, where $M$ is a Turing machine. In this case, we again obtain the standard notion of a computable numbering of computably enumerable sets (cf. [27]

and [36]), i.e., a numbering $\nu$ of a family of computably enumerable sets is a computable relative to $\underline{\text{int}}_{\text{r.e.}}$ if and only if the set $\{\langle x, y \rangle \mid y \in \nu(x)\}$ is computably enumerable. A numbering computable relative to $\underline{\text{int}}_{\text{r.e.}}$ is referred to as *computable*. A family $S$ is *computable* if there is a computable numbering of $S$ relative to $\underline{\text{int}}_{\text{r.e.}}$.

Let $S$ be a family of total functions on $\mathbb{N}$. Introduce the topology $\beta_S$ on $S$ as follows. For basis open sets we take sets of the family

$$B_S = \{V_g \mid g \text{ is a finite part of some function in } S\},$$

where $V_g \leftrightharpoons \{f \mid f \in S, g \subseteq f\}$. The family $S$ is *discrete* if the topology $\beta_S$ is discrete, i.e., for any $f \in S$ the finite part $g$ of $f$ is such that $V_g = \{f\}$. The family $S$ is *effectively discrete* if there exists a strictly computable sequence of finite sets $g_0, g_1, \ldots, g_n, \ldots$ such that $V_{g_n}$ contains only one element of $S$ for any $n$ and any element $f \in S$ belongs to some $V_{g_n}$, $n \in \mathbb{N}$. In this case, we say that the family $\{g_n \mid n \in \mathbb{N}\}$ *distinguishes* $S$ and $g_n$ *distinguishes* $f \supseteq g_n$. Note that an effectively discrete family is discrete. We say that a numbering $\nu$ of a set $S$ is *single-valued* (or is a *Friedberg numbering*) if $\nu$ is bijective, i.e., $\nu(x) \neq \nu(y)$ for any $x \neq y$ in $\mathbb{N}$. A numbering $\nu$ of a set $S$ is *positive* (*negative*) if the set $\eta_\nu = \{\langle x, y \rangle \mid \nu x = \nu y\}$ ($\overline{\eta}_\nu = \{\langle x, y \rangle \mid \nu x \neq \nu y\}$) is computably enumerable. A numbering $\nu$ is *solvable* if it is positive and negative, i.e., $\eta_\nu$ is computable. A single-valued numbering is solvable. A numbering $\nu$ of a set $S$ is *minimal* if $\nu/\equiv$ is a minimal element of **Num**$(S)$. Note that single-valued, solvable, and positive numberings are minimal.

One can prove that the Rogers semilattice of a computable nondiscrete family is infinite and, in the case of an effectively discrete family, consists of a single element [36]. There exists a discrete family with infinite Rogers semilattice (cf. [144]). As was shown in [87], the Rogers semilattice of any computable family of computably enumerable sets is infinite or has only one element. Selivanov [144] proved that the effective discreteness is not necessary for the Rogers semilattice to have only one element.

## 1.3. Models and Computability

Computability theory became play an important role in mathematics when the notion of computability was rigorously formulated and was applied by K. Gödel, A. Church, A. Turing, S. Kleene, E. Post, and A. Markov to the decidability of classical mathematical problems and to the proof of the Gödel theorem about the incompleteness of arithmetics.

In the XXth century, computability theory was rapidly developed. On the basis of results and methods of computability theory, new applications of mathematics have been formed, such as computer science, programing technology, automatization of various processes, etc. This can be explained by the fact that the computability approach suggests to represent an information in terms of natural numbers. Here, we briefly describe how numberings can be used for representation of mathematical objects and their structures.

Algorithmic properties of algebraic structures are naturally formulated and solved in numbering theory. Consider numberings of the basic sets of algebraic structures. Based on the standard algorithm theory, we can study the decidability of relations on elements with respect to numberings of such structures.

Consider a signature

$$\sigma = \langle P_0^{n_0}, \ldots, P_k^{n_k}, \ldots; F_0^{m_0}, \ldots, F_s^{m_s}, \ldots; c_0, \ldots, c_n, \ldots \rangle$$

such that there exist partial computable functions $[n]$ and $[m]$ defined as follows: $[n](i) = n_i$, where $n_i$ is the arity of the predicate symbol $P_i$, and $[m](i) = m_i$, where $m_i$ is the arity of the functional symbol $F_i$. We also consider the signature

$$\sigma_1 \leftrightharpoons \sigma \cup \langle a_0, a_1, \ldots \rangle$$

obtained from $\sigma$ by adding constant symbols.

Let $L$ and $L_1$ be families of all formulas of the first-order predicate calculus with equality $(P_0)$ of the signature $\sigma$ and $\sigma_1$ respectively. By a *Gödel numbering* of $L_1$ we mean any numbering $\gamma \colon L_1 \to \omega$ such that for a given $\gamma$-number we can effectively construct a formula with this number and for a given formula of $L_1$ we can effectively find its $\gamma$-number.

Now, define the Gödel numbering of formulas and terms of the signature $\sigma$. Note that the functions $[n]$ and $[m]$ from the definition of $\sigma$ exist if there are countably many predicate symbols or functional symbols. If there are countably many symbols with indices, it is required to recognize effectively the arity by the index.

We fix a set $V$ of variables $v_0, v_1, \ldots, v_n, \ldots$ and introduce the set $\text{Term}_\sigma(V)$ of terms of the signature $\sigma$ with variables in $V$ and the set $\text{Form}_\sigma(V)$ of formulas in variables of $V$. The Gödel numbering $\gamma$ is defined as the mapping $\gamma_\sigma \colon \text{Term}_\sigma(V) \cup \text{Form}_\sigma(V) @ > 1 - 1 >> \mathbb{N}$ such that we can effectively recognize a number of a formula or a term and obtain some information about the structure of formulas and terms. Then we construct $\gamma$ by induction on the complexity of formulas. We begin with $\text{Term}_\sigma(V)$:

(1) $\gamma(v_i) = c(0, c(0, i))$,

(2) $\gamma(c_i) = c(0, c(1, i))$ for $i$ such that $c_i \in \sigma$,

(3) if $t$ has the form $F_i(t_1, \ldots, t_{m_i})$, where $F_i$ is an $m_i$-ary predicate symbol, and $t_1, \ldots, t_{m_i}$ have the Gödel numbers $\gamma(t_1) = l_1, \ldots, \gamma(t_{m_i}) = l_{m_i}$, then $\gamma(t) = c(0, c((i + 2), c^{m_i}(l_1, \ldots, l_{m_i})))$.

It is obvious that the set of numbers of terms is computable. If the number of a term is known, we can recognize variables and their indices, as well as constants and their indices. Furthermore, we can find the index of the operation and the numbers of those subterms from which the term is constructed with the help of the symbol of this operation.

Define $\gamma$ on the set of formulas as follows:

(1) if $t$ and $q$ are terms and $\gamma(t) = n$, $\gamma(q) = m$, then $\gamma(t = q) \leftrightharpoons c(1, c(0, c(n, m)))$,

(2) if $P_i$ is an $n_i$-ary predicate symbol and $t_1, \ldots, t_{n_i}$ are terms with Gödel numbers $\gamma(t_1) = l_1, \ldots, \gamma(t_{n_i}) = l_{n_i}$, then $\gamma(P_i(t_1, \ldots, t_{n_i})) \leftrightharpoons c(1, c(1, c(i + 1, c^{n_i}(l_1, \ldots, l_{n_i}))))$ and $\gamma(t_1 = t_2) = c(1, c(1, c(0, c(l_1, l_2))))$,

(3) if $\varphi$ and $\psi$ are formulas with Gödel numbers $\gamma(\varphi) = n$ and $\gamma(\psi) = m$, then

$$\gamma((\varphi \,\&\, \psi)) \leftrightharpoons c(1, c(2, c(n, m))),$$
$$\gamma((\varphi \vee \psi)) \leftrightharpoons c(1, c(3, c(n, m))),$$
$$\gamma((\varphi \to \psi)) \leftrightharpoons c(1, c(4, c(n, m))),$$
$$\gamma(\neg\varphi) \leftrightharpoons c(1, c(5, n)),$$
$$\gamma((\exists v_i)\varphi) \leftrightharpoons c(1, c(6, c(i, n))),$$
$$\gamma((\forall v_i)\varphi) \leftrightharpoons c(1, c(7, c(i, n))).$$

By induction on the complexity of formulas, it is easy to show that every formula of $\mathrm{Form}_\sigma(V)$ has a Gödel number. Furthermore, we can recognize whether a given number is the Gödel number of a formula and obtain an information about the structure of this formula, for example, about free variables constants, the form of the formula, the presence of quantifiers, the complexity of the prefix formed by quantifiers, and the numbers of formulas that can be obtained by substitutions.

If the number of a formula is known, we can find the number of the equivalent formula in prenex normal form.

With every subset $S \subseteq L_1$ we associate the set $\gamma(S)$ of all numbers of formulas of $S$. A set $S$ is said to be *decidable* (*enumerable*) if $\gamma(S)$ is computable (computably enumerable).

Choosing some hierarchy of the complexity of subsets of $\mathbb{N}$ (for example, the arithmetic hierarchy, the analytic hierarchy [140], the Ershov hierarchy [25, 26, 29, 36, 37], etc.), we say that $X$ belongs to the *complexity class* $\Delta$ if $\gamma(X)$ belongs to $\Delta$.

For a given number $n$ we can recognize whether a formula with number $n$ is an axiom of the first-order predicate calculus $\mathrm{PC}^\sigma$. For a set of numbers we can recognize whether a given formula can be obtained from a finite set of formulas with the corresponding numbers by some of the rules of $\mathrm{PC}^\sigma$. Hence we can recognize whether a sequence of formulas with given Gödel numbers is a proof in $\mathrm{PC}^\sigma$. Thus, we arrive at the following assertion.

**Proposition 1.14.** *If a set of formulas is provable in* $PC^\sigma$ *from an enumerable set, then it is enumerable.*

Proposition 1.14 implies the following assertion.

**Proposition 1.15.** *If the set of axioms* $A$ *is enumerable, then the theory* $T_A \rightleftharpoons \{\varphi \mid A \vdash \varphi\}$ *is enumerable.*

We define the principal computable numbering $p_0(\overline{x}_0), \ldots, p_n(\overline{x}_n), \ldots$ of the set of all enumerable partial types consistent with a decidable theory $T$.

By a *partial type* $p(\overline{x})$ of a theory $T$ we mean the set of formulas in variables of $\overline{x}$ such that the set $p(\overline{x}) \cup T$ is consistent. A numbering $d_0(\overline{x}_0), \ldots, d_n(\overline{x}_n), \ldots$ of partial types of a theory $T$ is *computable* if $\overline{d}_0, \overline{d}_1, \ldots, \overline{d}_n, \ldots$, where $\overline{d}_i = \{n \mid n$ is the Gödel number of a formula in $d_i\}$, is a computable numbering of computably enumerable sets and there exists a computable function $v$ such that $v(n)$ is equal to the number of the tuple $\langle i_1, \ldots, i_{m_n} \rangle$ of indices such that $\overline{x}_n = (v_{i_1}, \ldots, v_{i_{m_n}})$. A numbering $p_0(\overline{x}_0), \ldots, p_n(\overline{x}_n), \ldots$ of partial types of a theory $T$ is *principal* if for every computable numbering $d_0(\overline{x}'_0), \ldots, d_n(\overline{x}'_n), \ldots$ of partial types of $T$ there is a computable function $f(n)$ such that $d_n(\overline{x}'_n) = p_{f(n)}(\overline{x}_{f(n)})$ for any $n$.

Consider the following sequence of finite sets:

$$\varnothing = p_n^0(\overline{x}_n) \subseteq p_n^1(\overline{x}_n) \subseteq \cdots \subseteq p_n^t(\overline{x}_n) \subseteq \cdots$$

Let $p_n(\overline{x}_n) \rightleftharpoons \cup_t p_n^t(\overline{x}_n)$. For $n$ we introduce $i$ and $k$ such that $c(i, k) = n$. We regard $i$ as the number of the $i$th computably enumerable set $W_i$ and $k$ as the number of the tuple $\langle i_1, \ldots, i_s \rangle$ relative to numberings of all tuples of finite length. We set

$$W_i^t \upharpoonright k \rightleftharpoons \{m \in W_i^t \mid m \text{ is the Gödel number}$$

$$\text{of a formula in free variables with}$$

$$\text{indices in } \{i_1, \ldots, i_s\} \text{ and number } k\},$$

$$p_n^t(\overline{x}_n) \rightleftharpoons \{\varphi \mid \varphi \text{ has the Gödel number in } W_{l(n)}^m \upharpoonright r(n)\},$$

where $m$ is the maximal number less than $t+1$ and such that the set

$$T \cup \{\varphi \mid \varphi \text{ has the Gödel number in } W_{l(n)}^m \upharpoonright r(n)\}$$

is consistent. Since $T$ is decidable, the consistency condition is decidable. Therefore, for $n$ and $t$ we can recognize whether a formula belongs to $p_n^t(\overline{x}_n)$ and indicate its number, i.e., we can list formulas in $p_n^t(\overline{x}_n)$. It is obvious that the Gödel numbers of such formulas are less than $t+1$ because of the assumption on $W_n^t$.

By the definition of $p_n$ on the basis of $W_{l(n)}$ and the possibility to compute exactly the set of free variables in a computable numbering of a family of finite types, as well as the fact that $\{W_n\}_{n\in\mathbb{N}}$ is a principal numbering, we conclude that $p_n$ is a principal numbering. Since the numbering $\{p_n\}_{n\in\mathbb{N}}$ is principal, we obtain the following assertion.

**Proposition 1.16.** *A family $S$ of partial types of a theory $T$ is computable, i.e., there is a computable numbering $d_0(\overline{x}_0'), \ldots, d_n(\overline{x}_n'), \ldots$ such that $S = \{d_0(\overline{x}_0'), \ldots, d_n(\overline{x}_n'), \ldots\}$ if and only if there exists a computably enumerable set $W$ such that $S = \{p_n(\overline{x}_n) \mid n \in W\}$.*

By a *numbered model* of the signature $\sigma$ without functional symbols we mean the pair $(\mathfrak{M}, \nu)$, where $\mathfrak{M} = \langle M, P_0, P_1, \ldots \rangle$ is a model of the signature $\sigma$ and $\nu$ is a numbering of the basic set $M$ of the model $\mathfrak{M}$. By a *homomorphism* from a numbered model $(\mathfrak{M}_0, \nu_0)$ into a numbered model $(\mathfrak{M}_1, \nu_1)$ we mean a mapping $\mu\colon M_0 \to M_1$ from the basic set $M_0$ of the model $\mathfrak{M}_0$ into the basic set $M_1$ of the model $\mathfrak{M}_1$, i.e., a homomorphism from $\mathfrak{M}_0$ into $\mathfrak{M}_1$ and a morphism from $(M_0, \nu_0)$ into $(M_1, \nu_1)$.

For a numbered model $(\mathfrak{M}, \nu)$ we can construct a $\sigma_1$-enrichment $\mathfrak{M}_\nu$ of $\mathfrak{M}$, i.e., a model of the signature $\sigma_1$ whose basic set is the basic set of $\mathfrak{M}$ and predicates of $\sigma$ in $\mathfrak{M}_\nu$ coincide with the corresponding predicates of $\mathfrak{M}$. Namely, for the value of the constant $a_k$, $k \in \omega$, we take $\nu k \in M$. We say that $\mathrm{Th}(\mathfrak{M}, \nu)$ is the *elementary theory* of $\mathfrak{M}_\nu$, i.e., the set of all closed formulas of the signature $\sigma_1$ that are true in $\mathfrak{M}_\nu$.

A numbered model $(\mathfrak{M}, \nu)$ is *constructive* if the set $\overline{D}(\mathfrak{M}, \nu) \leftrightharpoons \{\langle k, m_1, \ldots, m_{n_k}\rangle \mid \mathfrak{M} \vDash P_k(\nu m_1, \ldots, \nu m_{n_k})\}$ is computable.

Let $D(\mathfrak{M}, \nu) = \{\varphi(c_{m-1}, \ldots, c_{m_k} \mid \varphi(x_1, \ldots, x_{m_k})$ be a quantifier-free formula, and let $\mathfrak{M} \vDash \varphi(\nu m_1, \ldots, \nu m_k)\}$.

The following special class of constructive models plays an important role in the study of decidable theories. A numbered model $(\mathfrak{M}, \nu)$ is *strongly constructive* if $\mathrm{Th}(\mathfrak{M}, \nu)$ is a decidable theory. Models admitting strong constructivizations are said to be *decidable*.

The constructibility of a numbered model $(\mathfrak{M}, \nu)$ is equivalent to the decidability of the set of quantifier-free formulas in $\mathrm{Th}(\mathfrak{M}, \nu)$. Hence every strongly constructive model is constructive.

However, in the case of arbitrary numbered models and algebras, only numberings of algebras with effective operations are of interest. We consider this case in more detail. Let $\sigma = \langle f_0^{m_0}, f_1^{m_1}, \ldots\rangle$. If the signature $\sigma$ is infinite, we assume that the function $h: n \mapsto m_n$ is computable.

By a *computable numbering* of an algebra $\mathfrak{A} = \langle A, g_0, g_1, \ldots\rangle$ of the signature $\sigma$ we mean a numbering $\nu: \omega \to A$ of the basic set of $\mathfrak{A}$ such that there exists a binary computable function $G$ such that $g_n(\nu y_1, \ldots, \nu y_{m_n}) = \nu G\big(n, c^{m_n}(y_1, \ldots, y_{m_n})\big)$ for any $n \in \omega$ and $y_1, \ldots, y_{m_n}$.

The pair $(\mathfrak{A}, \nu)$ is referred to as a *computable numbered algebra* if $\nu: \omega \to A$ is a numbering of $\mathfrak{A}$. It turns out that any algebra admits a computable numbering.

**Theorem 1.17** ([37]). *Any at most countable algebra $\mathfrak{A}$ admits a computable numbering of this algebra.*

In this case, the complexity of this algebra depends only from numbering equivalence of that numbering.

A computable numbered algebra $(\mathfrak{A}, \nu)$ is constructive (i.e., it is a constructive model of the corresponding signature consisting of only functions) if and only if the numbering $\nu$ is solvable.

Let $\mathfrak{A} = \langle A; P_0, \ldots, P_n; F_0, \ldots, F_k; c_0, \ldots, c_s \rangle$ be an algebraic structure of a signature $\sigma$. If $\sigma$ is infinite, the functions $i \to m_i$ and $i \to n_i$ are assumed to be computable. A pair $(\mathfrak{A}, \nu)$, where $\nu$ is a mapping from $\mathbb{N}$ or from an initial interval of $\mathbb{N}$ to the basic set $A$ of $\mathfrak{A}$, is called a *numbered structure* and $\nu$ is called a *numbering* of $\mathfrak{A}$.

Let $\mathfrak{K}$ be a class of subsets and functions on $\mathbb{N}$. For $\mathfrak{K}$ we can take, for example, one of the following classes:

- the class $R$ of computable functions and relations,
- the class $R_A$ of computable relative to $A$ functions and relations,
- the class PRIM of primitive computable functions and relations,
- the class $P$ of relations and functions of the polynomial complexity,
- the class exp of relations and functions of the exponential complexity,
- the corresponding classes $\Delta_\alpha^0 (\Sigma_\alpha^{0,A}, \Pi_\alpha^{0,A})$ of relations and functions of the arithmetic hierarchy relative to $A$,
- the corresponding classes $\Delta_\alpha^1 (\Sigma_\alpha^{1,A}, \Pi_\alpha^{1,A})$ of relations and functions of the analytic hierarchy (relative to $A$),
- the corresponding classes $\Delta_\alpha^{m-1} (\Sigma_\alpha^{m-1,A}, \Sigma_\alpha^{m-1,A})$ of the Ershov hierarchy (relative to $A$) [**25, 26, 29**].

Let $B$ be a set or a family of sets, and let $\mathfrak{A}$ be an algebraic structure of a signature $\sigma$. A numbered structure $(\mathfrak{A}, \nu)$ is said to be *B-positive* if $\eta_\nu \leftrightharpoons \{(n, m) \mid \nu n = \nu m\}$ and $\nu^{-1}(P_i) = \{\langle l_1, \ldots, l_{m_i} \rangle \mid \langle \nu l_1, \ldots, \nu l_{m_i} \rangle \in P_i\}$, $i \leqslant n$, are computably enumerable with respect to a set in $B$ or with respect to the entire set $B$ and there exist $B$-computable functions $f_i$, $i \leqslant k$, such that $\nu f_i(l_1, \ldots, l_{n_i}) = F_i(\nu l_1, \ldots, \nu l_{n_i})$ for all $l_1, \ldots, l_{n_i} \in \mathbb{N}$. A $B$-positive structure $(\mathfrak{A}, \nu)$ is said to be *B-constructive* if $\eta_\nu$ and $\nu^{-1}(P_i)$ are $B$-computable. If the signature $\sigma$ is infinite, it is necessary to require the uniform computable numbering [**100**].

To study algorithmic properties of models, we need an algorithm checking the truth of formulas. We define the relative constructibility and strong constructibility. Let $\mathcal{B}$ be a class of subsets of $\mathbb{N}$. Suppose that a quantifier-free formula has no alternating groups of quantifiers and a formula $\Phi$ has $n$ alternating groups of quantifiers if the prenex normal form of $\Phi$ has $n$ alternating groups of quantifiers. Denote by $\mathfrak{F}_n$ the set of formulas possessing $n$ alternating groups of quantifiers and by $\mathfrak{F}_\omega$ the set of all formulas, called *fragments* (of the language). The sets $\mathfrak{F}_n$ are called *restricted fragments* (of the language). Let $\mathfrak{F}$ be a set of formulas of a signature $\sigma$. A numbered structure $(\mathfrak{A}, \nu)$ is said to be $\mathcal{B}\text{-}\mathfrak{F}\text{-}constructive$ or $\mathfrak{F}\text{-}constructive$ *relative to* $\mathcal{B}$ if the following set belongs to $\mathcal{B}$:

$$\{\langle s, l_1, \ldots, l_k \rangle \mid s \text{ is the number of a formula } \Phi(x_1, \ldots, x_k)$$
$$\text{in } \mathfrak{F} \text{ with } k \text{ free variables and } \mathfrak{A} \models \Phi(\nu l_1, \ldots, \nu l_k)\}.$$

It is easy to see that a structure is $\mathfrak{F}_0$-constructive relative to $\mathcal{B}$ if and only if it is $\mathcal{B}$-constructive. For the sake of brevity, we write $\mathfrak{F}$-constructive in the case of the $\mathfrak{F}$-constructibility relative to the class of computable relations and $\mathcal{B}$-constructive in the case $\mathfrak{F} = \mathfrak{F}_0$. If $\mathfrak{F} = \mathfrak{F}_0$ or $\mathcal{B} = \varnothing$, we omit $\mathfrak{F}$ or $\mathcal{B}$ in the notation.

$\mathcal{B}\text{-}\mathfrak{F}_\omega$-constructive structures are said to be *strongly $\mathcal{B}$-constructive* or *$\mathcal{B}$-$\omega$-constructive*, whereas $\mathcal{B}\text{-}\mathfrak{F}_n$-constructive structures are referred to as *$\mathcal{B}$-$n$-constructive* .

We describe the other approach. Let $\mathfrak{A}$ be an algebraic structure of a signature $\sigma$ such that the basic set $A$ is a subset of $\mathbb{N}$. Then it is reasonable to consider the effectiveness of different relations without any mention of numbers.

An algebraic structure $\mathfrak{A}$ is said to be $\mathcal{B}$-*computable* if the basic predicates and operations of $\mathfrak{A}$ belong to a class $\mathcal{B}$. For many computability classes $\mathcal{B}$ an abstract structure is $\mathcal{B}$-constructivizable if and only if it is isomorphic to a $\mathcal{B}$-computable structure. For a $\mathcal{B}$-constructive structure we can effectively construct a $\mathcal{B}$-computable structure relative to $\mathcal{B}$ provided that we can choose exactly one number in every set of the numbers of elements. We can pass from a $\mathcal{B}$-computable structure to a $\mathcal{B}$-constructive structure by using the $\mathcal{B}$-computable function that enumerates the basic set of the $\mathcal{B}$-computable model. Then we can define the $\mathcal{B}$-constructivization of this structure.

Similarly, for $\mathfrak{F}$-constructive structures relative to $\mathcal{B}$ we can define $\mathcal{B}$-$\mathfrak{F}$-computable models. If a model is isomorphic to a $\mathcal{B}$-$\mathfrak{F}_\omega$-computable model, then it is $\mathcal{B}$-decidable. In the case $B \subseteq \mathbb{N}$, the class of $B$-computable sets is denoted by $\mathcal{B}(B)$. We say that $\mathcal{B}(B)$-$\mathfrak{F}$-computable ($\mathcal{B}(B)$-$\mathfrak{F}$-constructive) structures are $\mathfrak{F}$-computable ($\mathfrak{F}$-constructive) relative to $B$.

We give the most important examples of relatively computable models. Assume that the language is computable and the basic set is a subset of $\omega$. We identify a structure $\mathcal{A}$ with its atomic diagram $D(\mathcal{A})$ and sentences with their Gödel numbers. In this case, we say that $\mathcal{A}$ is *computable* (*arithmetical* or *hyperarithmetical*) if $D(\mathcal{A})$, regarded as a subset of $\omega$, is computable (arithmetical or hyperarithmetical).

We say that a model has *constructivization* or admits a computable (arithmetical or hyperarithmetical) representation if there exists an isomorphic computable (arithmetical or hyperarithmetical) model. If for an abstract model there exists an isomorphic (arithmetical or hyperarithmetical) decidable model, then we say that this model has a decidable representation with respect to the class of (arithmetical or hyperarithmetical) sets.

Let $(\mathfrak{A}, \nu)$ and $(\mathfrak{B}, \mu)$ be numbered models, and let $\varphi \colon \mathfrak{A} \to \mathfrak{B}$ be a homomorphism. We say that $\varphi$ is $C$-*computable* if there exists a $C$-computable function $f$ such that $\varphi\nu = \mu f$, i.e., the following diagram is commutative:

$$
\begin{array}{ccc}
N & \xrightarrow{f} & N \\
\nu \downarrow & & \downarrow \mu \\
A & \xrightarrow{\varphi} & B
\end{array}
$$

In this case, the function $f$ represents $\varphi$ and $\varphi$ is called a $C$-*homomorphism*. If there exists a $C$-computable isomorphism $\varphi$ from $(\mathfrak{A}, \nu)$ into $(\mathfrak{B}, \mu)$, then $(\mathfrak{B}, \mu)$ is called a $C$-*extension* of

$(\mathfrak{A}, \nu)$ with respect to $\varphi$. If $\mathfrak{A} \subseteq \mathfrak{B}$ and the identity embedding of $\mathfrak{A}$ in $\mathfrak{B}$ is $C$-computable, then $(\mathfrak{B}, \mu)$ is a $C$-extension of $(\mathfrak{A}, \nu)$.

Let $(\mathfrak{N}, \nu)$ and $(\mathfrak{N}, \mu)$ be numbered algebraic structures. Recall that numberings $\nu$ and $\mu$ of $\mathfrak{N}$ are computably equivalent if there exist computable functions $f$ and $g$ such that $\nu = \mu f$ and $\mu = \nu g$. For constructivizations $\nu$ and $\mu$ it suffices to require the existence of only one computable function $f$ such that $\nu = \mu f$. We note that if there is the continual group of automorphisms of a constructivizable structure $\mathfrak{A}$, then there is the continuum of noncomputable equivalent constructivizations. Thus, for an atomless Boolean algebra we have the continuum of noncomputable equivalent constructivizations, although it has simple algorithmic structure. However, we consider abstract structures up to an isomorphism. The definition of the autoequivalence introduced by Mal'tsev [101] turns out to be more suitable in this situation. Two numberings $\nu$ and $\mu$ of an algebraic structure are *autoequivalent* if they are computably equivalent up to an automorphism, i.e., there exists an automorphism $\varphi$ of $\mathfrak{A}$ such that $\varphi \nu$ and $\mu$ are computably equivalent.

The questions on nonequivalent representations and their classification are important in the study of constructive structures. Within the framework of the above approaches, we can investigate the same properties by choosing a suitable language. In fact, the above approaches are equivalent. To demonstrate this fact, we show that the corresponding categories are equivalent.(for category theory we refer to [12] and [36]).

We consider the category Num of all numbered models with homomorphisms for morphisms and the category Nat of all models whose basic sets are subsets of $\mathbb{N}$ and morthisms are computable homomorphisms. Let $(\mathfrak{M}, \nu)$ be a numbered model. We define the value of the functor Com on $(\mathfrak{M}, \nu)$ by setting $\mathrm{Com}\,(\mathfrak{M}, \nu) \leftrightharpoons (N_{\mathfrak{M}}, \Sigma)$, where

$$N_{\mathfrak{M}} \leftrightharpoons \{n \mid n \text{ is the least number of } \nu n\},$$

$$P \leftrightharpoons \{\langle n_1, \ldots, n_k\rangle \in N_{\mathfrak{M}} \mid \mathfrak{M} \models P(\nu n_1, \ldots, \nu n_k)\},$$

$$F(n_1, \ldots, n_k) \leftrightharpoons \min\{m \mid F(\nu n_1, \ldots, \nu n_k) = \nu m\}.$$

If $\varphi$ is a homomorphism from $(\mathfrak{M}, \nu)$ into $(\mathfrak{N}, \mu)$, then

$$\mathrm{Com}\,(\varphi) \rightleftharpoons \{\langle n, m \rangle \mid n \in \mathbb{N}_{\mathfrak{M}}, m \in \mathbb{N}_{\mathfrak{N}}, \text{ and } \varphi(\nu n) = \mu m\}.$$

Note that $\mathrm{Com}\,(\varphi)$ is a homomorphism from $\mathrm{Com}\,(\mathfrak{M}, \nu)$ into $\mathrm{Com}\,(\mathfrak{N}, \mu)$.

REMARK. $\mathfrak{M}$ and $\mathrm{Com}\,(\mathfrak{M}, \nu)$ are isomorphic.

REMARK. If $\varphi$ is an isomorphism, then $\mathrm{Com}\,(\varphi)$ is also an isomorphism.

We define the functor K from Nat into Num by setting $\mathrm{K}\,(\mathfrak{M}) \rightleftharpoons (\mathfrak{M}, \nu)$, where $\nu$ is a numbering of $|\mathfrak{M}|$ in ascending order. If $|\mathfrak{M}|$ is finite, then all the numbers which do not appear in this numbering go to the last element (with respect to the numbering of $\mathfrak{M}$). As a result, we obtain a numbering $\nu$ which will be denoted by $\nu_{\mathrm{K}}$.

REMARK. The functor Com determines an equivalence between the categories Num and Nat.

Consider the subcategory $\mathrm{Con}^B$ of Num consisting of $B$-constructive models with $B$-computable homomorphisms for morphisms. We also consider the subcategory $\mathrm{Com}^B$ of Nat consisting of $B$-computable models with $B$-computable homomorphisms for morphisms.

**Theorem 1.18.** *The restriction $\mathrm{Com}^B$ of the functor $\mathrm{Com}$ to the subcategory $\mathrm{Con}^B$ determines an equivalence between $\mathrm{Con}^B$ and $\mathrm{Com}^B$.*

PROOF. It is easy to verify that if $(\mathfrak{M}, \nu)$ is $B$-constructive, then the model $\mathrm{Com}\,(\mathfrak{M}, \nu)$ is $B$-computable. Since the basic sets are $B$-computable and there exists a $B$-computable function $f$ such that $\varphi\nu = \mu f$, we conclude that $\mathrm{Com}\,\varphi$ is a partial $B$-computable function with $B$-computable graph. The restriction $\mathrm{Com}^B$ of Com to $\mathrm{Con}^B$ acts from $\mathrm{Con}^B$ into $\mathrm{Com}^B$. There exist isomorphisms $\varphi\colon 1_{\mathrm{Con}^B} \to \mathrm{Com}\,\mathrm{K}$ and $\psi\colon 1_{\mathrm{Com}^B} \to \mathrm{K}\,\mathrm{Com}$ such that $\mathrm{Com}\,\varphi = \psi\,\mathrm{Com}$. $\qquad\square$

**Corollary 1.19.** *Numbered models* $(\mathfrak{N}, \nu)$ *and* $(\mathfrak{M}, \mu)$ *are isomorphic if and only if* $\mathrm{Com}\,(\mathfrak{N}, \nu)$ *and* $\mathrm{Com}\,(\mathfrak{M}, \mu)$ *are isomorphic.*

**Corollary 1.20.** *B-constructive models* $(\mathfrak{N}, \nu)$ *and* $(\mathfrak{M}, \mu)$ *are B-isomorphic if and only if* $\mathrm{Com}\,(\mathfrak{M}, \mu)$ *is B-isomorphic to* $\mathrm{Com}\,(\mathfrak{N}, \nu)$.

Thus, the study of constructivizations of a model $\mathfrak{M}$, defined up to an autoequivalence, is equivalent to the study of computable models isomorphic up to a computable isomorphism to $\mathfrak{M}$. Consequently, if $\nu$ and $\mu$ are constructive models of $\mathfrak{M}$, then $\mathrm{Com}\,(\mathfrak{M}, \nu)$ and $\mathrm{Com}\,(\mathfrak{M}, \mu)$ are computable models isomorphic to $\mathfrak{M}$; moreover, $\nu$ and $\mu$ are autoequivalent if and only if $\mathrm{Com}\,(\mathfrak{M}, \nu)$ and $\mathrm{Com}\,(\mathfrak{M}, \mu)$ are computably isomorphic and for any computable model $\mathfrak{N}$ isomorphic to $\mathfrak{M}$ there exists a constructivization $\nu$ of $\mathfrak{M}$ such that $\mathrm{Com}\,(\mathfrak{M}, \nu)$ and $\mathfrak{N}$ are computably isomorphic.

Let $\mathfrak{M} = \langle M, P_0^{n_0}, \ldots, P_k^{n_k}, a_0, \ldots, a_s \rangle$ be a finite model of a finite signature $\sigma$ without functional symbols, and let $\nu$ be a mapping from $[0, n] = \{i \mid 0 \leqslant i \leqslant n\}$ onto $M$. The pair $(\mathfrak{M}, \nu)$ is called a *finitely numbered* (*n-numbered*) *model.*

With every finite signature $\sigma = \langle P_0^{n_0}, \ldots, P_k^{n_k}, a_0, \ldots, a_s \rangle$ we associate the number $\langle \langle \langle 0, n_0 \rangle, \ldots, \langle k, n_k \rangle \rangle, s \rangle$, $n_i \geqslant 1$. We extend $\sigma$ by constant symbols $c_0, \ldots, c_n, \ldots$ and define the $(n+1)$-diagram $\mathcal{D}(\mathfrak{M}, \nu)$ of $(\mathfrak{M}, \nu)$ by constructing the enrichment $\mathfrak{M}^n$ of $\mathfrak{M}$ to a model of the signature $\sigma_n = \sigma \cup \{c_0, \ldots, c_n\}$. For this purpose, assume that the value of $c_i$ is equal to $\nu i$, $i \leqslant n$, and $\mathcal{D}(\mathfrak{M}, \nu) = \{\varphi \mid \varphi$ is an atomic formula of the signature $\sigma_n$ without free variables or the negation of such a formula and $\mathfrak{M}^n \vDash \varphi\}$. Let $GD(\mathfrak{M}, \nu)$ be the set of the Gödel numbers of formulas in $\mathcal{D}(\mathfrak{M}, \nu)$. The number $\langle n, \langle \langle \langle 0, n_0 \rangle, \langle 1, n_1 \rangle, \ldots, \langle k, n_k \rangle \rangle, s \rangle, u \rangle$, where $u$ is the canonical number of the finite set $D_u = GD(\mathfrak{M}, \nu)$, is called the *Gödel number* of the numbered finite model of the finite signature $\sigma$. Such models are called *finitely numbered models* and their numbers are referred to as the *Gödel numbers.*

A numbering of finitely enumerable models possess the following obvious properties.

⟨1⟩ *The set of the Gödel numbers of finitely numbered models is computable.*

⟨2⟩ *For a given Gödel number of a finitely numbered model $(\mathfrak{M}, \nu)$ it is possible to compute the number of elements of $|\mathfrak{M}|$.*

⟨3⟩ *For a given Gödel number of a finitely numbered model it is possible to compute how many predicate symbols and constant symbols are contained in $\sigma$ and compute the arity of all predicate symbols.*

⟨4⟩ *For two Gödel numbers of finitely numbered models it is possible to recognize whether these models can be considered in the same signature.*

⟨5⟩ *The set of numbers of finite signatures is computable.*

⟨6⟩ *For numbers $n$ and $m$ it is possible to recognize whether a finitely numbered model with number $n$ is a model of the signature with number $m$.*

⟨7⟩ *For numbers $n$ and $m$ it is possible to recognize whether a finitely numbered model with number $n$ of the signature $\sigma = \langle P_0^{n_0}, \ldots, P_k^{n_k}, a_0, \ldots, a_s \rangle$ has an enrichment to a finitely numbered model of the signature $\sigma'$ with number $m$.*

A finite $n$-numbered model $(\mathfrak{M}, \nu)$ is called an *extension* of a $k$-numbered model $(\mathfrak{N}, \mu)$ if $k \leqslant n$, the models $\mathfrak{M}$ and $\mathfrak{N}$ are of the same signature, and the set $\{\langle \nu(i), \mu(i) \rangle \mid i \leqslant k\}$ is an isomorphic embedding of $\mathfrak{M}$ in $\mathfrak{N}$.

⟨8⟩ *For numbers $n$ and $m$ it is possible to recognize whether a finitely numbered model with Gödel number $m$ is an extension of a finitely numbered model with Gödel number $n$.*

An $n$-numbered model $(\mathfrak{M}, \nu)$ of the signature $\sigma = \langle P_0^{n_0}, \ldots, P_k^{n_k}, a_0, \ldots, a_s \rangle$ is called an *enriched extension* of a $k$-numbered model $(\mathfrak{N}, \mu)$ of the signature $\sigma' = (P_0^{m_0}, \ldots, P_{r'}^{m_{r'}}, a_0, \ldots, a_{s'})$ if $k \leqslant n$, $s' \leqslant s$, $r' \leqslant r$ and, for any $0 \leqslant i \leqslant r'$, the arity $m_i$ of the predicate $P_i$ is equal to the arity $n_i$ of the predicate $P_i$; moreover, $\{\langle \mu(i), \nu(i) \rangle \mid i \leqslant k\}$ is an isomorphism from $\mathfrak{N}$ into $\mathfrak{M} \upharpoonright \sigma$.

⟨9⟩ *For $n$ and $m$ it is possible to recognize whether a model with Gödel number $n$ is an enriched extension of a finitely numbered model with Gödel number $m$.*

For a numbered model $(\mathfrak{M}, \nu)$ of the signature $\sigma = \langle P_0^{n_0}, \ldots, P_k^{n_k}, \ldots, a_0, \ldots, a_s, \ldots \rangle$ we set $M_n = \{\nu i \mid i \leqslant n\}$, $n \in \mathbb{N}$, and consider $\sigma_n = \langle P_0^{n_0}, \ldots, P_r^{n_r}, a_{i_1}, \ldots, a_{i_k} \rangle$, where $r = n$ if $\sigma$ contains at least $n$ predicates and $r$ is the number of predicates of $\sigma$ otherwise. The set $\{i_1, \ldots, i_k\}$ consists of the indices $i$ of constants $c_i$ of $\sigma$ such that $i \leqslant n$ and the value of $c_i$ belongs to $M_n$.

Let $\mathfrak{M}_n$ be a submodel of the restriction $\mathfrak{M} \upharpoonright \sigma_n$ with the basic set $M_n$. Finitely numbered models $(\mathfrak{M}_n, \nu_n)$, where $\nu_n(k) \leftrightharpoons \nu(k)$, $k \leqslant n$, are called *finitely numbered submodels* of $(\mathfrak{M}, \nu)$. Denote by $(e\mathfrak{M}, \nu)$ the set of the Gödel numbers of finitely numbered submodels of $(\mathfrak{M}, \nu)$. The set $W(\mathfrak{M}, \nu)$ is called the *representation* of $(\mathfrak{M}, \nu)$.

**Proposition 1.21.** *A numbered model $(\mathfrak{M}, \nu)$ is constructive if and only if $W(\mathfrak{M}, \nu)$ is computably enumerable.*

By Proposition 1.21, it is possible to construct a universal computable numbering of all constructive and all computable models of a fixed signature without functional symbols.

The empty model of the empty signature with the empty numbering, as well as $n$-numbered models, is constructive. For a given set $W(\mathfrak{M}, \nu)$ we define a model $\mathfrak{M}$ and a numbering $\nu$ as follows. Let $M_W^0 = \{c_i \mid$ there exists the Gödel number of an $n$-numbered model in $W(\mathfrak{M}, \nu)$ and $i \leqslant n\}$. Introduce an equivalence relation on $M_W^0$ as $c_i \sim_W c_j$ if $c_i = c_j$ occurs in the diagram of some $n$-numbered model with number in $W(\mathfrak{M}, \nu)$. Let $M_W$ be the quotient set $M_W^0 / \sim_W$. We set $\nu_W(i) = c_i / \sim_W$, where $i \in M_W^0$, and $\nu_W(i) = a_j$ for a constant of the signature $\sigma$ of the model $\mathfrak{M}$ if $c_i = a_j$ occurs in the diagram of some $n$-numbered model with number in $W$. We set $P_i(\nu_W n_0, \ldots, \nu_W n_k)$ if $P_i(c_{n_0}, \ldots, c_{n_k})$ occurs in the diagram of some $n$-numbered model with number in $W$.

Thus, we obtain a model $\mathfrak{M}_W$ of the same signature as $\mathfrak{M}$. We also define the numbering $\nu_W$. Setting $\varphi(c_i/\sim) \rightleftharpoons \nu i$, we conclude that $\varphi$ is an isomorphism between $(\mathfrak{M}_W, \nu_W)$ and $(\mathfrak{M}, \nu)$; moreover, $\varphi$ is the identity mapping on numbers. If $W(\mathfrak{M}, \nu)$ is computable, then $(\mathfrak{M}_W, \nu_W)$ is constructive. The converse assertion is obvious.

Let us consider a finite signature $\sigma$ without functional symbols and define a numbering $\varkappa^\sigma$ of all constructive models of the signature $\sigma$. For this purpose, we consider the principal numbering $\{W_n\}_{n \in \mathbb{N}}$ of all computably enumerable subsets of $\mathbb{N}$. As usual, $W_n^t$ is the part of $W_n$ which was already numbered at the step $t$. We recall that we enumerate only $x < t$ in $W_n^t$. For $W_n$ we construct $V_n$ as follows. Let $V_n^0 = \varnothing$. At the step $t + 1$, we verify the following conditions:

(a) any element of $W_n^{t+1}$ is the Gödel number of some $k$-model of the signature $\sigma$,

(b) for any $x, y \in W_n^{t+1}$ one of finitely numbered models with the Gödel numbers $x$ and $y$ is an extension of the other.

We set $V_n^{t+1} \rightleftharpoons V_n^t$ if conditions (a) and (b) are not satisfied. Otherwise, we set $V_n^{t+1} \rightleftharpoons V_n^t \cup W_n^{t+1}$. The sequence $\{V_n\}_{n \in \mathbb{N}}$, where $V_n = \cup V_n^t$, is computable. Consequently, there is a computable function $\rho$ such that $V_n = W_{\rho(n)}$ for any $n$. Furthermore, $W_{\rho(\rho(n))} = W_{\rho(n)}$ for any $n$.

It is easy to see that every set $V_n$ is computably enumerable and represents some constructive model. By the above results, we can restore the constructive model $\mathfrak{M}_{V_n}$ and constructivization $\nu_{V_n}$.

We set $\varkappa^\sigma(n) \rightleftharpoons (\mathfrak{M}_{V_n}, \nu_{V_n})$ and write $\mathfrak{M}_n^\varkappa$ instead of $\mathfrak{M}_{V_n}$ and $\nu_n^\varkappa$ instead of $\nu_{V_n}$. It is easy to see that $\varkappa^\sigma(n)$ enumerates all constructive models of the signature $\sigma$ including finitely numbered models and the empty model as well. Assume that $\sigma$ is infinite and the function $i \to n_i$ is computable, where $n_i$ is the arity of the $i$th predicate symbol. Arguing as above, we obtain $W(\mathfrak{M}, \nu)$ and the numbering $\varkappa^\sigma$ of all constructive models of the signature $\sigma$ and finite constructive models of finite parts of $\sigma$ if, in the construction of $V_n$, finitely numbered models are models of the initial segments of the signature $\sigma$ and the enriched extension condition is used instead of the extension condition.

The notion of a computable sequence of constructive models is often used (cf., for example, [**22, 34, 37**]).

**Definition 1.22.** A sequence of constructive models $(\mathfrak{M}_0, \nu_0), \ldots, (\mathfrak{M}_n, \nu_n), \ldots$ is *computable* if the models are uniformly constructive, i.e., all computable functions of numbers of constants and indices of computable functions defining basic operations and predicates on numbers of elements can be computed for $(\mathfrak{M}_n, \nu_n)$ from computable functions by the number $n$.

Using the idea of numberings of sets, we can define, up to a recursive isomorphism, a numbering of any class $S$ of constructive models.

**Definition 1.23.** A numbering $\nu$ of models in $S$ is called a *computable numbering of class* $S$ if the sequence $(\mathfrak{M}_0, \nu_0), \ldots, (\mathfrak{M}_n, \nu_n), \ldots$ of constructive models in $S$ is computable, where $(\mathfrak{M}_n, \nu_n)$ is a model with number $n$ in the numbering $\nu$ $(\nu(n) = (\mathfrak{M}_n, \nu_n))$ and for any constructive model $(\mathfrak{M}, \mu)$ in $S$ there exists $n$ such that $(\mathfrak{M}, \mu)$ and $(\mathfrak{M}_n, \nu_n)$ are computably isomorphic.

A computable sequence of computable models is defined in a similar way.

**Definition 1.24.** A sequence $\mathfrak{M}_0, \ldots, \mathfrak{M}_n, \ldots$ of computable models is *computable* if the models are uniformly computable, i.e., all computable functions of constants and the indices of computable functions defining basic operations and predicates are computed for the models $\mathfrak{M}_n$ from computable functions by the number $n$.

Using again the ideas of numberings of sets, we can define, up to a recursive isomorphism, a numbering of any class $S$ of computable models.

**Definition 1.25.** A numbering $\nu$ of models in $S$ is called a *computable numbering of class $S$* if the sequence $\mathfrak{M}_0, \ldots, \mathfrak{M}_n, \ldots$ of computable models in $S$ is computable, where $\mathfrak{M}_n$ is a computable model with number $n$ in the numbering $\nu$ ($\nu(n) = \mathfrak{M}_n$) and for any computable model $\mathfrak{M}$ in $S$ there exists $n$ such that $\mathfrak{M}$ and $\mathfrak{M}_n$ are computably isomorphic.

**Proposition 1.26.** *A sequence* $(\mathfrak{M}_0, \nu_0), \ldots, (\mathfrak{M}_n, \nu_n), \ldots$ *of constructive models is computable if and only if there exist computable functions $f$ and $g$ such that $(\mathfrak{M}_n, \nu_n)$ and $\varkappa(f(n))$ are computably isomorphic for every $n$ and the number $g(n)$ of the computable function $\varkappa_{g(n)}$ defining this computable isomorphism is computed by $g$ from $n$, i.e., for $\varphi_n(\nu_n(m)) \leftrightharpoons \nu_{f(n)}^{\varkappa}(\varkappa_{g(n)}(m))$, $\varphi_n$ is an isomorphism from $\mathfrak{M}_n$ onto $\mathfrak{M}_{f(n)}^{\varkappa}$, where $\varkappa_n$ is the universal numbering of all partial computable functions.*

**Theorem 1.27.** *The sequence $(\mathfrak{M}_n^{\varkappa}, \nu_n^{\varkappa})$ of constructive models is computable.*

The proof is based on the construction and definition of a computable sequence. Indeed, for $(\mathfrak{M}_n^{\varkappa}, \nu_n^{\varkappa})$ and $n$ we can find the diagram $V_n = W(\mathfrak{M}_n^{\varkappa}, \nu_n^{\varkappa})$.

Let $\alpha = \{\langle \mathfrak{M}_n, \nu_n \rangle\}$ and $\beta = \{(\mathfrak{N}_n, \mu_n)\}$ be numberings of numbered models. We say that $\alpha$ is *reduced* to $\beta$ if there exists a computable function $f$ such that the constructive models $(\mathfrak{M}_n, \nu_n)$ and $(\mathfrak{N}_{f(n)}, \mu_{f(n)})$ are computably isomorphic (in the sense of constructive models). The reduction of $\alpha$ to $\beta$ is denoted by $\alpha \leqslant \beta$. We say that $\alpha$ is *effectively reduced* to $\beta$ if there exist computable functions $f$ and $g$ such that for any $n$ the function $\varkappa_{g(n)}$ defines an isomorphism from a constructive model $(\mathfrak{M}_n, \nu_n)$ onto the numbered model $(\mathfrak{N}_{f(n)}, \mu_{f(n)})$, i.e., the mapping $\varphi_n(\nu_n(m)) \leftrightharpoons \mu_{f(n)}(\varkappa_{g(n)}(m))$ is well defined and realizes an isomorphism between $\mathfrak{M}_n$ and $\mathfrak{N}_{f(n)}$.

Similarly, based on general ideas of numbering theory, we can define the reducibility for numbering of computable models.

Let $\alpha = \{\mathfrak{M}_n, n \in \omega\}$ and $\beta = \{\mathfrak{N}_n, n \in \omega\}$ be numberings of models. We say that $\alpha$ is *reduced* to $\beta$ if there exists a computable function $f$ such that the constructive models $(\mathfrak{M}_n, \nu_n)$ and $(\mathfrak{N}_{f(n)}, \mu_{f(n)})$ are computably isomorphic (in the sense of computable models). The reduction of $\alpha$ to $\beta$ is denoted by $\alpha \leqslant \beta$. We say that $\alpha$ is *effectively reduced* to $\beta$ if there exist computable functions $f$ and $g$ such that for any $n$ the function $\varkappa_{g(n)}$ is an isomorphism from a model $\mathfrak{M}_n$ onto the model $\mathfrak{N}_{f(n)}$, i.e., the mapping $\varphi_n(\nu_n(m)) \leftrightharpoons \mu_{f(n)}(\varkappa_{g(n)}(m))$ is well defined and realizes an isomorphism between $\mathfrak{M}_n$ and $\mathfrak{N}_{f(n)}$.

The following assertion follows from definitions.

**Proposition 1.28.** *If a numbering $(\mathfrak{M}_n, \nu_n)$, $n \in \mathbb{N}$, of numbered models is effectively reduced to a computable numbering of constructive models $(\mathfrak{N}_n, \mu_n)$, $n \in \mathbb{N}$, then $(\mathfrak{M}_n, \nu_n)$, $n \in \mathbb{N}$, is a computable numbering of constructive models.*

From the construction of numberings of constructive models $\varkappa^\sigma(n) \leftrightharpoons (\mathfrak{M}_{V_n}, \nu_{V_n})$ we obtain an important result due to Nurtazin about the existence of a universal computable numbering of constructive models and the existence of universal computable numbering of all computable models from the existence of a functor between categories.

**Theorem 1.29** ([130]). *There exists up to a recursive isomorphism a universal computable numbering of all constructive models of a computable signature without functional symbols i.e., a computable numbering $(\mathfrak{M}_n, \nu_n)$, $n \in \mathbb{N}$, of constructive models of the fixed structure such that any other computable numbering of constructive models of the same signature is reduced to this numbering.*

**Corollary 1.30** ([130]). *There exists up to recursive isomorphism a universal computable numbering of all computable models of a comuptable signature without functional symbols i.e., a computable numbering $\mathfrak{M}_n$, $n \in \mathbb{N}$, of computable models of this fixed signature to which is reduced any other computable numbering of computable models of the same signature.*

## 1.4. Perspective directions in the theory of computable models

We list the most important topics for the future development of the theory of computable models (cf. also [39]).

1. One of the main problem is connected with existence of computable representations. In particular, this approach is presented in [40, 37, 3, 5, 7, 13, 17, 28, 31, 32, 33, 35, 39, 43, 47, 53, 54, 57, 59, 65, 68, 69, 74, 76, 81, 83, 84, 85, 90, 92, 96, 100, 107, 108, 109, 110, 111, 112, 113, 114, 115, 116, 117, 118, 119, 120, 121, 122, 123, 124, 125, 129, 130, 132, 133, 134, 135, 136, 138, 150, 151, 152, 154].

2. The second approach is connected with the nonuniqueness of computable representations and algorithmic dimension (with some special properties). In particular, it is represented in [1, 5, 8, 10, 18, 23, 32, 35, 37, 39, 40, 44, 45, 46, 48, 49, 54, 57, 60, 67, 63, 64, 65, 72, 75, 101].

3. Interesting problems on the classification of computable models relative to structures connected with computable models. [40, 39].

4. Computable classes of models in the light of the above two approaches. The computability of families of computable representations and the computability of classes of computable models were studied in [40, 39, 1, 66, 14, 21, 22, 37, 45, 56, 58, 67, 85, 91, 92, 130].

5. Another class of problems connected with the classification of algorithmic problems with respect to complexity (cf., for example, [1, 2, 3, 5, 10, 13, 14, 37, 39, 40, 42, 43, 44, 54, 55, 56, 57, 62, 68, 72, 70, 75, 89, 82, 147, 148]).

6. There exists a closed connection between definability and complexity. An approach based on this fact was used in many papers, for example, [1, 5, 6, 10, 13, 14, 38, 39, 40, 42, 43, 58, 62, 66, 71, 70, 75, 77, 89, 127, 99, 147, 148, 154].

## 2. Bounds for Computable Models

Bounds for computable models are used for describing various mathematical constructions. We consider bounds for theories of computable models and the complexity of some models. We also examine the structure bounds from the point of view of the complexity of descriptions of computable models in a language with infinite disjunctions and conjunctions.

### 2.1. Bounds for the theory of computable models

By definition, the theory of a decidable model is decidable. We establish the existence of computable models satisfying a given specification in the language of the first-order predicate calculus.

**Theorem 2.1** ([32, 152]). *A decidable consistent theory $T$ possesses a decidable model.*

The situation is rather complicated if additional model-theoretic properties are required. Goncharov–Nurtazin and Harrington independently proved the following assertion for prime models.

**Theorem 2.2** ([65, 74]). *A decidable complete theory $T$ possesses a decidable prime model if and only if there exists an algorithm that for any formula consistent with $T$ produces a principal type of the theory containing this formula.*

Morley proved the existence theorem for saturated models and posed the decidability problem for homogeneous models.

**Theorem 2.3** ([125]). *A decidable complete theory $T$ possesses a decidable saturated model if and only if the set of all types of $T$ admits a computable numbering.*

Goncharov [52] and Peretyat'kin [136] independently found the decidability criteria for homogenous models. Goncharov [53] constructed an example of a totally transcendental decidable theory without decidable homogeneous models.

**Problem 1** (Goncharov). Whether there exists a decidable homogeneous model, determined up to an isomorphism, for an arbitrary decidable theory with countably many countable models?

The situation is quite simple in the case of countably categorical theories.

**Definition 2.4.** A theory $T$ is *countably categorical* if $T$ has a unique up to an isomorphism countable model.

Within the framework of model theory, countably categorical theories and models of such theories have been well studied. The following assertion is a simple consequence of the effective completeness theorem.

**Theorem 2.5.** *A countably categorical theory $T$ is decidable if and only if all models of $T$ are decidable, which holds if and only if $T$ has a decidable model.*

Thus, if we are interested in decidable models of countably categorical theories, an answer can be obtained in terms of decidability. However, the situation essentially changes for computable models.

If a theory $T$ possessing a computable model is computable in $\mathbf{0}^{\omega}$, then the degree of the $\omega$–jump of a computable set is $\mathbf{0}^{\omega}$. This bound is sharp because there exists a theory (for example, the theory of $(\omega, +, \times, \leqslant)$) possessing a computable model that is Turing equivalent to $\mathbf{0}^{\omega}$.

**Theorem 2.6.** *If $\mathbf{A}$ is a computable model, then the theory $\mathrm{Th}\,(\mathbf{A}^{*})$ is $\mathbf{0}^{\omega}$–decidable and the theory $\mathrm{Th}_{\Sigma_{n+1}}(\mathbf{A}^{*})$ is computably enumerable in $\mathbf{0}^{n}$ uniformly with respect to $n$, where $\mathbf{A}^{*}$ is an extension of $\mathbf{A}$ by constants.*

**Corollary 2.7.** *If $\mathbf{A}$ is a computable model, then the theory $\mathrm{Th}\,(\mathbf{A}^{*})$ is $\mathbf{0}^{\omega}$–decidable.*

These results suggest the following interesting problem.

**Problem 2.** Find necessary and sufficient conditions for the existence of computable models.

Our goal is to find natural sufficient conditions for the existence of computable models of given theories and to determine bounds for the complexity under different classifications of the complexity degrees of theories. Consider some special classes of theories.

## 2.1.1. Computable countably categorical models.

For countably categorical theories the question is trivial: All countable models of a countably categorical theory are decidable if and only if the theory is decidable. Naturally, the situation becomes much more complicated if we require the computability condition.

**Problem 3.** Characterize countably categorical theories possessing computable models.

Lerman and Schmerl [96] presented a sufficient condition for an arithmetic countably categorical theory to have a constructive model. More precisely, they proved that if $T$ is a countably categorical arithmetic theory such that the set of all sentences beginning with the existential quantifier and having $n + 1$ groups of quantifiers of the same type ($\Sigma_{n+1}$–formulas) is $\Sigma_n^0$ for every $n$, then $T$ has a constructive model.

It would be useful to weaken this condition, say, as follows: "the set of all sentences beginning with the existential quantifier and having $n + 1$ groups of quantifiers of the same type ($\Sigma_{n+1}$–formulas) is $\Sigma_{n+1}^0$ for every $n$."

**Problem 4.** Whether a countable model is 1-computable under the Lerman–Schmerl condition?

Knight [90] generalized the result to the case of non-arithmetical countably categorical theories. However, none of the mentioned results solves the problem. We do not even know any example of a theory satisfying this sufficient condition for sufficiently large $n$. Knight conjectured the existence of arithmetical and non-arithmetical countably categorical theories with computable models. An answer to this conjecture is contained in the following assertions which develop general methods for constructing computable models from arithmetical models with preserving some model-theoretical properties.

**Theorem 2.8** ([68]). *For every $n \geqslant 1$ there exists a countably categorical theory of Turing degree $0^n$ possessing a computable model.*

**Theorem 2.9** ([43]). *For every arithmetical Turing degree $d$ there exists a countably categorical theory of Turing degree $d$ possessing a computable model.*

The proof of the following assertion about the existence of a non-arithmetic countably categorical theory with computable models was based on the ideas of [68] and [43].

**Theorem 2.10** (Fokina, Goncharov, Khoussainov). *There exists a countably categorical theory $T$ with a computable model such that the Turing degree of $T$ is non-arithmetical.*

Having an answer to the question in Problem 5 below, it would be possible to obtain a complete description of the Turing degrees of countably categorical theories possessing computable models.

**Problem 5.** Is it true that for every Turing degree $d \leqslant 0^{(\omega)}$ there exists a countably categorical theory of Turing degree $d$ possessing a computable model?

### 2.1.2. Computable uncountably categorical models.

Here, we deal only with models of uncountably categorical theories.

Morley proved that a theory is categorical in uncountable power $\alpha$ if and only if the theory is categorical in uncountable power $\omega_1$. Among typical examples of uncountably categorical theories, there are the theory of algebraically closed fields of fixed charactcristic, the theory of vector spaces over a fixed countable field, the theory of the structure $(\omega, S)$, where $S$ is the successor function on $\omega$. Roughly speaking, all countable models of each of these theories can be listed into an $\omega + 1$ chain so that the first element is the prime model, the last element is the saturated model, and any two models are embedded each other. Apparently, it is one of the main structural properties of the class of models of an uncountably categorical theory.

Baldwin and Lachlan [9] showed that all models of an uncountably categorical theory $T$ can be listed in the following chain of elementary embeddings:

$$\text{chain}\,(T) : \mathbf{A}_0 \preccurlyeq \mathbf{A}_1 \preccurlyeq \mathbf{A}_2 \preccurlyeq \ldots \mathbf{A}_\omega,$$

where $\mathbf{A}_0$ is the prime model of $T$, $\mathbf{A}_\omega$ is the saturated model of $T$, and every $\mathbf{A}_{i+1}$ is prime over $\mathbf{A}_i$.

Assume that a theory $T$ is decidable. In the general case, the decidability of $T$ does not imply the decidability of all models of $T$. However, the following important result on decidable models of $T$ was established by Harrington and Khisamiev.

**Theorem 2.11** ([74, 83, 84]). *Let $T$ be an uncountably categorical theory. Then $T$ is decidable if and only if $T$ has a decidable model, which holds if and only if all models of $T$ admit decidable presentations.*

The situation is similar to that for countably categorical theories. Theorem 2.11 mainly answers to the question about the existence of decidable models of uncountably categorical decidable theories. However, it does not clarify how to build computable models of uncountably categorical theories if the decidability assumption is omitted. Correspondingly, the following problem is actual.

**Problem 6.** Characterize uncountably categorical theories possessing computable models.

The case of uncountably categorical theories is more complicated. In general, the existence of a computable model of an uncountably categorical theory $T$ does not imply that all models of $T$ admit constructivizations. As was shown by Goncharov [47], there exists an uncountably categorical theory $T$ such that only the prime model of $T$ is computable.

**Problem 7.** Is it true that any countable model of a strongly minimal theory possessing a computable prime model is $0^2$–computable?

**Problem 8.** Whether there exists an $\omega_1$–categorical theory $T$ such that the model $\mathbf{M_0}$ is computable, but any other model $\mathbf{M}_{n+1}$, $i = 2, 3, \ldots$, is not $0^i$-computable?

It is remarkable that all known uncountably categorical theories possessing computable models were regarded as computable in the double jump of $\mathbf{0}$ recently. But, at present, some of such theories are not viewed as computable owing to the lowering of the complexity of models preserving the basic model-theoretic properties. New theories with a given arithmetic complexity were constructed by the method suggested in [68].

**Theorem 2.12** ([68]). *There exist uncountably categorical theories $T_n$ of Turing degree $\mathbf{0}^n$, $n > 2$, such that all their models admit constructivizations.*

**Theorem 2.13** ([43]). *For every arithmetical Turing degree $d$ there exists an uncountably categorical theory $T$ of Turing degree $d$ such that any countable model of $T$ is isomorphic to a computable model.*

In the case of a special subclass of noncountably categorical theories (for example, strongly minimal theories with trivial pregeometry), an arithmetical bound holds for the complexity of such theories [**61**]. The sharpness of this bound was later proved by Khoussainov, Lempp, and Solomon. We consider this special case in more detail because of a new unexpected property allowing us to derive bounds for the complexity of such theories. A natural question arises: If the above results can be extended to the class of all strongly minimal theories or to some a sufficiently large subclass of such theories?

**Definition 2.14.** A formula $\varphi(x)$ is a *strongly minimal formula* of a complete theory $T$ if for any model $\mathcal{M}$ of $T$, elements $\overline{b}$ of $\mathcal{M}$, and a formula $\psi(x, \overline{y})$ one of the sets $\{a | \mathcal{M} \vDash \psi(a, \overline{b}) \& \varphi(a)\}$ or $\{a | \mathcal{M} \vDash \neg\psi(a, \overline{b}) \& \varphi(a)\}$ is finite.

If $\varphi(x)$ is strongly minimal formula of a complete theory $T$, then for any model $\mathcal{M}$ of the theory $T$ it is possible to define an operator of $cl(X)$ from the set $P(\varphi(\mathcal{M}))$ of all subsets $\varphi(\mathcal{M})$ to $P(\varphi(\mathcal{M}))$.

Let $\varphi(\mathcal{M}) \rightleftharpoons \{a | \mathcal{M}\varphi(a)\}$, and let $X$ be a subset of $\varphi(\mathcal{M})$. We set $cl(X) \rightleftharpoons \{a \mid$ there exists a formula $\theta(x)$ such that $\mathcal{M} \vDash \theta(a)$ and the set $\theta(\mathcal{M})$ is finite$\}$. Let $\varphi(x)$ be a strongly minimal formula of a complete theory $T$, and let $\mathcal{M}$ be a model of $T$. The cardinality of any maximal independent subset $Y$ of the model $\varphi(\mathcal{M})$ is called the *dimension* of the model $\mathcal{M}$ of $T$ and is denoted by $\dim(\mathcal{M})$.

Baldwin and Lachlan [**9**] found a remarkable property of theories categorical in uncountable power, owing to which it becomes possible to clarify globally the structure of all models of such theories.

**Theorem 2.15** ([**9**])**.** *Let $T$ be a complete uncountably categorical theory. Then there exists a complete formula $\rho(\overline{z})$ and constants $\overline{c}$ such that $T^* \rightleftharpoons T \cup \{\rho(\overline{c})\}$ is a complete theory (principal expansion of $T$) and there exists a strongly minimal formula $\varphi(x)$ of the theory $T^*$.*

**Theorem 2.16** ([9]). *Let $M_1$ and $M_1$ be models of a complete theory $T$ with strongly minimal formula. If $\dim(M_1) = \dim(M_2)$, then the models $M_1$ and $M_1$ are isomorphic.*

Consider a natural subclass of theories categorical in uncountable power.

**Definition 2.17.** A theory $T$ is *strongly minimal* if the formula $x = x$ is strongly minimal in $T$.

**Definition 2.18.** A model $M$ is *strongly minimal* if the theory $TH(M)$ is strongly minimal.

**Definition 2.19.** We say that a strongly minimal theory $T$ has *trivial pregeometry* if for any model $M$ of $T$ and any subset $X$ of the universe of $M$ the following equality holds: $\mathrm{cl}(X) = \cup_{a \in X} cl\{a\}$.

**Theorem 2.20** ([61]). *Let $M$ be a computable strongly minimal theory with trivial pregeometry. Then $\mathrm{Th}(M)$ forms a $\mathbf{0}''$-computable set of $\mathcal{L}$-sentences. Consequently, all countable models of $\mathrm{Th}(M)$ are $\mathbf{0}''$-decidable and, in particular, are $\mathbf{0}''$-computable.*

**Theorem 2.21** ([61]). *For any strongly minimal theory $T$ with trivial pregeometry the elementary diagram $FD(M)$ of any model $M$ of $T$ is a model complete $\mathcal{L}_M$-theory.*

Note that a model of a strongly minimal theory $T$ with trivial pregeometry is not necessarily model complete in the original language (for example, $\langle \omega S \rangle$ is not model complete).

PROOF OF THEOREM 2.21. Consider a model $M_0$ of $T$. To simplify the notation, we write $T^*$ instead of $\mathrm{Th}((M_0)_{M_0})$. Let $\mathcal{L}^*$ be the language of $T^*$ (i.e., $\mathcal{L}^* = \mathcal{L}_{M_0}$). Consider two models $M \subseteq N$ of $T^*$ of size $\varkappa$, where $\varkappa > |M_0|$ is fixed. Since $M$ and $N$ are models of $T^*$, we can assume that $M_0 \preceq M$ and $M_0 \preceq N$. We need to show that $M \preceq N$. For this purpose, we use two standard facts, the so-called *non-finite cover property* and *finite satisfiability*.

The non-finite cover property of an uncountably categorical theory means that for all $\mathcal{L}^*$-formulas $\varphi(\overline{x}, \overline{y})$ there is a number $k$ such that for any $M^* \models T^*$ and $\overline{b}$ in $M^*$ either $\varphi(\overline{b}, M^*)$ is infinite or has size at most $k$. The number $k$ depends only on $\varphi$ and the partition of free variables in $(\overline{x}, \overline{y})$. Thus, one can use the quantifiers $\exists^{<\infty}$ and $\exists^{\infty}$, where $\exists^{<\infty}\overline{y}\varphi(\overline{x}, \overline{y})$ denotes $\exists^{\leqslant k}\overline{y}\varphi(\overline{x}, \overline{y})$ and $\exists^{\infty}\overline{y}\varphi(\overline{x}, \overline{y})$ denotes $\neg\exists^{<\infty}\overline{y}\varphi(\overline{x}, \overline{y})$.

The following assertion is an immediate consequence of the pigeon-hole principle.

**Lemma 2.22.** *If* $\mathcal{N} \models \exists^{\infty}\overline{y}\varphi(\overline{b}, \overline{y})$ *and* $\lg(\overline{y}) = k + 1$, *then there is a partition of* $\overline{y}$ *in* $w\overline{z}$ *with* $\lg(w) = 1$ *and* $\lg(\overline{z}) = k$ *such that* $\mathcal{N} \models \exists^{\infty}w\exists\overline{z}\varphi(\overline{b}, w, \overline{z})$.

The following general fact, referred to as the *finite satisfiability*, asserts that if $\mathcal{M}_0 \preceq \mathcal{N}$ are models of a stable theory and $\mathcal{N} \models \varphi(\overline{b}, \overline{c})$ for some $\mathcal{L}_{M_0}$-formula and some $\overline{b}, \overline{c}$ in $N$ that are independent (i.e., do not fork over $M_0$), then there is $\overline{a}$ in $M_0$ such that $\mathcal{N} \models \varphi(\overline{a}, \overline{c})$. This fact is obvious because, in a stable theory, every complete type over a model is definable. We formulate this assertion in a special case of strongly minimal theories.

**Lemma 2.23.** *Suppose that* $\mathcal{M}_0 \preceq \mathcal{N}$ *are models of a strongly minimal theory and* $\overline{b}$, $\overline{c}$ *are tuples in* $N$ *such that* $\mathrm{acl}(M_0\overline{b}) \cap \mathrm{acl}(M_0\overline{c}) = M_0$. *If* $\mathcal{N} \models \varphi(\overline{b}, \overline{c})$ *for any* $\mathcal{L}_{M_0}$-*formula* $\varphi$, *then there is* $\overline{a}$ *in* $M_0$ *such that* $\mathcal{N} \models \varphi(\overline{a}, \overline{c})$.

We also need the following notion.

**Definition 2.24.** *An* $\mathcal{L}^*$-*formula* $\varphi(\overline{x})$ *is* *absolute* *if for all* $\overline{b}$ *in* $M$ *we have* $M \models \varphi(\overline{b})$ *if and only if* $\mathcal{N} \models \varphi(\overline{b})$.

To complete the proof of $M \preceq \mathcal{N}$, it suffices to show that any $\mathcal{L}^*$-formula is absolute. It is obvious that every quantifier-free $\mathcal{L}^*$-formula is absolute and a family of absolute formulas is closed under the Boolean operations. Thus, to obtain the model completeness of $T^*$, it suffices to show that if an $\mathcal{L}^*$-formula $\varphi(\overline{x}, y)$ is absolute, then $\exists y \varphi(\overline{x}, y)$ is also absolute.

**Definition 2.25.** An $\mathcal{L}^*$-formula $\varphi(\overline{x}, \overline{y})$ is said to be an $(n, m)$-*formula* if $\lg(\overline{x}) = n$ and $\lg(\overline{y}) = m$. We identify three interrelated families of statements:

- $A_{n,m}$ is the statement that for all absolute $(n, m)$-formulas $\varphi(\overline{x}, \overline{y})$ the formula $\exists^{<\infty} \overline{y} \varphi(\overline{x}, \overline{y})$ is absolute,

- $B_{n,m}$ is the statement that for all absolute $(n, m)$-formulas $\varphi(\overline{x}, \overline{y})$, if $\overline{b} \in M^n$ and $\mathcal{N} \models \exists^{<\infty} \overline{y} \varphi(\overline{b}, \overline{y})$, then $\varphi(\overline{b}, \mathcal{N}) = \varphi(\overline{b}, \mathcal{M})$, i.e., every realization of $\varphi(\overline{b}, \overline{y})$ in $N^m$ is an element of $M^m$,

- $C_{n,m}$ is the statement that for all absolute $(n, m)$-formulas $\varphi(\overline{x}, \overline{y})$ the formula $\exists \overline{y} \varphi(\overline{x}, \overline{y})$ is absolute.

By the above arguments, to prove the model completeness of $T^*$, it suffices to show that $C_{n,1}$ holds for all $n \in \omega$.

It is obvious that each of three statements in Definition 2.25 is preserved if subscripts decrease (for example, $B_{n,m}$ implies $B_{n',m'}$ for all $n' \leqslant n$ and all $m' \leqslant m$).

**Lemma 2.26.** *The following assertions hold:*

(a) $B_{n,m}$ *implies* $C_{n,m}$ *for all* $n, m \in \omega$,

(b) $B_{n,m}$ *implies* $A_{n,m+1}$ *for all* $n, m \in \omega$,

(c) $B_{1,m}$ *(consequently, $B_{0,m}$) holds for all* $m \in \omega$.

**Proposition 2.27.** $B_{n,m+1}$ *and* $A_{n+1,m}$ *imply* $B_{n+1,m}$ *for all* $n, m \in \omega$.

As was already noted, $T^*$ is model complete if $\mathcal{M} \preceq \mathcal{N}$.

We show that $B_{n,m}$ holds for all $n, m \in \omega$. For this purpose, we show by induction on $n$ that $B_{n,m}$ holds for all $n$. Note that $B_{1,m}$ holds for all $m \in \omega$. We fix $n \geqslant 1$ and assume that $B_{n,m}$ holds for all $m$. Let us prove that $B_{n+1,m}$ holds for all $m$ by induction on $m$. It is obvious that $B_{n+1,0}$ holds. Assume that $B_{n+1,m}$ holds for some $m$. Then $B_{n,m+2}$ holds by the induction assumption and $A_{n+1,m+1}$ holds since $B_{n+1,m}$ holds. Thus, $B_{n+1,m+1}$ holds by Proposition 2.27, and the induction procedure is complete.

By Lemma 2.26 (a), $C_{n,m}$ holds for all $n, m \in \omega$. In particular, $C_{n,1}$ holds for all $n \in \omega$. This means that the family of absolute $\mathcal{L}^*$-formulas is closed under the existential quantification. As is known, the family of absolute $\mathcal{L}^*$-formulas contains quantifier-free formulas and is closed under Boolean connectives. Hence every $\mathcal{L}^*$-formula is absolute. Thus, $\mathcal{M} \preceq \mathcal{N}$, as required.                $\square$

For a structure $\mathcal{M}$ we denote by $\mathrm{Th}_{\forall\exists}(\mathcal{M}_M)$ the set of all $\forall\exists$-sentences $\sigma \in \mathrm{Th}(\mathcal{M}_M)$ (in the language $\mathcal{L}_M$).

**Lemma 2.28.** *If the elementary diagram of a structure $\mathcal{M}$ is model complete, then $\mathrm{Th}_{\forall\exists}(\mathcal{M}_M)$ and $\mathrm{Th}(\mathcal{M}_M)$ are equivalent $\mathcal{L}_M$-theories.*

PROOF. It is obvious that $\mathrm{Th}_{\forall\exists}(\mathcal{M}_M)$ is a subset of $\mathrm{Th}(\mathcal{M}_M)$. On the other hand, if $\mathrm{Th}(\mathcal{M}_M)$ is model complete, then it is $\forall\exists$-axiomatizable in the language $\mathcal{L}_M$. But any $\forall\exists$-axiomatization of $\mathrm{Th}(\mathcal{M}_M)$ is a subset of $\mathrm{Th}_{\forall\exists}(\mathcal{M}_M)$.                $\square$

It turns out that the model completeness of the elementary diagram of a structure $\mathcal{M}$ is a property of the theory of $\mathcal{M}$. To prove this fact, we introduce the following definition.

**Definition 2.29.** An existential $\mathcal{L}$-formula $\psi(\overline{x}, \overline{y})$ and an $\forall\exists$-formula of $\mathcal{L}$ form a *linked pair* (for $T$) if $T \models \exists \overline{y}\theta(\overline{y})$ and $T \models \forall\overline{y}\forall\overline{y}'\forall\overline{x}(\theta(\overline{y}) \wedge \theta(\overline{y}') \wedge \psi(\overline{x}, \overline{y}) \rightarrow \psi(\overline{x}, \overline{y}'))$.

**Proposition 2.30 ([67]).** *The elementary diagram of an $\mathcal{L}$-structure $\mathcal{M}$ is model complete if and only if for every $\mathcal{L}$-formula $\varphi(\overline{x})$ there is a linked pair $(\theta, \psi)$ such that $\mathcal{M} \models \exists \overline{y}\theta(\overline{y})$ and*

$$\mathcal{M} \models \forall\overline{y}(\theta(\overline{y}) \rightarrow \forall\overline{x}[\varphi(\overline{x}) \leftrightarrow \psi(\overline{x}, \overline{y})]). \qquad (*)$$

PROOF. Assume that the elementary diagram of $\mathcal{M}$ is model complete. Fix an $\mathcal{L}$-formula $\varphi(\overline{x})$. Since $\mathrm{Th}(\mathcal{M}_M)$ is model complete, there is an existential $\mathcal{L}$-formula $\psi(\overline{x}, \overline{y})$ and a tuple $\overline{b}$ in $M$

such that $\mathcal{M} \models \delta(\overline{b})$, where $\delta(\overline{y}) := \forall \overline{x}[\varphi(\overline{x}) \leftrightarrow \psi(\overline{x}, \overline{y})]$. Hence $\text{Th}_{\forall \exists}(\mathcal{M}_M) \models \delta(\overline{b})$ in view of Lemma 2.28. By compactness, there is an $\forall \exists$-formula $\theta(\overline{y})$ in $\mathcal{L}$ such that $\theta(\overline{b}) \in \text{Th}_{\forall \exists}(\mathcal{M}_M)$ and $\{\theta(\overline{b})\} \models \delta(\overline{b})$. (Without loss of generality, by padding $\delta$, we can assume that any constant symbol appearing in $\theta$ also appears in $\delta$.)

Conversely, assume that the right-hand side of $(*)$ holds. Fix an $\mathcal{L}_M$-formula $\varphi(\overline{x}, \overline{a})$, where $\varphi(\overline{x}, \overline{z})$ is an $\mathcal{L}$-formula and $\overline{a}$ belongs to $M$. Choose $\theta(\overline{y})$ and $\psi(\overline{x}, \overline{z}, \overline{y})$ corresponding to $\varphi(\overline{x}, \overline{z})$. Let $\overline{b}$ be any realization of $\theta(\overline{y})$ in $M$. Then

$$\mathcal{M} \models \forall \overline{x} \forall \overline{z}[\varphi(\overline{x}, \overline{z}) \leftrightarrow \psi(\overline{x}, \overline{z}, \overline{b})].$$

In particular, $\mathcal{M} \models \forall \overline{x}[\varphi(\overline{x}, \overline{a}) \leftrightarrow \psi(\overline{x}, \overline{a}, \overline{b})]$. Thus, every $\mathcal{L}_M$-formula is $\text{Th}(\mathcal{M}_M)$-equivalent to an existential $\mathcal{L}_M$-formula, which implies the model completeness of $\text{Th}(\mathcal{M}_M)$. $\qquad \square$

**Corollary 2.31** ([61]). *If $\mathcal{M}$ and $\mathcal{N}$ are elementarily equivalent $\mathcal{L}$-structures then the elementary diagram of $\mathcal{M}$ is model complete if and only if the elementary diagram of $\mathcal{N}$ is model complete. In particular, if $T$ is a complete theory and the elementary diagram of some model of $T$ is model complete, then the elementary diagram of every model of $T$ is model complete.*

**Proposition 2.32** ([61]). *Let $T$ be an $\mathcal{L}$-theory such that the elementary diagram of every model of $T$ is model complete. Then $T$ is $\exists \forall \exists$-axiomatizable.*

PROOF. Let $\mathcal{M}$ be an arbitrary model of $T$. Then $\text{Th}_{\forall \exists}(\mathcal{M}_M)$ implies $\sigma$. Therefore, there is a conjunction $\psi$ of $\forall \exists$-sentences of $\mathcal{L}_M$ that logically implies $\sigma$. Since none of the extra constant symbols in $M$ appears in $\sigma$, we can existentially quantify out these constant symbols and obtain a formula of the desired complexity which logically implies $\sigma$. $\qquad \square$

The following assertion immediately follows from Theorem 2.21 and Proposition 2.32.

**Corollary 2.33** ([61]). *Every strongly minimal theory with trivial pregeometry is $\exists\forall\exists$-axiomatizable.*

PROOF. By Theorem 2.21, $\mathrm{Th}(\mathcal{M}_M)$ is model complete and, consequently, $\forall\exists$-axiomatizable. Then $\mathrm{Th}_{\forall\exists}(\mathcal{M}_M)$ is a $\mathbf{0}''$-computable set of formulas which axiomatizes $\mathrm{Th}(\mathcal{M}_M)$, and so $\mathrm{Th}(\mathcal{M}_M)$ and its reduct $\mathrm{Th}(\mathcal{M})$ are $\mathbf{0}''$-computable sets of formulas as well. By relativisation theorem due to Harrington [74] and Khisamiev [83], any countable model of the theory $\mathrm{Th}(\mathcal{M})$ decidable relative to $\mathbf{0}''$, is $\mathbf{0}''$-computable. □

The following question still remains open.

**Problem 9.** Is the assertion of Corollary 2.33 remains valid for an arbitrary strongly minimal theory?

Recently, Khoussainov, Lempp, and Solomon proved the following result.

**Theorem 2.34** ([86]). *There exists an uncountably categorical strongly minimal theory $T$ with trivial pregeometry possessing a computable prime model such that all other models has the complexity of Turing degree $\mathbf{0}^2$.*

It is of interest to generalize the result of [86].

**Problem 10.** Whether there are examples of uncountably categorical strongly minimal theories $T_n$ possessing a computable models such that other models have the complexity of Turing degree $\mathbf{0}^{n+3}$, $n \geqslant 0$?

The following conjecture was suggested by S. Lempp.

**Conjecture 2.35.** *An uncountably categorical theory possessing a computable model is arithmetical.*

Note that the above result of Harrington [74] and Khisamiev [83, 84] can be relativized to show that if $T$ is uncountably categorical and arithmetic, then all models of $T$ admit arithmetic numbering. If Conjecture 2.35 could be confirmed, this would mean that all models of an uncountably categorical arithmetic theory admit arithmetic numberings. To confute Conjecture 2.35, it suffices to construct a theory with the properties listed in the following problem.

**Problem 11.** Whether there exists an uncountably categorical theory $T$ with models $\mathbf{A}_0 \preceq \mathbf{A}_1 \preceq \ldots \preceq \mathbf{A}_\omega$ such that $\mathbf{A}_0$ has a constructivization, and every $\mathbf{A}_{i+1}$, $i \in \omega$, has $\mathbf{0}^{i+1}$-constructivization, but not $\mathbf{0}^i$-constructivization?

### 2.1.3. Computable models of Ehrenfeucht theories.

In the case of countably categorical theories, the question about bounds for theories with decidable models is trivial. All countable models of such a theory are decidable if and only if the theory is decidable. The Ehrenfeucht theories are close to the countably categorical theories. Recall that a theory is a called an *Ehrenfeucht theory* if it has finitely many countable models. Naturally, the question is much more complicated if the computablity condition is required.

Peretyat'kin [135] proved that a prime model of an Ehrenfeucht theory is decidable. Lachlan constructed the first example of an Ehrenfeucht theory possessing six countable models such that only the prime model is decidable. Later, such examples for any $n \geqslant 3$ were constructed by Peretyat'kin.

However, there are still many open questions concerning theories possessing decidable models. First of all, we recall the well-known Morley problem.

**Problem 12** ([125]). Is it true that any countable model of any Ehrenfeucht theory with computable types is decidable?

The following weakened version of the Morley problem is also of interest.

**Problem 13.** Is it true that any countable models of any Ehrenfeucht theory with computable types is arithmetical?

Ash and Millar [7] proved that all models of hyperarithmetical Ehrenfeucht theories are decidable in hyperarithmetical degrees. Millar [107] and Reed [138] constructed examples of decidable Ehrenfeucht theories with a given complexity for some nonprincipal type and, consequently, with a given hyperarithmetical complexity of their countably saturated model and some other models. The following problem arises in a natural way.

**Problem 14.** Is it true that all countable models of any Ehrenfeucht theory with arithmetical types are decidable (computable) relative to some arithmetical Turing degree?

Note that all homogeneous models of an Ehrenfeucht theory with decidable (arithmetic) types are decidable (relative to some arithmetic Turing degree) [40]. This fact immediately follows from the decidablity theorem for homogeneous models with countable family of types realized there and countable family of all decidable types of theories of this model [40].

The above discussion suggests the following strategy.

**Problem 15.** Show that all almost homogeneous models of an Ehrenfeucht theory with decidable (arithmetical) types are decidable (relative to some arithmetical type).

The following weaker property is also of interest.

**Problem 16.** Show that all almost homogeneous models of an Ehrenfeucht theory with decidable (arithmetical) types are computable (relative to some arithmetical type).

Recall that a model is said to be *almost homogeneous* if it is homogeneous in some enrichment by constants for a finite collection of its elements.

It is remarkable that, in all known examples of Ehrenfeucht theories, all countable models are almost homogeneous. The assertion that all countable models of any Ehrenfeucht theory are almost homogeneous (if it is true) could be helpful for resolving the problems. On the other hand, a counterexample could open a door to the negative solution of the Morley problem.

To consider computable models of such theories, we start with the following open question.

**Problem 17.** Characterize Ehrenfeucht theories possessing computable models.

It is of interest to generalize the result of Lerman and Schmerl [96] for countably categorical *arithmetic* theories to the case of Ehrenfeucht theories. The same question can be considered regarding the Knight theorem [90] for non-arithmetic Ehrenfeucht theories.

The above results on the complexity for countably categorical theories yield the following assertions.

**Corollary 2.36** ([68]). *For every $n \geqslant 1$ there exists am Ehrenfeucht theory of Turing degree $0^n$ that has a computable model.*

**Corollary 2.37** ([43]). *For every arithmetical Turing degree $d$ there exists an Ehrenfeucht theory of Turing degree $d$ that has a computable model.*

We complete this section with the following result asserting the existence of computable models of non-arithmetic countably categorical theories.

**Theorem 2.38** (Fokina, Goncharov, Khoussainov). *There exists an Ehrenfeucht theory $T$ with a computable model and the Turing degree $T$ is non-arithmetical.*

## 3. Structure Complexity of Computable Models

In this section, we discuss necessary conditions on the structure of computable models from the point of view of the model-theoretic complexity. For this purpose, we choose a language with infinite disjunctions and conjunctions. Then every countable model can be described up to an isomorphism and the number of necessary infinite disjunctions and conjunctions determines the ordinal level of the structure complexity of the model. Using the theory of admissible sets, it is possible to obtain an upper bound for the complexity of a computable model. The sharpness of the bound and the realizability of all less complexities play an important role for describing structural properties of computable models. We present two methods based on the Scott rank and on the Barwise rank.

## 3.1. Definability of computable models

Recall that the Scott rank is a measure of the model-theoretic complexity. This term came from the Scott isomorphism theorem [144].

**Theorem 3.1** (the Scott isomorphism theorem). *For every countable structure $\mathcal{A}$ (for a countable language $L$) there is an $L_{\omega_1\omega}$ sentence whose countable models are isomorphic copies of $\mathcal{A}$.*

To prove this assertion, Scott assigned countable ordinals to tuples in $\mathcal{A}$ and to $\mathcal{A}$ itself. There are several different definitions of the *Scott rank*.

Let $\overline{a}$ and $\overline{b}$ be tuples in $\mathcal{A}$.

- We write $\overline{a} \equiv^0 \overline{b}$ if $\overline{a}$ and $\overline{b}$ satisfy the same quantifier-free formulas.
- Let $\alpha > 0$. We write $\overline{a} \equiv^\alpha \overline{b}$ if for all $\beta < \alpha$ and $\overline{c}$ there exists $\overline{d}$ and for every $\overline{d}$ there exists $\overline{c}$ such that $\overline{a}, \overline{c} \equiv^\beta \overline{b}, \overline{d}$.

**Definition 3.2.** The *Scott rank of a tuple $\overline{a}$* in $\mathcal{A}$ is the least $\beta$ such that for all $\overline{b}$ from $\overline{a} \equiv^\beta \overline{b}$ it follows that $(\mathcal{A}, \overline{a}) \cong (\mathcal{A}, \overline{b})$.

**Definition 3.3.** The *Scott rank* $\mathrm{SR}(\mathcal{A})$ *of $\mathcal{A}$* is the least ordinal $\alpha$ greater than the ranks of all tuples in $\mathcal{A}$.

**Example 3.4.** If $\mathcal{A}$ is an ordering of type $\omega$, then SR $(\mathcal{A}) = 2$. We have $\bar{a} \equiv^0 \bar{b}$ if $\bar{a}$ and $\bar{b}$ are ordered in the same way. We have $\bar{a} \equiv^1 \bar{b}$ if the corresponding intervals (before the first element and between successive elements) are of the same size, and this fact is enough to assure an isomorphism. Hence the tuples have Scott rank 1 and the ordering has Scott rank 2.

## 3.2. The Kleene notation system $\mathcal{O}$

As in the general algorithm theory, for constructing models of given complexity and estimating the complexity an important role is played by computable ordinals and the Kleene notation system $\mathcal{O}$ (cf. [140]) for all computable ordinals. The least ordinal having no notation in the Kleene system is referred to as the *Church–Kleene ordinal* and is denoted by $\omega_1^{CK}$. It is easy to check that it is the least noncomputable ordinal.

Recall that the Kleene notation system consists of a set $\mathcal{O}$ of notations equipped with a partial ordering $<_\mathcal{O}$. The ordinal 0 has notation 1. If $a$ is the notation of $\alpha$, then $2^a$ is the notation of $\alpha + 1$. Then $a <_\mathcal{O} 2^a$, and $b <_\mathcal{O} a$ implies $b <_\mathcal{O} 2^a$.

Suppose that $\alpha$ is a limit ordinal. If $\varphi_e$ is a total function providing notations for an increasing sequence of ordinals with limit $\alpha$, then $3 \cdot 5^e$ is the notation of $\alpha$. For all $n$ we have $\varphi_e(n) <_\mathcal{O} 3 \cdot 5^e$, and $b <_\mathcal{O} \varphi_e(n)$ implies $b <_\mathcal{O} 3 \cdot 5^e$. The set $\mathcal{O}$ is $\Pi_1^1$ complete.

## 3.3. Computable infinitary formulas

For any notation from the Kleene notation system $\mathcal{O}$ it is possible to introduce infinitary formulas which are used to describe computable structures. Roughly speaking, we will define infinitary formulas on a fixed level where the disjunctions and conjunctions of computable formulas from previous levels are computable. They are essentially the same as the formulas in the least admissible fragment of $L_{\omega_1\omega}$.

We may classify computable infinitary formulas as *computable* $\Sigma_\alpha$, or *computable* $\Pi_\alpha$, for various computable ordinals $\alpha$. We have the useful fact that in a computable structure, a relation defined by a computable $\Sigma_\alpha$ (or computable $\Pi_\alpha$) formula will be $\Sigma_\alpha^0$ (or $\Pi_\alpha^0$). To illustrate the expressive power of computable infinitary formulas, we note that there is a natural computable $\Pi_2$ sentence characterizing the class of Abelian $p$-groups. For every computable ordinal $\alpha$ there is a computable $\Pi_{2\alpha}$ formula saying that the height is at least $\omega \cdot \alpha$ for an element of an Abelian $p$-group.

The following theorem presents a well-known useful version of the compactness theorem for computable infinitary formulas.

**Theorem 3.5** (the Barwise–Kreisel compactness theorem). *Let $\Gamma$ be a $\Pi_1^1$ set of computable infinitary sentences. If every $\Delta_1^1$ subset of $\Gamma$ has a model, then $\Gamma$ also has a model.*

Theorem 3.5 can be used for obtaining computable structures and special computable sequences of computable structures.

**Corollary 3.6.** *Let $\Gamma$ be a $\Pi_1^1$ set of computable infinitary sentences. If every $\Delta_1^1$ subset has a computable model, then $\Gamma$ also has a computable model.*

Corollary 3.6 can be applied uniformly to $\Pi_1^1$ sets of computable infinitary sentences.

The following two assertions demonstrate the expressive power of computable infinitary formulas.

**Corollary 3.7.** *If $\mathcal{A}$ and $\mathcal{B}$ are computable structures satisfying the same computable infinitary sentences, then $\mathcal{A} \cong \mathcal{B}$.*

**Corollary 3.8.** *Suppose that $\overline{a}$ and $\overline{b}$ are tuples satisfying the same computable infinitary formulas in a computable structure $\mathcal{A}$. Then there is an automorphism of $\mathcal{A}$ sending $\overline{a}$ to $\overline{b}$.*

Theorem 3.5 and Corollaries 3.6–3.8 are well known and may be found, for example, in [5] (cf. also [66]).

## 3.4. Computable rank

**Definition 3.9.** The *computable rank* $R^c(\mathcal{A})$ of a structure $\mathcal{A}$ is the first ordinal $\alpha$ such that for all tuples $\bar{a}$, $\bar{b}$ in $\mathcal{A}$ of the same length the following holds: if for all $\beta < \alpha$ all computable $\Pi_\beta$ formulas true of $\bar{a}$ are also true of $\bar{b}$, then there is an automorphism sending $\bar{a}$ to $\bar{b}$.

By Corollary 3.7, if $\mathcal{A}$ is a hyperarithmetical structure, then $R^c(\mathcal{A}) \leqslant \omega_1^{CK}$. In this case, Definition 3.9 can be formulated as follows: The *computable rank* is the first ordinal $\alpha$ such that for all tuples $\bar{a}$ and $\bar{b}$ of the same length the following holds: if $\bar{a}$ and $\bar{b}$ satisfy the same computable $\Pi_\beta$ formulas for $\beta < \alpha$, then they satisfy the same computable $\Pi_\alpha$ formulas.

**Proposition 3.10.** *For any computable language $L$ and computable ordinal $\alpha$ (or any notation) there exists a computable infinitary sentence saying that $R^c(\mathcal{A}) \geqslant \alpha$ for an $L$-structure $\mathcal{A}$.*

Note that the notion of computable rank essentially differs from that of Scott rank. Nevertheless, in the case of a hyperarithmetical structure $\mathcal{A}$, the computable rank is a computable ordinal just as the Scott rank is computable. If $R^c(\mathcal{A})$ is computable, then $\mathcal{A}$ has a computable Scott sentence. The converse assertion is also true.

**Proposition 3.11** (J. Millar[1]). *Suppose that $\mathcal{A}$ is a hyperarithmetical and $R^c(\mathcal{A}) = \omega_1^{CK}$. If $\psi$ is a computable infinitary sentence true in $\mathcal{A}$, then $\psi$ is also true in some hyperarithmetical $\mathcal{B} \not\cong \mathcal{A}$.*

SKETCH OF PROOF. Let $\mathcal{A}^*$ be an expansion of $\mathcal{A}$ with a predicate $R_\varphi$ for every computable infinitary formula $\varphi$, up to complexity $\alpha$. Since the rank of $\mathcal{A}$ is not computable, $\mathcal{A}^*$ is not homogeneous. Therefore, there is some tuple $\bar{a}$ realizing a non-principal type in $\mathcal{A}^*$. We produce a hyperarithmetical model $\mathcal{B}^*$ of the

---

[1]Private communication.

elementary first order theory of $\mathcal{A}^*$ omitting the type of $\overline{a}$ and satisfying $\psi$. To guarantee that $\psi$ is true, we make sure that for all subformulas $\varphi(\overline{u})$

$$\mathcal{B}^* \models \forall \overline{u}\, [\varphi(\overline{u}) \leftrightarrow R_\varphi(\overline{u})].$$

If $\varphi(\overline{u})$ is the disjunction of $\varphi_i(\overline{u})$, we need to omit the type consisting of $R_\varphi(\overline{u})$ and the formulas $\neg R_{\varphi_i}(\overline{u})$. If $\varphi(\overline{u})$ is the conjunction of $\varphi_i(\overline{u})$, we need to omit the type consisting of $\neg R_\varphi(\overline{u})$ and the formulas $R_{\varphi_i}(\overline{u})$. □

## 3.5. Rank and isomorphisms

We revise the Scott isomorphism theorem by looking for isomorphisms of bounded complexity.

**Definition 3.12.** Let $\alpha$ be a computable ordinal. A *formally* $\Sigma_\alpha^0$ *Scott family* is a c.e. Scott family $\Phi$ made up of computable $\Sigma_\alpha$ formulas, possibly with a fixed tuple of parameters.

**Definition 3.13.** A computable structure $\mathcal{A}$ is $\Delta_\alpha^0$ *categorical* if $\mathcal{A} \cong_{\Delta_\alpha^0} \mathcal{B}$ for every computable copy $\mathcal{B}$.

**Theorem 3.14** (Ash, Goncharov). *Suppose that $\mathcal{A}$ is computable. If $\mathcal{A}$ has a formally $\Sigma_\alpha^0$ Scott family, then it is $\Delta_\alpha^0$ categorical. With some added effectiveness on one copy of $\mathcal{A}$, the converse holds.*

This assertion was proved in [45, 46] in the computable case and in [1] in the general case.

**Proposition 3.15.** *Let $\alpha$ be a computable ordinal. For a given index of a computable structure $\mathcal{A}$ such that $R^c(\mathcal{A}) = \alpha$ there is an index of a formally $\Sigma_{\alpha+2}^0$ Scott family for $\mathcal{A}$ without parameters.*

Suppose that $K$ is a class of structures such that there is a computable bound on $R^c(\mathcal{A})$ for $\mathcal{A} \in K^c$. Proposition 3.15 asserts that for a given index of $\mathcal{A} \in K^c$ we can find an index of a Scott family consisting of formulas of bounded complexity. Then we can pass to a computable infinitary Scott sentence.

These results yield a bound on the Scott ranks for computable structures [**127**]. There are examples of computable structures having various computable Scott ranks and familiar structures (for example, the Harrison ordering) with Scott rank $\omega_1^{CK} + 1$ [**70**]. Makkai [**99**] constructed a structure of Scott rank $\omega_1^{CK}$ which can be made computable [**68**] and simplified it so that it is just a tree [**14**]. As was shown in [**13**], it is possible to construct further computable structures of Scott rank $\omega_1^{CK}$ in the classes of undirected graphs, fields of any characteristic, and linear orderings. These results give us interesting examples of computable structures with different complexity of the isomorphism problem for different computable representations.

**Proposition 3.16.** *Let $\mathcal{A}$ be a computable structure. Then* $\mathrm{SR}\,(\mathcal{A}) \leqslant \omega_1^{CK} + 1$.

The further properties of computable structures are listed in the following assertion.

**Proposition 3.17.** *Let $\mathcal{A}$ be a computable structure. Then*

(1) $\mathrm{SR}\,(\mathcal{A}) < \omega_1^{CK}$ *if there is a computable ordinal $\beta$ such that the orbits of all tuples are defined by computable $\Pi_\beta$ formulas,*

(2) $\mathrm{SR}\,(\mathcal{A}) = \omega_1^{CK}$ *if the orbits of all tuples are defined by computable infinitary formulas, but there is no computable bound on the complexity of these formulas,*

(3) $\mathrm{SR}\,(\mathcal{A}) = \omega_1^{CK} + 1$ *if there is some tuple whose orbit is not defined by any computable infinitary formula.*

The low Scott rank is associated with simple Scott sentences. Recall that a *Scott sentence* for $\mathcal{A}$ is a sentence whose countable models are just the isomorphic copies of $\mathcal{A}$ (as in the Scott isomorphism theorem).

**Theorem 3.18** ([127, 128]). *Let $\mathcal{A}$ be a computable structure. Then* SR $(\mathcal{A})$ *is computable if and only if $\mathcal{A}$ has a computable infinitary Scott sentence.*

**Corollary 3.19.** *Let $\Gamma$ be a $\Pi_1^1$-set of computable infinitary sentences. If every $\Delta_1^1$-set $\Gamma' \subseteq \Gamma$ has a computable model, then $\Gamma$ has a computable model.*

**Proposition 3.20** ([139]). *Suppose that $\mathcal{A}$ is a hyperarithmetical structure. Let $\Gamma$ be a $\Pi_1^1$-set of computable infinitary sentences in a finite expansion of the language of $\mathcal{A}$. Suppose that for each $\Delta_1^1$-set $\Gamma' \subseteq \Gamma$ the structure $\mathcal{A}$ can be expanded to a model of $\Gamma'$. Then $\mathcal{A}$ can be expanded to a model of $\Gamma$.*

**Corollary 3.21.** *Let $\mathcal{A}$ be a hyperarithmetical structure. If $\bar{a}$ and $\bar{b}$ are tuples in $\mathcal{A}$ satisfying the same computable infinitary formulas, then there is an automorphism of $\mathcal{A}$ sending $\bar{a}$ to $\bar{b}$.*

Consider three different types of Scott rank for computable models described in Proposition 3.17 that are realized in classical algebras and models.

In the case SR $(\mathcal{A}) < \omega_1^{CK}$, the following structural property of computable models holds.

**Proposition 3.22.** *All computable members of the following structures have a computable Scott rank:*

- *well orderings,*
- *superatomic Boolean algebras,*
- *reduced Abelian p-groups.*

An interesting class of models is formed by computable models with SR $(\mathcal{A}) = \omega_1^{CK} + 1$. There are well-known examples of computable structures of Scott rank $\omega_1^{CK} + 1$. Harrison showed that there is a computable ordering of type $\omega_1^{CK}(1 + \eta)$, called the *Harrison ordering*, which gives rise to some other computable structures with similar properties. The *Harrison Boolean algebra* is the interval algebra of the Harrison ordering. The *Harrison*

*Abelian p-group* has length $\omega_1^{CK}$, with all infinite Ulm invariants and a divisible part of infinite dimension.

**Proposition 3.23.** *The Harrison ordering, Harrison Boolean algebra, and Harrison Abelian p-groups have Scott rank* $\omega_1^{CK} + 1$.

It was unexpected that there exist models with $\mathrm{SR}\,(\mathcal{A}) = \omega_1^{CK}$.

In the case of the Scott rank $\omega_1^{CK}$, it is not easy to find computable examples. An arithmetical example was constructed by Makkai.

**Theorem 3.24 ([99]).** *There is an arithmetical structure* $\mathcal{A}$ *of rank* $\omega_1^{CK}$.

Models $\mathcal{A}$ of Scott rank $\mathrm{SR}\,(\mathcal{A}) = \omega_1^{CK}$ will be referred to as *Makkai models*.

In the Makkai example, in contrast to the Harrison ordering, the set of computable infinitary sentences that are true in the structure is $\aleph_0$ categorical. Hence the conjunction of these sentences is a Scott sentence for the structure. The following assertion can be proved on the basis of the results of [68] and [92].

**Theorem 3.25.** *There exists a computable structure of Scott rank* $\omega_1^{CK}$.

As was proved in [14], there exists a computable tree of Scott rank $\omega_1^{CK}$. This construction may be employed in other situations. The authors of [14] used the idea to take trees as Knight–Millar Trees and add a homogeneity property. In more detail, let $\mathcal{T}$ be a subtree of $\omega^{<\omega}$. We have a top node $\varnothing$. We will define the *tree rank* for $\sigma \in \mathcal{T}$ and then for $\mathcal{T}$. Below, we use the notation $rk(\sigma)$, $rk(\mathcal{T})$.

- $rk(\sigma) = 0$ if $\sigma$ is terminal.
- For $\alpha > 0$, $rk(\sigma) = \alpha$ if all successors of $\sigma$ have ordinal rank, and $\alpha$ is the first ordinal greater than these ordinals.

- $rk(\sigma) = \infty$ if $\sigma$ does not have ordinal rank.

We set $rk(T) = rk(\varnothing)$.

REMARK. $rk(\sigma) = \infty$ if and only if $\sigma$ extends to a path.

If $T$ is a tree, we denote by $T_n$ the set of elements at level $n$ in the tree, i.e., $T_n = T \cap \omega^n$.

**Definition 3.26.** A tree $T$ is *thin* if for all $n$ the set of ordinal ranks of elements of $T_n$ has order type at most $\omega \cdot n$.

This definition is used as follows. If $T$ is a computable thin tree, then for every $n$ there is a computable $\alpha_n$ such that for all $\sigma \in T_n$ from $rk(\sigma) \geqslant \alpha_n$ it follows that $rk(\sigma) = \infty$.

**Theorem 3.27 ([92]).** *The following assertions hold.*

(1) *There exists a computable thin tree $T$ with a path but no hyperarithmetical path.*

(2) *If $T$ is a computable thin tree with a path but no hyperarithmetical path, then $\mathcal{A}(T)$ is a computable structure of Scott rank $\omega_1^{CK}$.*

A computable tree of Scott rank $\omega_1^{CK}$ was constructed in [14]. This tree satisfies some conditions from [92] and the following homogeneity property.

**Definition 3.28.** A tree $T$ is *rank-homogeneous* if for all $n$ the following conditions are satisfied:

- for all $\sigma \in T_n$ and computable $\alpha$, if there exists $\tau \in T_{n+1}$ such that $rk(\tau) = \alpha < rk(\sigma)$, then $\sigma$ has infinitely many successors $\sigma'$ with $rk(\sigma') = \alpha$,
- for all $\sigma \in T_n$, if $rk(\sigma) = \infty$, then $\sigma$ has infinitely many successors $\sigma'$ with $rk(\sigma') = \infty$.

REMARK. Suppose that $T$ and $T'$ are rank-homogeneous trees and for all $n$ there is an element in $T_n$ of rank $\alpha \in \mathrm{Ord} \cup \{\infty\}$ if and only if there is an element in $T'_n$ of rank $\alpha$. Then $T \cong T'$.

In [14], the construction of a tree of Scott rank $\omega_1^{CK}$ is based on the following result.

**Theorem 3.29** ([92]). *The following assertions hold.*

(1) *There is a computable thin rank-homogeneous tree $T$ such that $rk(T) = \infty$ but $T$ has no hyperarithmetical path.*

(2) *If $T$ is a computable thin rank-homogeneous tree such that $rk(T) = \infty$ but $T$ has no hyperarithmetical path, then $\mathrm{SR}(T) = \omega_1^{CK}$.*

As in the case of group-trees, the computable infinitary theory is $\aleph_0$ categorical for the trees considered in [14]. But, unlike group-trees, there are many nontrivial hyperarithmetical automorphisms. It is possible to produce a tree as above, with the property of strong computable approximability [14].

**Definition 3.30.** A structure $\mathcal{A}$ is *strongly computably approximable* if for any $\Sigma_1^1$ set $S$ there exists a uniformly computable sequence $(\mathcal{C}_n)_{n\in\omega}$ such that $n \in S$ if and only if $\mathcal{C}_n \cong \mathcal{A}$. The structures $\mathcal{C}_n$ with $n \notin S$ are said to be *approximating*.

For example, it is well known that the Harrison ordering is strongly computably approximable by computable well orderings.

**Theorem 3.31** ([14]). *There is a computable tree $T$ of Scott rank $\omega_1^{CK}$ such that $T$ is strongly computably approximable. Moreover, the approximating structures are trees of computable Scott rank.*

Using these trees, it is possible to construct many new examples of Makkai models.

**Theorem 3.32** ([13]). *Each of the following classes contains computable structures of Scott rank $\omega_1^{CK}$:*

- *undirected graphs,*
- *linear orderings,*
- *Boolean algebras,*
- *fields of any characteristic.*

Thus, it is of interest to clarify how to determine the nonconstructive Scott rank on a computable model from its computable representation.

**Problem 18.** What is the complexity of the index set of Makkai models in universal computable numberings of computable models of a fixed signature?

**Problem 19.** What is the complexity of the index set of computable models of Scott rank $\omega_1^{CK} + 1$ in universal computable numberings of computable models of a fixed signature?

### 3.5.1. Barwise rank.

Recall the definition of the *quantifier rank* of a formula (we assume that the implication $\Rightarrow$ is expressed in terms of $\neg$ and $\wedge$ and thereby it does not occur directly in the formulas under consideration):

$$
qr(\varphi) = \begin{cases}
0 & \text{if } \varphi \text{ is quantifier-free;} \\
qr(\psi) & \text{if } \varphi \text{ is } \neg\psi; \\
qr(\psi) + 1 & \text{if } \varphi \text{ is } \exists v\psi \text{ or } \forall v\psi; \\
sup\{qr(\psi) \mid \psi \in \Phi\} & \text{if } \varphi \text{ is } \bigwedge \Phi \text{ or } \bigvee \Phi.
\end{cases}
$$

Show that for computable models we have $\mathrm{SR}\,(\mathcal{A}) \leqslant \omega_1^{CK}$ for the complexity of Barwise rank. Let $\alpha$ be an ordinal.

Models $\mathfrak{M}$ and $\mathfrak{N}$ are $\alpha$-*equivalent* ($\mathfrak{M} \equiv^\alpha \mathfrak{N}$) if they satisfy the same sentences with quantifier rank at most $\alpha$. Two tuples $\bar{a}, \bar{b} \in \mathfrak{M}^{<\omega}$ are $\alpha$-*equivalent* ($\bar{a} \equiv^\alpha \bar{b}$) if they satisfy the same formulas with quantifier rank at most $\alpha$.

We say that a tuple $\bar{a} \in \mathfrak{M}^{<\omega}$ has *quantifier rank* $\alpha$ in $\mathfrak{M}$ if $(\bar{a} \equiv^\alpha \bar{b} \Rightarrow \bar{a} \equiv \bar{b})$ for all tuples $\bar{b} \in \mathfrak{M}^{<\omega}$.

The *Barwise rank* $\mathrm{br}\,(\mathfrak{M})$ of a model $\mathfrak{M}$ is the minimal ordinal $\alpha$ such that $(\bar{a} \equiv^\alpha \bar{b} \Rightarrow \bar{a} \equiv^{\alpha+1} \bar{b})$ for all $\bar{a}, \bar{b} \in \mathfrak{M}^{<\omega}$.

As is known, the Barwise rank of a model $\mathfrak{M} \in \mathbb{HYP}_\omega$ does not exceed $\omega_1^{\mathrm{CK}}$.

The following assertion concerning the existence of hyperarithmetical isomorphisms for different computable representations of models shows a close connection between this problem and the $\Pi_1^1\Pi$ definability of relations on computable models.

**Theorem 3.33** ([71]). *Let $\mathfrak{M}$ be a hyperarithmetical model. The following assertions are equivalent.*

(1) *There exist tuples $\bar{a}, \bar{b} \in \mathfrak{M}^{<\omega}$ such that $\langle \mathfrak{M}, \bar{a} \rangle \cong \langle \mathfrak{M}, \bar{b} \rangle$, but $\langle \mathfrak{M}, \bar{a} \rangle \not\cong_h \langle \mathfrak{M}, \bar{b} \rangle$.*

(2) *There is a tuple $\bar{a} \in \mathfrak{M}^{<\omega}$ such that there exists an infinite family $(\bar{a}_i)_{i<\omega}$ of tuples in $\mathfrak{M}^{<\omega}$ with the following properties:*
   (a) *$\langle \mathfrak{M}, \bar{a} \rangle \cong \langle \mathfrak{M}, \bar{a}_i \rangle$ for all $i < \omega$,*

   (b) *$\langle \mathfrak{M}, \bar{a}_i \rangle \not\cong_h \langle \mathfrak{M}, \bar{a}_j \rangle$ for all $i < j < \omega$.*

(3) *The Barwise rank of $\mathfrak{M}$ is equal to $\omega_1^{\mathrm{CK}}$.*

(4) *$I_{\mathfrak{M}} \notin \Pi_1^1$, where $I_{\mathfrak{M}} = \{ \langle \bar{a}, \bar{b} \rangle \in \mathfrak{M}^{<\omega} \times \mathfrak{M}^{<\omega} \mid \bar{a} \cong \bar{b} \}$.*

Assume that there exist two isomorphic hyperarithmetical models $\mathfrak{M}$ and $\mathfrak{N}$ that are not hyperarithmetically isomorphic.

**Problem 20.** Is it true that there exists a computable sequence of hyperarithmetical models $\mathfrak{M}_n$, $n \in \omega$, such that every $\mathfrak{M}_n$ is isomorphic, but not hyperarithmetically isomorphic to $\mathfrak{M}$?

### 3.5.2. Intrinsically $\Pi_1^1$ relations.

In view of Theorem 3.33, it is important to have a description for relations with $\Pi_1^1$ complexity in computable hyperarithmetical models. The first results on analytic complexity were obtained by Soskov [147, 148].

**Proposition 3.34** ([148]). *Suppose that $\mathcal{A}$ is computable and $R$ is a $\Delta_1^1$ relation invariant under automorphisms of $\mathcal{A}$. Then $R$ is definable in $\mathcal{A}$ by a computable infinitary formula without parameters.*

**Corollary 3.35.** *For a computable structure $\mathcal{A}$ and a relation $R$ on $\mathcal{A}$ the following assertions are equivalent:*

(1) *$R$ is intrinsically $\Delta^1_1$ on $\mathcal{A}$,*

(2) *$R$ is relatively intrinsically $\Delta^1_1$ on $\mathcal{A}$,*

(3) *$R$ is definable in $\mathcal{A}$ by a computable infinitary formula with finitely many parameters.*

**Definition 3.36.** A relation $R$ on $\mathcal{A}$ is *formally $\Pi^1_1$ on $\mathcal{A}$* if it is defined in $\mathcal{A}$ by the $\Pi^1_1$ disjunction of computable infinitary formulas with finitely many parameters.

We formulate the result of [147] in the following form.

**Proposition 3.37.** *For a computable (hyperarithmetical) structure $\mathcal{A}$ and a relation $R$ on $\mathcal{A}$ the following assertions are equivalent:*

(1) *$R$ is relatively intrinsically $\Pi^1_1$ on $\mathcal{A}$,*

(2) *$R$ is formally $\Pi^1_1$ on $\mathcal{A}$.*

**Theorem 3.38** ([70])**.** *Suppose that $\mathcal{A}$ is a computable structure and $R$ is a relation on $\mathcal{A}$ such that it is $\Pi^1_1$ and is invariant under automorphisms of $\mathcal{A}$. Then $R$ is formally $\Pi^1_1$. Moreover, it is possible to define it without parameters.*

**Corollary 3.39** ([70])**.** *For a computable structure $\mathcal{A}$ and a relation $R$ the following assertions are equivalent:*

(1) *$R$ is intrinsically $\Pi^1_1$ on $\mathcal{A}$,*

(2) *$R$ is relatively intrinsically $\Pi^1_1$ on $\mathcal{A}$,*

(3) *$R$ is formally $\Pi^1_1$ on $\mathcal{A}$.*

We say that a relation is *properly $\Pi^1_1$* if it is $\Pi^1_1$, but not $\Sigma^1_1$.

**Corollary 3.40** ([70])**.** *If a relation $R$ on a computable structure $\mathcal{A}$ is invariant and properly $\Pi^1_1$, then the image of $R$ in any computable copy is also properly $\Pi^1_1$.*

There are several examples of computable structures with intrinsically $\Pi_1^1$ relations.

**Example 3.41.** The *Harrison ordering* is a computable ordering of type $\omega_1^{CK}(1 + \eta)$. The existence of such an ordering was proved by Harrison who showed that for any computable tree $T \subseteq \omega^{<\omega}$ such that $T$ has paths, but no hyperarithmetical paths, the Kleene–Brouwer ordering on $T$ is a computable ordering of type $\omega_1^{CK}(1 + \eta) + \alpha$ with some computable ordinal $\alpha$.

Let $\mathcal{A}$ be the Harrison ordering, and let $R$ be the initial segment of type $\omega_1^{CK}$. This set is intrinsically $\Pi_1^1$ since it is defined by the disjunction of computable infinitary formulas saying that the interval to the left of $x$ has order type $\beta$ for computable ordinals $\beta$.

**Example 3.42.** The *Harrison Boolean algebra* is the interval algebra of the Harrison ordering.

Let $\mathcal{A}$ be the Harrison Boolean algebra, and let $R$ be the set of superatomic elements containing in some of the Frechet ideals. This set is intrinsically $\Pi_1^1$ since it is defined by the disjunction of computable infinitary formulas saying that $x$ is a finite join of $\alpha$-atoms, where $\alpha$ is a computable ordinal.

**Example 3.43.** Recall that a countable Abelian $p$-group $\mathcal{G}$ is determined up to an isomorphism by its Ulm sequence $(u_\alpha(\mathcal{G}))_{\alpha < \lambda(\mathcal{G})}$ and the dimension of the divisible part. The *Harrison $p$-group* is a computable Abelian $p$-group $\mathcal{G}$ such that $\lambda(\mathcal{G}) = \omega_1^{CK}$, $u_\mathcal{G}(\alpha) = \infty$ for all $\alpha < \omega_1^{CK}$ and the divisible part $D$ has infinite dimension.

By a *Harrison group* we mean the Harrison $p$-group for some $p$. Let $\mathcal{A}$ be a Harrison group, and let $R$ be the set of elements with computable ordinal height, the complement of the divisible part. Then $R$ is intrinsically $\Pi_1^1$ on $\mathcal{A}$ since it is defined by the disjunction of computable infinitary formulas saying that $x$ has height $\alpha$, where $\alpha$ is a computable ordinal.

**Theorem 3.44.** *For the Harrison groups, Harrison Boolean algebra, and Harrison ordering there are computable representations without hyperarithmetical isomorphisms.*

**Problem 21.** Characterize $\Pi_1^1$ relations for other classes of analytic hierarchy.

## 4. Isomorphism Problem

In this section, we consider isomorphisms of constructive and computable models. Some of the results described below are taken from [**67**].

### 4.1. Isomorphisms of countably categorical models

Owing to the fundamental concept of a computable isomorphism, it is possible to recognize whether or not two constructivizations of a model have the same computability–theoretic properties.

**Definition 4.1.** Constructive algebraic systems $(\mathbf{A}, \nu)$ and $(\mathbf{A}, \mu)$ are *computably isomorphic* if there exists an automorphism $\alpha$ of $\mathbf{A}$ and a computable function $f$ such that $\alpha\nu(n) = \mu(f(n))$ for all $n \in \omega$. In this case, $\nu$ and $\mu$ are said to be *autoequivalent*.

A similar definition can be introduced for computable models.

**Definition 4.2.** Let $\mathcal{A}$ be a computable structure. We say that $\mathcal{A}$ is *computably categorical* if for all computable $\mathcal{B} \cong \mathcal{A}$ there is a computable isomorphism from $\mathcal{A}$ onto $\mathcal{B}$.

Computably isomorphic structures cannot be distinguished in terms of computability–theoretic properties of definable relations. This means that for any definable relation $R$ in $\mathbf{A}$ (or even for $R$ invariant under automorphisms of $\mathbf{A}$) the Turing degrees of $R$

under the constructivizations $\nu$ and $\mu$ are equivalent, i.e., $\nu^{-1}(R)$
and $\mu^{-1}(R)$ have the same Turing degree. In addition, if $\nu$ and
$\mu$ are bijections, then $\nu^{-1}(R)$ and $\mu^{-1}(R)$ are computably invariant. Within the study of computable isomorphisms, the following
important notion was introduced by Goncharov.

**Definition 4.3.** The *dimension* dim $(\mathbf{A})$ of an algebraic system $\mathbf{A}$ is the maximal number of its nonautoequivalent constructivizations of $\mathbf{A}$.

It is easy to see that the algebraic dimension can be expressed
in terms of computable models. Namely, the dimension of an algebraic system $\mathbf{A}$ is equivalent to the maximal number of computable
models that are not computably isomorphic each other, but they
are isomorphic to $\mathbf{A}$. Informally, if we know the dimension of an
algebraic system $\mathbf{A}$, we know the number of effective realizations
of $\mathbf{A}$. The dimension of an algebraic system $\mathbf{A}$ can be represented
in computability–theoretic terms as the number of computable isomorphism types of $\mathbf{A}$. Thus, if dim $\mathbf{A} = 1$, the algebraic system
$\mathbf{A}$ has exactly one effective realization. We single out algebraic
systems of dimension 1.

**Definition 4.4** ([100]). An algebraic system $\mathbf{A}$ is said to
be *autostable* if dim$(\mathbf{A}) = 1$ and *strongly autostable* if all strong
constructivizations of $\mathbf{A}$ are autoequivalent.

The notion of an effectively infinite algebraic system, introduced by Goncharov, is used in the study of computable isomorphisms. A sequence $(\mathbf{A}_0, \nu_0), (\mathbf{A}_1, \nu_1), \ldots$ of constructive models is
*effective* if the set $\{(i, \varphi) | \varphi \in AD_{\nu_i}(\mathbf{A}_i)\}$ is uniformly computable.

**Definition 4.5.** An algebraic system $\mathbf{A}$ is said to be *effectively infinite* if there is an algorithm such that, applying it
to any index of an effective sequence of constructive systems
$(\mathbf{A}, \nu_0), (\mathbf{A}, \nu_1), \ldots$, we obtain a constructive model $(\mathbf{A}, \nu)$ such
that $(\mathbf{A}, \nu)$ is not computably isomorphic to $(\mathbf{A}, \nu_i)$ for any $i \in \omega$.

Thus, an effectively infinite algebraic system $\mathbf{A}$ has infinite
dimension.

The following characterization of strongly autostable algebraic systems was one of the first important results of the theory of autostable models.

**Theorem 4.6** ([129]). *A strongly constructive algebraic system* $(\mathbf{A}, \nu)$ *is strongly autostable if and only if there exists finitely many elements* $a_0, \ldots, a_n \in A$ *such that*

(1) *the set of all complete formulas of the theory* $T$ *of the algebraic system* $(\mathbf{A}, a_0, \ldots, a_n)$ *is computable,*

(2) *the algebraic system* $(\mathbf{A}, a_0, \ldots, a_n)$ *is the prime model of the theory* $T$.

Furthermore, if $(\mathbf{A}, \nu)$ is not strongly autostable, then there exists an algorithm such that, applying it to any index of an effective sequence of strongly constructive systems $(\mathbf{A}, \nu_0), (\mathbf{A}, \nu_1), \ldots$, we obtain a strongly constructive algebraic system $(\mathbf{A}, \nu)$ such that $(\mathbf{A}, \nu)$ is not computably isomorphic to $(\mathbf{A}, \nu_i)$ for all $i \in \omega$. Thus, the dimension of a strongly constructive algebraic system that is not strongly autostable is infinite.

Similar questions are considered for other classes of algebraic structures, for example, linearly ordered sets, Boolean algebras, Abelian groups, rings, groups, partially ordered sets, fields, vector spaces, etc. The first results were obtained for linearly ordered sets, Boolean algebras, and torsion-free Abelian groups.

Together with the result of Nurtazin [130], the following theorem provides a characterization of all strongly autostable countably categorical models.

The known Ryll–Nardzewski theorem (cf. [16]) characterizes countably categorical theories in terms of types. It asserts that a theory $T$ is countably categorical if and only if for every $n$ the number of $n$-types of $T$ is finite. This theorem suggests to introduce a Ryll–Nardzewski type function $\mathrm{type}_T$ associating with every $n \geqslant 1$ the number of $n$-types of $T$. For a decidable theory $T$ the function $\mathrm{type}_T$ is a $\Delta_2^0$-function.

**Theorem 4.7.** *A strongly constructive model* $(\mathbf{A}, \nu)$ *of a countably categorical theory* $T$ *is strongly autostable if and only if the type function* $\text{type}_T$ *is computable.*

**Corollary 4.8.** *Let* $\mathbf{A}$ *be a model of a countably categorical theory* $T$ *that admits the effective elimination of quantifiers. Then the following assertions are equivalent.*

(1) *The dimension of* $\mathbf{A}$ *is* 1.

(2) *There exists a finite sequence* $a_0, \ldots, a_n$ *of elements of* $\mathbf{A}$ *such that* $(\mathbf{A}, a_0, \ldots, a_n)$ *is the prime model of the theory* $T'$ *of* $(\mathbf{A}, a_0, \ldots, a_n)$ *and the set of atoms of* $T'$ *is computable.*

(3) *The type function* $\text{type}_T$ *is computable.*

A natural question arises: What can be said about the computability–theoretic complexity of $\text{type}_T$ if $T$ is decidable? An answer is contained in the following assertion proved independently by Venning [**153**].

**Theorem 4.9.** *For any c.e. degree* $\mathbf{x}$ *there exists a decidable countably categorical theory* $T$ *such that* $\text{type}_T$ *has degree* $x$.

Note that there exists a strongly autostable, but not autostable countably categorical model.

At the first glance, it seems that, if $\text{type}_T$ of a countably categorical theory is not computable, the dimension of the model of $T$ is greater than 1. However, there exists a counterexample that can be obtained from the following result due to Khoussainov, Lempp, and Solomon.

**Theorem 4.10** ([**86**]). *There exists a countably categorical theory* $T$ *such that the type function* $\text{type}_T$ *is not computable, whereas the model of* $T$ *is autostable.*

If a countably categorical theory $T$ has a computable model, then the type function of $T$ is computable in $\mathbf{O}^\omega$. Together with the above results, this remark leads to the following open question.

**Problem 22.** Whether there exists a countably categorical theory $T$ such that the type function $type_T$ is not arithmetical, whereas $T$ has a constructive autostable model?

Note that the results concerning the construction of nonautostable algebraic systems of finite dimension do not control the model–theoretic properties of structures. For example, all the structures constructed in [**48, 49, 40, 18, 85**] have theories without prime models. Moreover, all known countably categorical models have dimensions equal to either 1 or $\omega$. So, it is reasonable to put the following questions.

**Problem 23.** Whether a countably categorical model is effectively infinite if it is not autostable?

**Problem 24.** Assume that a countably categorical theory $T$ has a computable model. Is it true that the model of $T$ is not autostable if $T$ is computable in $\mathbf{0}^n$ and $type_T$ is not computable in $\mathbf{0}^n$?

## 4.2. Isomorphisms of uncountably categorical models

Consider the algebraic system $(\omega, S)$. The theory $T$ of $(\omega, S)$ is uncountably categorical. The isomorphism type of a model $\mathbf{A}$ of $T$ is determined by the number of its components. The saturated model of $T$ has infinitely many components. All nonsaturated models of $T$ are autostable. One can prove that the saturated model of $T$ is not autostable; moreover, it is effectively infinite.

Let $V$ be a vector space over an infinite computable field $F$. Then the theory $T$ of $V$ (in the language consisting of $+$ for vector addition and unary operation $f$, $f \in F$, for multiplication by $f$) is uncountably categorical. As is known, the isomorphism type of a model $\mathbf{A}$ of $T$ is characterized by the dimension of $\mathbf{A}$.

The saturated model of $T$ has infinite dimension. As above, every finite dimensional vector space over $F$ is autostable, the saturated model of $T$ is not autostable and; moreover, is effectively infinite.

**Theorem 4.11.** *Let $T$ be the theory of algebraically closed fields of a fixed characteristic. Then a model **A** of $T$ is autostable if and only if it has a finite transcendence degree over its prime field.*

In all these examples, all the theories are decidable and admit the elimination of quantifiers; moreover, non-saturated models are autostable. At the same time, there exists a decidable uncountably categorical theory $T$ admitting the elimination of quantifiers such that the prime model of $T$ is not autostable.

Let $T$ be a decidable uncountably categorical theory with strongly autostable prime model.

**Problem 25.** Is it true that every nonsaturated model of $T$ is strongly autostable?

**Conjecture 4.12.** *There exists an uncountably categorical theory such that the countably saturated model is autostable.*

**Problem 26.** Is it true that any field with infinite basis is not autostable?

Without the requirement of decidability of an uncountably categorical theory, the situation becomes much more complicated. No results are known for computable isomorphisms and dimensions of computable models of uncountably categorical theories. For example, we do not know the spectra of dimensions of uncountably categorical models. Recall that all models of an uncountably categorical theory $T$ can be listed in the $\omega + 1$ chain of models chain $(T)$: $\mathbf{A}_0 \preceq \mathbf{A}_1 \preceq \mathbf{A}_2 \preceq \ldots \preceq \mathbf{A}_\omega$, where $\mathbf{A}_i$ is the prime model over $\bar{\mathbf{A}}_i$ and $\mathbf{A}_\omega$ is the saturated model.

**Problem 27.** Let $\mathbf{A}_i$ be a model of an uncountably categorical theory $T$ in chain $(T)$. What sufficient and necessary conditions for $\mathbf{A}_i$ to be autostable?

In Problem 27, it is also of interest to control the dimension of uncountably categorical models. In particular, the following open question can be suggested.

**Problem 28.** Whether there exists an uncountably categorical nonautostable model of finite dimension?

As was already mentioned, Goncharov constructed a nonautostable algebraic system of finite dimension. Thus, it is reasonable to formulate the following problem.

**Problem 29.** Whether it is possible to construct an algebraic system of finite dimension greater than 1 whose theories belong to some class of well–studied theories, for example, countable or uncountably categorical theories, Erenfeucht theories, etc.

The following problem is of general character.

**Problem 30.** Characterize uncountably categorical models of dimension 1.

There are known examples of computable structures of computable Scott rank. At the same time, there are known structures (for example, the Harrison ordering) of Scott rank $\omega_1^{CK} + 1$. Makkai [99] constructed a structure of Scott rank $\omega_1^{CK}$ which can be made computable [68]."" Then he simplified it in such a way that it becomes just a tree [14]. As was shown in [13], there are other computable structures of Scott rank $\omega_1^{CK}$ among undirected graphs, fields of any characteristic, and linear orderings. The new examples share a strong approximability property with the Harrison ordering and the tree in [14]. These results provide us with examples of computable structures with different complexity of the isomorphism problem for different computable representations.

## 4.3. Computable categoricity

Let $\mathcal{A}$ be a computable structure. We say that $\mathcal{A}$ is *computably categorical* if for all computable $\mathcal{B} \cong \mathcal{A}$ there is a computable isomorphism from $\mathcal{A}$ onto $\mathcal{B}$. Similarly, $\mathcal{A}$ is $\Delta^0_\alpha$ *categorical* if for all computable $\mathcal{B} \cong \mathcal{A}$ there is a $\Delta^0_\alpha$ isomorphism. We say that $\mathcal{A}$ is *relatively computably categorical* if for all $\mathcal{B} \cong \mathcal{A}$ there is an isomorphism that is computable relative to $\mathcal{B}$, and we say that $\mathcal{A}$ is *relatively $\Delta^0_\alpha$ categorical* if for all $\mathcal{B} \cong \mathcal{A}$ there is a $\Delta^0_\alpha(\mathcal{B})$ isomorphism.

**Definition 4.13.** A *Scott family for* $\mathcal{A}$ is a set $\Phi$ of formulas with a fixed tuple of parameters $\bar{c}$ in $\mathcal{A}$ such that

- every tuple in $\mathcal{A}$ satisfies some $\varphi \in \Phi$,
- if $\bar{a}$, $\bar{b}$ are tuples in $\mathcal{A}$ satisfying the same formula $\varphi \in \Phi$, then there is an automorphism of $\mathcal{A}$ sending $\bar{a}$ to $\bar{b}$.

A *formally c.e. Scott family* is a c.e. Scott family made up of finitary existential formulas.

A *formally $\Sigma^0_\alpha$ Scott family* is a $\Sigma^0_\alpha$ Scott family made up of "computable $\Sigma_\alpha$" formulas.

**Proposition 4.14.** *For a structure $\mathcal{A}$ the set $\{\bar{a} : \mathcal{A} \models \varphi(\bar{a})\}$ is $\Sigma^0_\alpha(\mathcal{A})$ if $\varphi$ is computable $\Sigma_\alpha$, and $\Pi^0_\alpha(\mathcal{A})$ if $\varphi$ is computable $\Pi_\alpha$. Moreover, this assertion remains valid with all imaginable uniformity over structures and formulas.*

It is easy to see that if $\mathcal{A}$ has a formally c.e. Scott family, then it is relatively computably categorical, so it is computably categorical. More generally, if $\mathcal{A}$ has a formally $\Sigma^0_\alpha$ Scott family, then it is relatively $\Delta^0_\alpha$ categorical and, consequently, $\Delta^0_\alpha$ categorical.

Goncharov showed that, under some additional effectiveness conditions (on a single copy), if $\mathcal{A}$ is computably categorical, then it has a formally c.e. Scott family.

Ash showed that, under some effectiveness conditions (on a single copy), if $\mathcal{A}$ is $\Delta^0_\alpha$ categorical, then it has a formally $\Sigma^0_\alpha$ Scott family.

For the relative notions, we do not have the effectiveness conditions. The following assertion was proved in [**6**] and [**17**].

**Proposition 4.15.** *A computable structure $\mathcal{A}$ is relatively $\Delta_\alpha^0$ categorical if and only if it has a formally $\Sigma_\alpha^0$ Scott family. In particular, $\mathcal{A}$ is relatively computably categorical if and only if it has a formally c.e. Scott family.*

## 4.4. Basic results in numbering theory

We present some basic results in numbering theory [72] and applications to computable models. For $\mathcal{S} \subseteq P(\omega)$ a *numbering* is a binary relation $\nu$ such that $\mathcal{S} = \{\nu(i) : i \in \omega\}$, where $\nu(i) = \{x : (i, x) \in \nu\}$. A numbering $\nu$ of $\mathcal{S}$ is called a *Friedberg numbering* if it is a bijection in the sense that $i \neq j$ implies $\nu(i) \neq \nu(j)$.

Suppose that $\nu$ and $\mu$ are two numberings of a family $\mathcal{S}$. We write $\nu \leqslant \mu$ if there is a computable function $f$ such that $\nu(i) = \mu(f(i))$ for all $i$, i.e., we can effectively pass from a $\nu$-index to a $\mu$-index for the same set. We say that $\nu$ and $\mu$ are *computably equivalent* if $\mu \leqslant \nu$ and $\nu \leqslant \mu$. Note that if $\mu$ and $\nu$ are Friedberg numberings of $\mathcal{S}$, then $\mu \leqslant \nu$ implies $\nu \leqslant \mu$.

**Definition 4.16.** A family $\mathcal{S} \subseteq P(\omega)$ is *discrete* if for every $A \in \mathcal{S}$ there exists $\sigma \in 2^{<\omega}$ such that for all $B \in \mathcal{S}$ the following holds: $\sigma \subseteq \chi_B$ if and only if $B = A$.

**Definition 4.17.** A family is *effectively discrete* if there is a c.e. set $E \subseteq 2^{<\omega}$ such that

(a) for every $A \in \mathcal{S}$ there is $\sigma \in E$ such that $\sigma \subseteq \chi_A$,

(b) for all $\sigma \in E$ and $A, B \in \mathcal{S}$ from $\sigma \subseteq \chi_A, \chi_B$ it follows that $A = B$.

**Proposition 4.18** ([144]). *There exists a unique up to a computable equivalence family $\mathcal{S} \subseteq P(\omega)$ with computable Friedberg numbering such that it is discrete, but not effectively discrete.*

**Proposition 4.19** ([**50**]). *For every finite $n \geqslant 1$ there is a family of sets with just $n$ computable Friedberg numberings determined up to a computable equivalence.*

**Proposition 4.20** ([**154, 145**]). *There is a family $S \subseteq P(\omega)$ with numberings in all noncomputable degrees but not a computable numbering.*

The numbering results of Selivanov, Goncharov, and Wehner can be relativized. In [**40, 72**], one can find a general method of constructing a model from any computable family of c.e. sets with computable numberings. Owing to this method, problems in the theory of computable models are reduced to some problems in numbering theory.

Let $S$ be a family of sets. For every $A \in S$ we can construct a *daisy graph* $\mathcal{G}_A$ such that

(a) $\mathcal{G}(S)$ is a rigid graph,

(b) if $S$ has a unique computable Friedberg numbering, then $\mathcal{G}(S)$ is computably categorical,

(c) if $S$ has just $n$ computable Friedberg numberings determined up to a computable equivalence, then $\mathcal{G}(S)$ has computable dimension $n$,

(d) if $S$ is discrete, then every element of $\mathcal{G}(S)$ has a finitary existential definition without parameters,

(e) if $S$ has a computable Friedberg numbering, and is discrete but not effectively discrete, then $\mathcal{G}(S)$ does not have a formally c.e. defining family.

For lifting the basic results of Goncharov, Manasse, Slaman, and Wehner, we formulate them in the following form.

**Proposition 4.21** ([**46, 40**]). *There is a rigid graph structure $\mathcal{G}$ that is computably categorical without a formally c.e. defining family.*

**Proposition 4.22** ([**126**]). *There is a computable structure $\mathcal{A}$ with a relation $R$ that is intrinsically c.e. but not relatively intrinsically c.e.*

Consider the cardinal sum of disjoint computable copies of the graph structure $\mathcal{G}$ from Proposition 4.21. Let $R$ be a unique isomorphism.

**Proposition 4.23** ([**48, 49, 40, 39**]). *For every finite $n$ there is a rigid graph structure $\mathcal{G}$ with computable dimension $n$.*

**Proposition 4.24** ([**145, 154**]). *There is a structure $\mathcal{A}$ with copies in just the noncomputable degrees.*

A coding of a $\Delta_\alpha^0$ structure in a computable structure was suggested in [**72**] to preserve some complexity of algorithmic properties.

To lift the basic results of Goncharov and Manasse, we relativize by producing a $\Delta_\alpha^0$ graph. To pass to a computable structure, we use a pair of structures for coding the arrow relation.

For a graph $\mathcal{G}$, a pair of structures $\mathcal{B}_1$, $\mathcal{B}_2$, and a relational language we set $\mathcal{G}^* = (G \cup U, G, U, Q, \dots)$, where $G$ is the basic set of $\mathcal{G}$, $G$ and $U$ are disjoint, $Q$ is a ternary relation assigning to every pair $a, b \in G$ an infinite set $U_{(a,b)}$, the sets $U_{(a,b)}$ form a partition of $U$, every relation in the notation "$\dots$" is the union of the bounds to $U_{(a,b)}$, and for every pair $a, b \in G$

$$(U_{(a,b)}, \dots) \cong \begin{cases} \mathcal{B}_1 & \text{if } \mathcal{G} \models a \to b, \\ \mathcal{B}_2 & \text{otherwise.} \end{cases}$$

**Theorem 4.25** ([**72**]). *Suppose that $\mathcal{G}$ is a graph structure and $\mathcal{G}^*$ is constructed from $\mathcal{G}$, $\mathcal{B}_i$ in the same way as above. In this case, $\mathcal{G}$ has a $\Delta_\alpha^0$ copy if and only if $\mathcal{G}^*$ has a computable copy. More generally, for any $X$ the structure $\mathcal{G}$ has a $\Delta_\alpha^0(X)$ copy if and only if $\mathcal{G}^*$ has an $X$-computable copy. In addition,*

(a) *if $\mathcal{G}$ has a unique up to a $\Delta_\alpha^0$ isomorphism $\Delta_\alpha^0$ copy, then $\mathcal{G}^*$ is $\Delta_\alpha^0$ categorical,*

(b) *if $\mathcal{G}$ has just $n$ $\Delta_\alpha^0$ copies, determined up to a $\Delta_\alpha^0$ isomorphism, then $\mathcal{G}^*$ has $\Delta_\alpha^0$ dimension $n$,*

(c) *if $\mathcal{G}$ does not have a $\Sigma_\alpha^0$ Scott family made up of finitary existential formulas, then $\mathcal{G}^*$ does not have a formally $\Sigma_\alpha^0$ Scott family.*

The following construction allows us to reduce the above consideration to graph structure and other algebraic structures.

**Theorem 4.26** ([58]). *Suppose that $\mathfrak{M}$ is a countable structure of a signature $\sigma$ such that the arity of all predicate and functional symbols in $\sigma$ is bounded by a number $k$. There exists a partial ordering (graph) $\mathfrak{M}^*$ with the following properties: The model $\mathfrak{M}$ has a computable copy if and only if $\mathfrak{M}^*$ has a computable copy. More generally, for any $X$ the model $\mathfrak{M}$ has an $X$-computable copy if and only if $\mathfrak{M}^*$ has an $X$-computable copy. In addition,*

(a) *if $\mathfrak{M}$ is $\Delta_\alpha^0$ categorical, then $\mathfrak{M}^*$ is $\Delta_\alpha^0$ categorical,*

(b) *if $\mathfrak{M}$ has $\Delta_\alpha^0$ dimension $n$, then $\mathfrak{M}^*$ has $\Delta_\alpha^0$ dimension $n$,*

(c) *if $\mathfrak{M}$ does not have a formally $\Sigma_\alpha^0$ Scott family, then $\mathfrak{M}^*$ does not have a formally $\Sigma_\alpha^0$ Scott family.*

The proof is based on the following constructions [48] of categories of computable algebraic systems. Consider a computable signature $\sigma = \langle P_0^{n_0}, P_1^{n_1}, \ldots, P_k^{n_k}, \ldots \rangle$ such that $\sigma$ is countable or finite. Denote by $\mathrm{Mod}^\sigma$ the category whose objects are models of the signature $\sigma$ and morphisms are their isomorphisms. Introduce the subcategory $\mathrm{Mod}_{\mathrm{com}}^\sigma$ of $\mathrm{Mod}^\sigma$. It is easy to see that a model $\mathfrak{M}'$ is computable if it is computably isomorphic to a computable model $\mathfrak{M}$ with computable basic set $|\mathfrak{M}|$ which is a computable subset of some sets of words of finite alphabet and the set $\{\langle i, m_1, \ldots, m_{n_i} \rangle / \mathfrak{M} \models P_i(m_1, \ldots, m_{n_i})\}$ is computable. If $\mathfrak{M}_1$ and $\mathfrak{M}_2$ are computable models, then the isomorphism

$\varphi : \mathfrak{M}_1 \underset{\text{onto}}{\to} \mathfrak{M}_2$ is computable provided that $\varphi$ is partially computable. The objects of $\text{Mod}\,^{\sigma}_{\text{com}}$ are computable models of the signature $\sigma$ and the morphisms are computable isomorphisms. If $\mathfrak{M}$ is a model of the signature $\sigma$, then $\text{Mod}\,^{\sigma}_{\text{com}}(\mathfrak{M})$ is the complete subcategory of $\text{Mod}\,^{\sigma}_{\text{com}}$ whose objects are computable models isomorphic to $\mathfrak{M}$. If $K_0$ is a subcategory of $K$ and $F$ is a function from $K$ into $K_1$, then denote by $F \upharpoonright K_0$ the restriction of $F$ to $K_0$. A signature $\sigma$ is *bounded* if there exists $k$ such that $m_i \leqslant k$ for every $i$.

**Proposition 4.27** ([58]). *For an arbitrary bounded signature $\sigma$ there exists a finite signature $\sigma_0$ and a completely univalent functor $F_1$ from $\text{Mod}\,^{\sigma}$ to $\text{Mod}\,^{\sigma_0}$ such that the following assertions hold.*

(i) *$F_1 \upharpoonright \text{Mod}\,^{\sigma}_{\text{com}}$ is a completely univalent functor from $\text{Mod}\,^{\sigma}_{\text{com}}$ to $\text{Mod}\,^{\sigma_0}_{\text{com}}$.*

(ii) *For an arbitrary model $\mathfrak{M}$ of the signature $\sigma$ the functor $F_1 \upharpoonright \text{Mod}\,^{\sigma}_{\text{com}}(\mathfrak{M})$ realizes an equivalence of the categories $\text{Mod}\,^{\sigma}_{\text{com}}(\mathfrak{M})$ and $\text{Mod}\,^{\sigma_0}_{\text{com}}(F_1(\mathfrak{M}))$. In addition,*

  (a) *if $\mathfrak{M}$ is $\Delta^0_\alpha$ categorical, then $F_1(\mathfrak{M})$ is $\Delta^0_\alpha$ categorical,*

  (b) *if $\mathfrak{M}$ has $\Delta^0_\alpha$ dimension $n$, then $F_1(\mathfrak{M})$ has $\Delta^0_\alpha$ dimension $n$,*

  (c) *if $\mathfrak{M}$ does not have a formally $\Sigma^0_\alpha$ Scott family, then $F_1(\mathfrak{M})$ does not have a formally $\Sigma^0_\alpha$ Scott family.*

PROOF. Let $\sigma = \langle P_0^{n_0}, P_1^{n_1}, \ldots, P_k^{n_k}, \ldots \rangle$. Suppose that the set $\langle n_i | i \in \mathfrak{N} \rangle$ is bounded by $k$. For every $k \leqslant K$ we consider all predicates $P_{i^{k_0}}, P_{i^{k_1}}, \ldots, P_{i^{k_l}}, \ldots,\ l \in N'_k$, from $\sigma$ of arity $k$, where $N'_k$ is equal to $N$ or is an initial segment of $N$. We set $\sigma_0 = \{=, P_0^1, P_1^2, \ldots, P_k^{k+1}, \ldots, P_k^{k+1}, A^{12}, \lhd\}$ and define a functor $F_1$ on objects of $\text{Mod}\,^{\sigma}$. Let $\mathfrak{M}$ be an arbitrary model of the signature $\sigma$. If $M$ is the basic set of $\mathfrak{M}$, then for the basic set of $\mathfrak{M}_0 \rightleftharpoons F_1(\mathfrak{M})$ we take $M_0 = M \cup \{a_0, a_1, \ldots, a_n, \ldots\}$, where $\{a_0, a_1, \ldots, a_n, \ldots\} \cap M = \varnothing$ and $a_i \neq a_j$ for $i \neq j$.

Introduce predicates as follows:

1) $A_{\mathfrak{M}_0} \rightleftharpoons \{a_0, a_1, \ldots, a_n, \ldots\}$,

2) $x \lhd y$ if $x = a_n$ and $y = a_{n+1}$ for some $n$,

3) $(x_0, x_1, \ldots, x_s) \in (P_s)_{\mathfrak{M}_0}$ if $x_0 = a_d$, $\bigwedge\limits_{j=1}^{i} x_j \in M$ and $\mathfrak{M} \models P^s_{i^s d}(x_1, \ldots, x_s)$.

It is easy to see that $\mathfrak{M}_0$ is a computable model if $\mathfrak{M}$ is computable. If $\mathfrak{M}$ and $\mathfrak{M}^0$ are objects of the category $\mathrm{Mod}^0$ and $\varphi$ is an isomorphism of $\mathfrak{M}$ onto $\mathfrak{M}^0$, then we define $F(\mathfrak{M}, \mathfrak{M}^0)(\varphi)$. We define it only in the case where the basic sets of both models are subsets of $N$. The remaining cases are treated in a similar way. Thus,

$$[F(\mathfrak{M}, \mathfrak{M}^0)(\varphi)](x) \rightleftharpoons \begin{cases} x & \text{if } x \in \{a_0, a_1, \ldots, a_n, \ldots\}, \\ \varphi(x) & \text{if } x \in M\}. \end{cases}$$

It is clear that $F_1(\mathfrak{M}, \mathfrak{M}^0)(\varphi)$ is a computable isomorphism relative to a Turing degree $a$ if $\varphi$ is a computable isomorphism relative to $a$.

To prove that $F_1$ from $K_1$ into $K_2$ is completely univalent, it suffices to show that $F_*(A, B)$ : $\mathrm{Hom}\,(A, B) \longrightarrow \mathrm{Hom}\,(F_*(A), F_*(B))$ is a bijection for every pair $A, B$ of objects of $K_1$. Thus, $F_1$ and $F_1 \upharpoonright \mathrm{Mod}^{\sigma}_{\mathrm{com}}(\mathfrak{M})$ are completely univalent functors. We can prove that $F_1 \upharpoonright \mathrm{Mod}^{\sigma}_{\mathrm{com}}(\mathfrak{M})$ realizes an equivalence by showing that for every object $\mathfrak{M}'$ of the category $\mathrm{Mod}^{\sigma_0}_{\mathrm{com}}(F_1(\mathfrak{M}))$ there exists an object $\mathfrak{M}'_0$ of the category $\mathrm{Mod}_{\mathrm{com}}(F_1)(\mathfrak{M})$ such that $\mathfrak{M}'$ and $F_1(\mathfrak{M}'_0)$ are isomorphic in the category $\mathrm{Mod}_{\mathrm{com}}(F_1(\mathfrak{M}))$. Let $\mathfrak{M}' \in \mathrm{Mod}^{\sigma}_{\mathrm{com}}(F_1)(\mathfrak{M})$. The case of a finite model is trivial.

Let $\mathfrak{M}$ be an infinite model. We can consider a computable function $f : N \xrightarrow[\text{onto}]{1-1} M' \backslash A_{\mathfrak{M}'}$. Since $\mathfrak{M}'$ is a computable model, it follows that $N' \backslash A_{\mathfrak{M}'}$ is computable and the function exists. Let $a$ be an element of $A_{\mathfrak{M}'}$ that does not have a $\lhd$–predecessor. Let us now define predicates of the signature $\sigma$ on $N$: $(n_1, \ldots, n_{m_k}) \in P^{m_k}_{i_k}$ if and only if $(l, f(n_1), \ldots, f(n_{m_i})) \in P^{m_k+1}_{m_k}$, where the elements $l_0, l_1, l_2, \ldots, l_k$ are such that $l_i \lhd l_{i+1}$ for $0 \leqslant i < k$, $a = l_0$, and

$l_k = l$. It is easy to see that such a model $\mathfrak{M}''$ of the signature $\sigma$ is computable. We show that $F(\mathfrak{M}'')$ is computably isomorphic to $\mathfrak{M}'$. For this purpose, consider a function $g$ defined as follows:

$$
g(m) = \begin{cases} f(m) & \text{if } m \epsilon N, \\ l & \text{if } m = a_k \text{ and there exist } l_0, l_1, \ldots, l_k \text{ such} \\ & \text{that } \mathfrak{M}' \models \bigwedge_{i=0}^{k-1} l_i \lhd l_{i+1} \text{ and } l_0 = a \& l_k = l. \end{cases}
$$

It is clear that $g$ is an isomorphism and a computable function.

Properties (a)–(c) of th Scott families can be derived from the definability of basic predicates and their negations by $\exists$–formulas. $\qquad\square$

**Proposition 4.28** ([58]). *For an arbitrary finite signature $\sigma_0$ there exist a signature $\sigma_1$ consisting of a single predicate symbol $P$ and a completely univalent functor $F_2$ from $\mathrm{Mod}^{\sigma_0}$ into $\mathrm{Mod}^{\sigma_1}$ such that the following assertions hold.*

(i) *$F_2 \upharpoonright \mathrm{Mod}^{\sigma_0}_{\mathrm{com}}$ is a completely univalent functor from $\mathrm{Mod}^{\sigma_0}_{\mathrm{com}}$ into $\mathrm{Mod}^{\sigma_1}_{\mathrm{com}}$.*

(ii) *For every model $\mathfrak{M}$ of the signature $\sigma_0$ the functor $F_2 \upharpoonright \mathrm{Mod}^{\sigma_0}_{\mathrm{com}}(\mathfrak{M})$ realizes an equivalence of the categories $\mathrm{Mod}^{\sigma_0}_{\mathrm{com}}(\mathfrak{M})$ and $\mathrm{Mod}^{\sigma_1}_{\mathrm{com}}(F_2(\mathfrak{M}))$. In addition,*

    (a) *if $\mathfrak{M}$ is $\Delta^0_\alpha$ categorical, then $F_2(\mathfrak{M})$ is $\Delta^0_\alpha$ categorical,*

    (b) *if $\mathfrak{M}$ has $\Delta^0_\alpha$ dimension $n$, then $F_2(\mathfrak{M})$ has $\Delta^0_\alpha$ dimension $n$,*

    (c) *if $\mathfrak{M}$ does not have a formally $\Sigma^0_\alpha$ Scott family, then $F_2(\mathfrak{M})$ does not have a formally $\Sigma^0_\alpha$ Scott family.*

PROOF. Our goal is to define the functor $F_2$. Let $\sigma_0 = \langle P_0^{n_0}, P_1^{n_1}, \ldots, P_k^{n_k} \rangle$. Suppose that $\mathfrak{M}$ is a model of the finite signature $\sigma_0$. Consider a predicate symbol $P$ of arity $n = \sum_{i=0}^{k} n_i$ and a signature $\sigma_1 = \langle P^n \rangle$. We begin by defining $F_2$ on the objects of $\mathrm{Mod}^{\sigma_0}$.

Let $\mathfrak{M}$ be a model of the signature $\sigma_0$ with the basic set $M$. For the basic set $M_0$ of the model $F_2(\mathfrak{M})$ we take $\{\infty\} \cup M$. We define $P$ on $M_0$ as follows: $\langle x_1, \ldots, x_n \rangle \in P$ if and only if one of the following conditions is satisfied:

(a) $x_1 = x_2 = \ldots = x_n = 0$,

(b) there exist $i \leqslant k$ and $y_1, \ldots, y_{n_i}$ such that $x_j = 0$ and $y_j = x_{j+m_i}$ for any $j$ such that $1 \leqslant j \leqslant n_i$ and $\mathfrak{M} \models P_i(y_1, \ldots, y_{n_i})$.

But $x_j = 0$ for any $j$ such that $1 \leqslant j \leqslant m_i$ or $m_i + n_i + 1 \leqslant j \leqslant n$. We put $m_0 = 0$ and $m_i = \sum_{l=0}^{i-1} n_l$ for $i > 1$.

If $\mathfrak{M}$ is a computable model, the model $F_2(\mathfrak{M})$ is also computable. Let $\mathfrak{M}$ and $\mathfrak{N}$ be two models of the signature $\sigma_0$. We define a mapping $F_2(\mathfrak{M}, \mathfrak{N}) : \mathrm{Hom}\,(\mathfrak{M}, \mathfrak{N}) \to \mathrm{Hom}\,(F_2(\mathfrak{M}), F_2(\mathfrak{N}))$ as follows:

$$[F_2(\mathfrak{M}, \mathfrak{N})(\varphi)](x) \rightleftharpoons \begin{cases} \infty & \text{if } x = \infty, \\ \varphi(x) & \text{if } x \neq \infty. \end{cases}$$

It is easy to see that $F_2(\mathfrak{M}, \mathfrak{N})$ is an isomorphism if $\varphi$ is an isomorphism, and it is computable if $\varphi$ is computable. The remaining assertions can be proved in the same way as in Proposition 4.27. The additional properties (a)–(c) can be proved by induction.                                                              $\square$

## 4.5. Categories of graphs and partial orders

Consider a signature $\sigma^*$ consisting of a single binary predicate $Q$. The category $\mathrm{Mod}^{\sigma^*}$ is called the *category of graphs* and is denoted by Graph. Denote by Ord the complete subcategory of $\mathrm{Mod}^{\sigma^*}$ whose objects are the models $\langle M, Q \rangle$, where $Q$ is a partial order on $M$.

**Proposition 4.29** ([58]). *For every signature $\sigma_1$ consisting of a single predicate of arity $n \geqslant 3$ there exists a completely univalent functor $F_3$ from* $\mathrm{Mod}^{\sigma_1}$ *into* Graph *with binary predicate $R$ such that the following assertions hold.*

(i) $F_3 \restriction \mathrm{Mod}^{\sigma_1}_{\mathrm{com}}$ *is a completely univalent functor from* $\mathrm{Mod}^{\sigma_1}_{\mathrm{com}}$ *into* $\mathrm{Graph}_{\mathrm{com}} = \mathrm{Mod}^{\sigma^*}_{\mathrm{com}}$.

(ii) *For every model $\mathfrak{M} \in \mathrm{Ob}_{\mathrm{Mod}^{\sigma_1}}$ the functor $F_3 \restriction \mathrm{Mod}^{\sigma_1}_{\mathrm{com}}(\mathfrak{M})$ realizes an equivalence between the categories* $\mathrm{Mod}^{\sigma_1}_{\mathrm{com}}(\mathfrak{M})$ *and* $\mathrm{Mod}^{\sigma^*}_{\mathrm{com}}(F_3(\mathfrak{M}))$. *In addition,*

   (a) *if $\mathfrak{M}$ is $\Delta^0_\alpha$ categorical, then $F_3(\mathfrak{M})$ is $\Delta^0_\alpha$ categorical,*

   (b) *if $\mathfrak{M}$ has $\Delta^0_\alpha$ dimension $n$, then $F_3(\mathfrak{M})$ has $\Delta^0_\alpha$ dimension $n$,*

   (c) *if $\mathfrak{M}$ does not have a formally $\Sigma^0_\alpha$ Scott family, then $F_3(\mathfrak{M})$ does not have a formally $\Sigma^0_\alpha$ Scott family.*

PROOF. (i) We construct directly the functor $F_3$ from $\mathrm{Mod}^{\sigma_1}$ into Graph. Let $\langle M, P \rangle$ be a model of the signature $\sigma_1$, where $P$ is a predicate of arity $n$. Consider $I = \{0, 1, \ldots, n\}$ and $M' = I \times M^n \cup M$. For the basic set $|F_3(\mathfrak{M})|$ we take the set $M_0 \leftrightharpoons M' \cup \{a_0, a_1, a_2, b_0, b_1, b_2, c_0, c_1, c_2, c_3, c_4, c_5, c_6, c_7, c_8\}$. Suppose that all the elements in $\{a_0, a_1, a_2, b_0, b_1, b_2, c_0, c_1, c_2, c_3, c_4, c_5, c_6, c_7, c_8\}$ are different and new. Fix $a_0, a_1, a_2, b_0, b_1, b_2, c_0, c_1, c_2, c_3, c_4, c_5, c_6, c_7, c_8$. These elements will be referred to as basic elements for the definability of $F_3$ on $\mathfrak{M}$. We define a predicate $R$ on $M_0$ as follows. Let $x, y, \in M_0$. We set $\langle x, y \rangle \in R$ if one of the following conditions is satisfied:

(a) $x = a_i \& y = c_j$, and $1 \leqslant i \leqslant 3$ and $(i = 0 \& j \in \{0, 1\}) \vee (i = 1 \& j \in \{2, 3, 4\}) \vee (i = 2 \& j \in \{5, 6, 7, 8\})$,

(b) $x = c_j \& y = b_i$, and $1 \leqslant i \leqslant 3$ and $(i = 0 \& j \in \{0, 1\}) \vee (i = 1 \& j \in \{2, 3, 4\}) \vee (i = 2 \& j \in \{5, 6, 7, 8\})$,

(c) $x \in M \& y \in I \times M^n \& y = \langle i, x_1, \ldots, x_n \rangle \& x = x_i$ and $n \geqslant i \geqslant 1$,

(d) $x, y \in I \times M^n \& x = \langle i, x_1, \ldots, x_n \rangle \& y = \langle i+1, x_1, \ldots, x_n \rangle$ and $i \geqslant 1$,

(e) $x = a_1 \& y = \langle 0, y_1, \ldots, y_n \rangle \in I \times M^n \& \mathfrak{M} \nvDash P(y_1, \ldots, y_n)$,

(f) $x = a_0 \& y = \langle 0, y_1, \ldots, y_n \rangle \in I \times M^n \& \mathfrak{M} \vDash P(y_1, \ldots, y_n)$,

(g) $x = a_2 \& y \in M$.

Thus, we constructed a graph on $M_0$. Let $\mathfrak{M}$ and $\mathfrak{M}^0$ be two models of the signature $\sigma_1$, and let $\varphi$ be an isomorphism from $\mathfrak{M}$ onto $\mathfrak{M}^0$. We set

$$[F'(\mathfrak{M}, \mathfrak{M}^0)(\varphi)](x)$$

$$\leftrightharpoons \begin{cases} \varphi(x), & \text{if } x \in M, \\ \langle i, \varphi(x_1), \ldots, \varphi(x_n) \rangle & \text{if } x = \langle i, x_1, \ldots, x_n \rangle \in I \times M^n, \\ x & \text{otherwise.} \end{cases}$$

Successively considering all the cases, we can show that $F_3'(\mathfrak{M}, \mathfrak{M}^0)(\varphi)$ is an isomorphism; moreover, it is computable if $\varphi$ is computable.

It remains to prove that the functor $F_3$ is completely univalent. Let $\Psi$ be an isomorphism from $F_3(\mathfrak{M})$ onto $F_3(\mathfrak{M}^0)$. Then the restriction of $\Psi$ to the definable by an existential formula subset $M$ in $F_3(\mathfrak{M})$, equal to $\{2n \mid n \in N\}$, induces an isomorphism $\Psi_0$ between the models $\mathfrak{M}$ and $\mathfrak{M}^0$. Since all the elements of $\langle M_0, P \rangle$ are of the type $\langle i, x_1, \ldots, x_n \rangle$ and are definable over elements of $M$ by existential formulas, it is easy to show that $F_3(\mathfrak{M}, \mathfrak{M}^0)(\Psi_0) = \Psi$.

(ii) Consider a model $\mathfrak{M}$ of the signature $\sigma_1$ and $\mathfrak{M}' \in \mathrm{Mod}\,\sigma^*_{\mathrm{com}}(F_3(\mathfrak{M}))$. Since elements among $a_0$, $a_1$, $a_2$, $b_0$, $b_1$, $b_2$, $c_0$, $c_1$, $c_2$, $c_3$, $c_4$, $c_5$, $c_6$, $c_7$, $c_8$ are definable by existential formulas over elements in $F_3(\mathfrak{M})$, we select them in $\mathfrak{M}'$. Suppose that these elements are the following: $a_0^0$, $a_1^0$, $a_2^0$, $b_0^0$, $b_1^0$, $b_2^0$, $c_0^0$, $c_1^0$, $c_2^0$, $c_3^0$, $c_4^0$, $c_5^0$, $c_6^0$, $c_7^0$, $c_8^0$. Choosing elements connected with $a_2^0$ by the basic binary predicate, we obtain exactly the definable set $X_0$ which is isomorphic to $M$ in $F_3(\mathfrak{M})$.

Define the predicate $P^n$ on $X_0$ as follows:

$$\langle x_1, \ldots, x_n \rangle \in P^n \Leftrightarrow \mathfrak{M}' \models (\exists y_1, \ldots, y_n)(y_1 R \ldots R y_n$$
$$\& \left( \bigwedge_{1 \leqslant i \leqslant j \leqslant n} x_i R y_j \right) \& a_0' R y_1 )$$

It is easy to see that

$$\langle x_1, \ldots, x_n \rangle \notin P^n \Leftrightarrow \mathfrak{M}' \models (\exists y_1 \ldots y_n)(y_1 P^2 \ldots P^2 y_n$$
$$\& ( \bigwedge_{1 \leqslant i \leqslant j \leqslant n} x_i P^2 y_j ) \& a_1' P^2 y_1 ).$$

Therefore, $\langle X_0, P^n \rangle$ is a computable model of the signature $\sigma_1$. A direct verification shows that the model of $F'(\langle X_0, P \rangle)$ is computably isomorphic to $\mathfrak{M}'$.

It remains to prove the additional properties (a)–(c). All the elements of $M' = I \times M^n \cup \{a_0, a_1, a_2, b_0, b_1, b_2, c_0, c_1, c_2, c_3, c_4, c_5,$ $c_6, c_7, c_8\}$ are definable in $F_3(\mathfrak{M})$ over elements of $M$ by existential formulas from the computable set of these formulas. Thus, we can construct a formally $\Sigma_\alpha^0$ Scott family for $F_3(\mathfrak{M})$ from the formally $\Sigma_\alpha^0$ Scott family for $\mathfrak{M}$. If we have a formally $\Sigma_\alpha^0$ Scott family for the model $F_3(\mathfrak{M})$, we can see that $F_3(\mathfrak{M})$ is $\Delta$–definable in $\mathfrak{M}$ with the basic set $M \bigcup \bigcup_{i=1}^{n+1} M^{n+i} / \Theta_i \bigcup \bigcup_{i=1}^{15} M^{2n+1+i} / \Delta_i$. Here, we put $\langle X, Y \rangle \in \Theta_i$ if $X = \langle x_1, \ldots, x_{n+i} \rangle$, $Y = \langle y_1, \ldots, y_{n+i} \rangle$ and $x_j = y_j$ for any $1 \leqslant j \leqslant n$. For the other equivalence relation we put $\langle X, Y \rangle \in \Delta_i$ for any elements $X$, $Y$ of $M^{2n+1+i}$. Since this model is $\Delta$-definable, we can define a formally $\Sigma_\alpha^0$ Scott family for $\mathfrak{M}$.                                                            $\square$

**Proposition 4.30** ([58]). *For every signature $\sigma_1$ consisting of a single binary predicate $R$ there exists a completely univalent functor $F_4$ from $\mathrm{Mod}^{\sigma_1}$, into $\mathrm{Ord}$ such that the following assertions hold.*

(i) *$F_4 \upharpoonright \mathrm{Mod}_{\mathrm{com}}^{\sigma_1}$ is a completely univalent functor from $\mathrm{Mod}_{\mathrm{com}}^{\sigma_1}$ into $\mathrm{Mod}_{\mathrm{com}}^{\sigma^*}$.*

(ii) *For every model $\mathfrak{M} \in \mathrm{Ob}_{\mathrm{Mod}^{\sigma_1}}$ the functor $F_4 \upharpoonright \mathrm{Mod}_{\mathrm{com}}^{\sigma_1}(\mathfrak{M})$ realizes an equivalence between the categories $\mathrm{Mod}_{\mathrm{com}}^{\sigma_1}(\mathfrak{M})$ and $\mathrm{Mod}_{\mathrm{com}}^{\sigma^*}(F_4(\mathfrak{M}))$. In addition,*

(a) *if $\mathfrak{M}$ is $\Delta_\alpha^0$ categorical, then $F_4(\mathfrak{M})$ is $\Delta_\alpha^0$ categorical,*

(b) *if $\mathfrak{M}$ has $\Delta_\alpha^0$ dimension $n$, then $F_4(\mathfrak{M})$ has $\Delta_\alpha^0$ dimension $n$,*

(c) *if $\mathfrak{M}$ does not have a formally $\Sigma_\alpha^0$ Scott family, then $F_4(\mathfrak{M})$ does not have a formally $\Sigma_\alpha^0$ Scott family.*

PROOF. We construct the functor $F_4$ from $\mathrm{Mod}^{\{R\}}$ into $\mathrm{Ord}$ satisfying the requirement conditions. Let $\mathfrak{M} = \langle M, R \rangle$ be a model with a single binary predicate $R$. We define the partially ordered set $\langle M_0, \leqslant \rangle$, where the basic set $M_0$ is the image of $M$ under the functor $F_4$. Then we set $M_0 = M \cup M^2 \times \{0, 1\} \cup \{a_1, a_2, a_3, a_4, a_5\} \cup \{b_1, \ldots, b_7, b_8\}$, where elements of the set $\{a_1, a_2, a_3, a_4, a_5\} \cup \{b_1, \ldots, b_7, b_8\}$ are new.

Introduce a partial order $\leqslant$ on $M_0$ such that its transitive closure is the desired partial order on $M_0$:

1) $a_1 \leqslant a_2, a_2 \leqslant a_4, a_2 \leqslant a_3, a_4 \leqslant a_5$,

2) $b_1 \leqslant b_2, b_2 \leqslant b_3, b_3 \leqslant b_4, b_4 \leqslant b_5, b_5 \leqslant b_6, b_5 \leqslant b_7, b_7 \leqslant b_8$,

3) if $x_1, x_2 \in M$ and $x_1 \neq x_2$, then $\langle \langle x_1, x_2 \rangle, 0 \rangle \leqslant x_1$, and $\langle \langle x_1, x_2 \rangle, i \rangle \leqslant x_2$ for $i \in \{0, 1\}$,

4) if $x_1 \neq x_2 \in M$ and $\mathfrak{M} \models P(x_1, x_2)$, then $a_5 \leqslant \langle \langle x_1, x_2 \rangle, 0 \rangle$,

5) if $x_1 \neq x_2 \in M$ and $\mathfrak{M} \nvDash P(x_1, x_2)$, then $a_3 \leqslant \langle \langle x_1, x_2 \rangle, 0 \rangle$,

6) if $x_1 \in M$ and $\mathfrak{M} \models P(x_1, x_1)$, then $b_6 \leqslant \langle \langle x_1, x_2 \rangle, 0 \rangle$,

7) if $x_1 \in M$ and $\mathfrak{M} \nvDash P(x_1, x_1)$, then $b_8 \leqslant \langle \langle x_1, x_2 \rangle, 0 \rangle$.

We define $F_4(\mathfrak{M}, \mathfrak{M}')$ on isomorphisms $\varphi$ in the same way as in the case of the functor $F_3$. The proof of the properties of this functor is similar to that in Proposition 4.29.                        $\square$

Using the idea of the proof of Proposition 4.29, it is easy to construct a functor from the category of an arbitrary signature into the category of a bounded signature.

**Proposition 4.31** ([58]). *For every signature $\Sigma$ there exists a bounded signature $\Sigma_0$ and a completely univalent functor $F_6$ from* $\mathrm{Mod}^{\Sigma}$ *into* $\mathrm{Mod}^{\Sigma_0}$ *such that the following assertions hold.*

(i) *$F_5 \upharpoonright \mathrm{Mod}^{\Sigma}_{\mathrm{com}}$ is a completely univalent functor from $\mathrm{Mod}^{\Sigma}_{\mathrm{com}}$ into $\mathrm{Mod}^{\Sigma_0}_{\mathrm{com}}$.*

(ii) *For every model $\mathfrak{M}$ of the signature $\Sigma$, the functor $F_5 \upharpoonright \mathrm{Mod}^{\Sigma}_{\mathrm{com}}(\mathfrak{M})$ realizes an equivalence of the categories $\mathrm{Mod}^{\Sigma}_{\mathrm{com}}(\mathfrak{M})$ and $\mathrm{Mod}^{\Sigma_0}_{\mathrm{com}}(F_5(\mathfrak{M}))$. In addition,*

   (a) *if $\mathfrak{M}$ is $\Delta^0_\alpha$ categorical, then $F_5(\mathfrak{M})$ is $\Delta^0_\alpha$ categorical,*

   (b) *if $\mathfrak{M}$ has $\Delta^0_\alpha$ dimension $n$, then $F_5(\mathfrak{M})$ has $\Delta^0_\alpha$ dimension $n$,*

   (c) *if $\mathfrak{M}$ does not have a formally $\Sigma^0_\alpha$ Scott family, then $F_5(\mathfrak{M})$ does not have a formally $\Sigma^0_\alpha$ Scott family.*

PROOF. Consider a new signature $\sigma^*$. We put in $\sigma^*$ all predicates from $\sigma$ with arity $n \leqslant 2$. If a predicate symbol $P_n$ has arity $m_n \geqslant 3$, then we add three new predicate symbols in $\sigma^*$: the binary predicate symbol $R_n$ and two unary predicate symbols $A_n$ and $B_n$. We also add one new unary predicate symbol $U$. Then we consider the impoverishment $\mathfrak{M}_n$ of the model $\mathfrak{M}$ of the signature $\Sigma_n = \langle P_n^{m_n} \rangle$ for every $m_n \geqslant 3$. We consider a model $\mathfrak{L}_n$ with $M \subseteq |\mathfrak{L}_n|$ that is isomorphic to the model $F_3(\mathfrak{M}_n)$ from Proposition 4.29 with isomorphism $\varphi_n$ from $F_3(\mathfrak{M}_n)$ on this model $\mathfrak{L}_n$ such that for any $m \in M$ we have $\varphi(m) = m$, but $|\mathfrak{L}_n| \cap |\mathfrak{L}_k| = M$ for any $n \neq k$. The basic set $|F_5(\mathfrak{M})|$ of the model $F_5(\mathfrak{M})$ is $\bigcup_n |\mathfrak{L}_n|$. We set $U = M$. Define a predicate symbol $P$ from $\sigma$ with arity $n \leqslant 2$ as the interpretation of this predicate in $\mathfrak{M}$. Now, define the remaining symbols of the signature $\sigma^*$. Let $R_n$ on $|F_5(\mathfrak{M})|$ be equal to the binary predicate from $\mathfrak{L}_n$. But $A_n$ is the set $|\mathfrak{L}_n| \setminus M$ and $B_n$ is the set $\{\varphi(a_0), \varphi(a_1), \varphi(a_2), \varphi(b_0), \varphi(b_1), \varphi(b_2), \varphi(c_0), \varphi(c_1), \varphi(c_2), \varphi(c_3), \varphi(c_4), \varphi(c_5), \varphi(c_6), \varphi(c_7), \varphi(c_8)\}$, where $\{a_0, a_1, a_2, b_0, b_1, b_2, c_0, c_1, c_2, c_3, c_4, c_5, c_6, c_7, c_8\}$ is the set of basic elements for the definability of $F_3$ on $\mathfrak{M}_n$.

Thus, we get the desired functor $F_5$ on the objects of the category. Using the construction of Proposition 4.29, we can define it on the morphisms of our category. The proof of the remaining assertions is similar to that of Proposition 4.29.                                  □

REMARK. If the signature contains functional symbols, we can pass to a new signature with predicates for graphs of this functions.

Thus, we proved the following assertion.

**Theorem 4.32** ([58]). *For every signature $\Sigma$ there exist a signature $\Sigma_0$ containing only one binary predicate $R$ and a completely univalent functor $F$ from $\mathrm{Mod}^{\Sigma}$ to $\mathrm{Mod}^{\Sigma_0}$ such that the following assertions hold.*

(i) *$F \upharpoonright \mathrm{Mod}^{\Sigma}_{\mathrm{com}}$ is a completely univalent functor from $\mathrm{Mod}^{\Sigma}_{\mathrm{com}}$ to $\mathrm{Mod}^{\Sigma_0}_{\mathrm{com}}$.*

(ii) *For every model $\mathfrak{M}$ of the signature $\Sigma$ the functor $F \upharpoonright \mathrm{Mod}^{\Sigma}_{\mathrm{com}}(\mathfrak{M})$ realizes an equivalence of the categories $\mathrm{Mod}^{\Sigma}_{\mathrm{com}}(\mathfrak{M})$ and $\mathrm{Mod}^{\Sigma_0}_{\mathrm{com}}(F_5(\mathfrak{M}))$. In addition,*

(a) *if $\mathfrak{M}$ is $\Delta^0_\alpha$ categorical, then $F(\mathfrak{M})$ is $\Delta^0_\alpha$ categorical,*

(b) *if $\mathfrak{M}$ has $\Delta^0_\alpha$ dimension $n$, then $F(\mathfrak{M})$ has $\Delta^0_\alpha$ dimension $n$,*

(c) *if $\mathfrak{M}$ does not have a formally $\Sigma^0_\alpha$ Scott family, then $F(\mathfrak{M})$ does not have a formally $\Sigma^0_\alpha$ Scott family.*

By Theorem 4.32, it suffices to consider only problems connected to computable equivalence and self-equivalence on partially ordered sets or graphs since there are no essential difficulties arise in the case of a more complicated signature.

The above results lead to the following assertion.

**Theorem 4.33** (Goncharov–Tusupov, [58]). *Suppose that $\mathcal{G}$ is a graph structure and the partial ordering (graph) $\Delta(\mathcal{G})$ is constructed from $\mathcal{G}$, $\mathcal{B}_i$ in the same way as in Theorems 4.25 and 4.26. Then $\mathcal{G}$ has a $\Delta^0_\alpha$ copy if and only if $\Delta(\mathcal{G})$ has a computable* *copy. In general, for any $X$ the structure $\mathcal{G}$ has a $\Delta_\alpha^0(X)$ copy if and only if $\Delta(\mathcal{G})$ has an $X$-computable copy. In addition,*

(a) *if $\mathcal{G}$ has a unique up to a $\Delta_\alpha^0$ isomorphism $\Delta_\alpha^0$ copy, then $\Delta(\mathcal{G})$ is $\Delta_\alpha^0$ categorical,*

(b) *if $\mathcal{G}$ has just $n$ $\Delta_\alpha^0$ copies, determined up to a $\Delta_\alpha^0$ isomorphism, then $\Delta(\mathcal{G})$ has $\Delta_\alpha^0$ dimension $n$,*

(c) *if $\mathcal{G}$ does not have a $\Sigma_\alpha^0$-computable Scott family made up of finitary existential formulas, then $\Delta(\mathcal{G})$ does not have a formally $\Sigma_\alpha^0$-Scott family.*

## 4.6. Lift of basic results

The following assertion lifts the result of Goncharov about computably categorical structures that are not relatively computably categorical.

**Theorem 4.34** ([**72**]). *For every computable successor ordinal $\alpha$ there is a structure that is $\Delta_\alpha^0$ categorical, but not relatively $\Delta_\alpha^0$ categorical (and does not have a $\Sigma_\alpha^0$-Scott family).*

**Corollary 4.35** (Goncharov–Tusupov, [**58**]). *For every computable successor ordinal $\alpha$ there is a partial ordering (graph) that is $\Delta_\alpha^0$ categorical, but not relatively $\Delta_\alpha^0$ categorical (and does not have a $\Sigma_\alpha^0$-Scott family).*

The following assertion lifts the result of Manasse [**106**] about relations that are intrinsically c.e., but not relatively intrinsically c.e.

**Theorem 4.36** ([**72**]). *For every computable successor ordinal $\alpha$ there is a computable structure with a relation that is intrinsically $\Sigma_\alpha^0$, but not relatively intrinsically $\Sigma_\alpha^0$.*

**Corollary 4.37** (Goncharov–Tusupov, [**58**]). *For every computable successor ordinal $\alpha$ there is a computable partial ordering* (graph) with a relation that is intrinsically $\Sigma^0_\alpha$ but not relatively intrinsically $\Sigma^0_\alpha$.

The following assertion lifts the result of Goncharov about structures with finite computable dimension.

**Theorem 4.38** ( [72]). *For any computable successor ordinal $\alpha$ and a finite number $n$ there is a computable structure with $\Delta^0_\alpha$ dimension $n$.*

**Corollary 4.39** (Goncharov–Tusupov, [58]). *For any computable successor ordinal $\alpha$ and a finite number $n$ there is a computable partial ordering (graph) with $\Delta^0_\alpha$ dimension $n$.*

The following assertion lifts the result of Slaman and Wehner.

**Theorem 4.40** ([72]). *For every computable successor ordinal $\alpha$ there is a structure with copies in just the degrees of sets $X$ such that $\Delta^0_\alpha(X)$ is not $\Delta^0_\alpha$. In particular, for every finite $n$ there is a structure with copies in just the non-low$_n$ degrees.*

**Corollary 4.41** (Goncharov–Tusupov, [58]). *For every computable successor ordinal $\alpha$ there is a partial ordering (graph) with copies in just the degrees of sets $X$ such that $\Delta^0_\alpha(X)$ is not $\Delta^0_\alpha$. In particular, for every finite $n$ there is a structure with copies in just the non-low$_n$ degrees.*

Based on examples of computable graphs and the construction of [77], one can construct many other algebraic structures with the same properties as in Theorems 4.34, 4.36, 4.38, 4.40.


# 5. Classes of Computable Models and Index Sets

In the study of computable models, it is important to consider not only individual models, but also classes of models defined by certain properties and to find relationships between the definability problems and algorithmic complexity expressed in terms of their index sets. As was shown by Nurtazin, [**130**], for a predicate signature there exists a computable numbering of all computable models of a given signature which is universal computable numbering of this class and is unique up to a recursive permutation. This fact provides us with a good tool for studying the algorithmic complexity of different classes of models of this signature.

One of questions in this direction is to express the complexity of definability of a class of mathematical structures in terms of the complexity of definability of the corresponding index sets in a universal numbering of all computable models of a given signature. This question is close to the investigations of Goncharov and Knight [**66**] on the structural properties of classes of computable models.

## 5.1. Computable classification or structure theorem

If $K$ is a class, we denote by $K^c$ the set of computable members of $K$. A *computable characterization* for $K$ should separate computable members of $K$ from other structures that either are outside $K$ or belong to $K$, but are not computable. A computable classification (or a structure theorem) should describe up to an isomorphism (or up to a some other equivalence relation) every member of $K^c$, in terms of relatively simple invariants. On the other hand, a computable non-structure theorem should assert the absence of a computable structure theorem.

We consider three different approaches from [**66**]. Each of them gives a "correct" answer in the case of vector spaces over $Q$ and linear orderings. Under each of these three approaches, both classes have computable characterization and there is a computable classification for vector spaces, but not for linear orderings.

In the first approach, $K$ has a computable characterization if $K^c$ is the set of computable models of some "computable" infinitary sentence. There is a computable classification for $K$ if there is a computable bound on the "ranks" of elements of $K^c$.

In the second approach, $K$ has a computable characterization if the set $I(K)$ of computable indices for elements of $K^c$ is hyperarithmetical. There is a computable classification for $K$ if the set $E(K)$ of pairs of indices corresponding to isomorphic structures is hyperarithmetical. (We also consider computable isomorphisms or $\Delta_\alpha^0$ isomorphisms.)

In the third approach, $K$ has a computable characterization if there is a hyperarithmetical list (an *enumeration*) of elements of $K^c$ representing all isomorphism types. A computable classification theorem holds for $K$ if there is an enumeration such that every computable isomorphism type is represented only once. (Again, we consider computable isomorphisms or $\Delta_\alpha^0$ isomorphisms.)

Uncountable and countable structures are of great interest in model theory, The compactness theorem is a central result, so it is natural to use elementary first order formulas. In model theory, classes are normally characterized by elementary first order theories. In computable structure theory, we are interested in computable structures. Within the framework of computable structure theory, the compactness theorem does not play an essential role since it does not yield computable structures. If the compactness is established, we can deal with such classes as the Abelian $p$-groups which are not characterized by the elementary first order theory.

### 5.1.1. First approach.

We discuss characterization and classification in the following sense.

*Computable characterization.* There is a computable infinitary sentence whose computable models are just elements of $K^c$.

*Computable classification.* In addition to a computable infinitary sentence characterizing the computable members of $K$, there is a computable bound on "ranks" of elements of $K^c$.

We beging with a definition of rank and then indicate how it is connected with the complexity of isomorphisms. Then we consider applications of the characterization and classification statements to some well-known classes of structures.

Let us clarify how the above computable characterization and classification statements can be applied to some well-known classes of structures.

*Computable characterization.* Linear orderings, Boolean algebras, and equivalence structures can be characterized by a single elementary first order sentence. Vector spaces over $Q$ and algebraically closed fields of a given characteristic can be characterized by either an infinite set of elementary first order sentences or a single computable $\Pi_2$ sentence. The class of Abelian $p$-groups is not characterized by any set of elementary first order sentences, but it is characterized by a single computable $\Pi_2$ sentence.

Some classes, for example, well orderings, superatomic Boolean algebras, and reduced Abelian $p$-groups cannot be characterized by a computable infinitary sentence. In fact, they cannot be characterized by any $L_{\omega_1\omega}$ sentence. The case of well orderings was considered by Lopez–Escobar [98].

*Computable classification.* For vector spaces over $Q$ and algebraically closed fields of a given characteristic the computable rank is 1; we have the elimination of quantifiers. For equivalence structures the rank is at most 3.

The following assertion is well known.

**Proposition 5.1** ([66]). *There is no computable bound on the ranks for the following classes $K$ :*

(a) *linear orderings,*

(b) *Boolean algebras,*

(c) *Abelian p-groups,*

(d) *structures for language with at least one binary relation symbol.*

Each of the classes listed in Proposition 5.1 contains a structure of a noncomputble rank. The following assertion shows that it is a general fact.

**Proposition 5.2** ([66]). *Let $K^c$ be a set of computable models of a computable infinitary sentence $\psi$. If there is no computable bound on $R^c(\mathcal{A})$, for $\mathcal{A}$ in $K^c$, then there exists $\mathcal{A}$ in $K^c$ such that $R^c(\mathcal{A}) = \omega_1^{CK}$.*

### 5.1.2. Second approach.

We consider the characterization and classification in terms of indices.

**Definition 5.3.** The *computable index* of a structure $\mathcal{A}$ is a number $e$ such that $D(\mathcal{A}) = W_e$. The *index set* $I(K)$ of a class $K$ is the set of computable indices of elements of $K^c$.

We assume that $\mathcal{A}_e$ is a structure with computable index $e$. The *isomorphism problem* for a class $K$ is stated as follows:

$$E(K) = \{(a, b) : a, b \in I(K) \,\&\, \mathcal{A}_a \cong \mathcal{A}_b\}.$$

We write $\mathcal{A} \cong_{\Delta_\alpha^0} \mathcal{B}$ if $\mathcal{A}$ and $\mathcal{B}$ are isomorphic by a $\Delta_\alpha^0$ isomorphism. The $\Delta_\alpha^0$ *isomorphism problem* is stated as follows:

$$E_{\Delta_\alpha^0}(K) = \{(a, b) : a, b \in I(K) \,\&\, \mathcal{A}_a \cong_{\Delta_\alpha^0} \mathcal{A}_b\}.$$

*Computable characterization*: $I(K)$ is hyperarithmetical.

*Computable classification*: $E(K)$ is hyperarithmetical.

If we consider $\Delta_\alpha^0$ isomorphisms, the classification result means that $E_{\Delta_\alpha^0}(K)$ is hyperarithmetical.

For many classes the index set is at a low level in the hyperarithmetical hierarchy.

**Proposition 5.4** ([66]). $I(K)$ *is* $\Pi_2^0$ *for the following classes* $K$ :

(a) *linear orderings,*

(b) *Boolean algebras,*

(c) *Abelian p-groups,*

(d) *equivalence structures,*

(e) *vector spaces over* $Q$,

(f) *structures for a fixed computable language.*

For some well-known classes, the index set is not hyperarithmetical.

**Proposition 5.5** ([140]). $I(K)$ *is* $\Pi_1^1$ *complete for the following classes* $K$:

(a) *well orderings,*

(b) *superatomic Boolean algebras,*

(c) *reduced Abelian p-groups.*

We refer to [139] or [5] for the proof of (a).

We turn to the isomorphism problems. If $I(K)$ is hyperarithmetical, then $E(K)$ is at least $\Sigma_1^1$. For vector spaces over $Q$ and algebraically closed fields of a given characteristic the isomorphism problem is at a low level of the hyperarithmetical hierarchy.

**Proposition 5.6** (Calvert). $E(K)$ *is* $\Pi_3^0$ *complete for the following classes* $K$:

(a) *vector spaces over* $Q$ *(or other infinite computable field),*

(b) *algebraically closed fields of a given characteristic.*

Below, we list several classes for which the isomorphism problem is $\Sigma_1^1$ complete (maximum complexity). These results are firmly established in folklore and are seemed to be known since the 1960's. However, I am not able to say exactly who was the first who proved them. In [44], there are related results in descriptive set theory concerning the Borel completeness of the isomorphism problem for various classes of structures with a fixed countable basic set. Note that the arguments in [44] can serve as the proof of the assertions formulated below.

Let $E(K)$ be $\Sigma_1^1$ complete for the following classes $K$:

(a) Abelian $p$-groups,

(b) trees,

(c) Boolean algebras,

(d) linear orderings,

(e) arbitrary structures for language with at least one binary relation symbol.

## 5.2. Special isomorphisms

We considered the set $I(K)$ with the equivalence relation $E(K)$. Now, we replace $E(K)$ with a computable isomorphism.

**Proposition 5.7.** *If $I(K)$ is $\Delta_3^0$, then $E_{\Delta_1^0}(K)$ is at least $\Sigma_3^0$.*

**Theorem 5.8** ([66]). *$E_{\Delta_1^0}(K)$ is $\Sigma_3^0$ complete (maximum complexity) for the following classes $K$ :*

(a) *linear orderings,*

(b) *arbitrary structures for language with at least one binary relation symbol,*

(c) *Boolean algebras,*

(d) *Abelian p-groups,*

(e) *equivalence structures.*

We consider $\Delta_\alpha^0$ isomorphisms instead of computable ismorphisms and generalize Proposition 5.7 and Theorem 5.8.

**Proposition 5.9** ([66]). *If $I(K)$ is $\Delta_{\alpha+2}^0$, then $E_{\Delta_\alpha^0}(K)$ is $\Sigma_{\alpha+2}^0$.*

**Theorem 5.10.** *Let $\alpha > 1$ be computable. Then $E_{\Delta_\alpha^0}(K)$ is $\Sigma_{\alpha+2}^0$ complete (maximum complexity) for the following classes $K$:*

(a) *linear orderings,*

(b) *arbitrary structures for a computable language with at least one binary relation symbol,*

(c) *Boolean algebras,*

(d) *Abelian p-groups.*

The following assertion about linear orderings is very useful. The construction is based on the method from the Ash metatheorem [5, 1]. This method has many applications and, possibly, can serve as a metaconstruction for new computable models.

**Theorem 5.11** ([66]). *There is a fixed computable linear ordering $\mathcal{B}$ such that for any $\Sigma_{\alpha+2}^0$ set $S$ there is a uniformly computable sequence of linear orderings $(\mathcal{C}_n)_{n\in\omega}$ such that*

$$\begin{cases} \mathcal{C}_n \cong_{\Delta_\alpha^0} \mathcal{B} & \text{if } n \in S, \\ \mathcal{C}_n \ncong \mathcal{B} & \text{otherwise.} \end{cases}$$

To prove this theorem, we need the following lemma.

**Lemma 5.12.** *If $\mathcal{A}$ is a $\Delta_\alpha^0$ ordering, then there is a computable $\mathcal{B} \cong \omega^\alpha \cdot \mathcal{A}$ with a $\Delta_\alpha^0$ function sending every element of $\mathcal{A}$ to the first element of the corresponding copy of $\omega^\alpha$ in $\mathcal{B}$ and there is a $\Delta_\alpha^0$ procedure associating with every $b \in \mathcal{B}$ the position of $b$ in the copy of $\omega^\alpha$. Moreover, it is possible to pass effectively from a $\Delta_\alpha^0$ index for $\mathcal{A}$ to a computable index for $\mathcal{B}$, $\Delta_\alpha^0$ indices for the rest.*

PROOF. There are known related results (cf., for example, [4, 5]), but it seems that none of these results provides us with the desired assertion. Namely, the mentioned results can yield a $\Delta_3^0$ embedding of a $\Delta_3^0$ ordering $\mathcal{A}$ in a computable ordering of type

$\omega\mathcal{A}$, but not a $\Delta_2^0$ embedding of a $\Delta_2^0$ ordering $\mathcal{A}$ in a computable ordering of type $\omega\mathcal{A}$.

We use the metatheorem of Ash [1]. However, the general formulation is too large and restrict ourselves with some definitions and verify one nontrivial condition.

We define an $\alpha$-system $(L, U, \widehat{\ell}, P, E, (\leqslant_\beta)_{\beta<\alpha})$ and a $\Delta_\alpha^0$ instruction function $q$ such that $E(\pi)$ is the diagram of the desired $\mathcal{C}$, whereas $\pi$ yields the rest. Without loss of generality, we assume that $\mathcal{A}$ has the first element. Suppose that the basic set $A$ of $\mathcal{A}$ is an infinite computable set of constants and the first element in the ordering is also the first constant. Let $U$ be the set of linear orderings on initial segments of $A$, including the first element. For every $u \in U$ we denote by $\mathcal{O}_u$ an ordering of type $\omega^\alpha u$. Assume that the following assertions hold.

(i) If $u \subseteq v$, then $\mathcal{O}_u \subseteq \mathcal{O}_v$.

(ii) The orderings $\mathcal{O}_u$ are computable uniformly in $u$ and it is possible to determine effectively the Cantor normal form of intervals.

Let $B$ be an infinite computable set of constants. Suppose that $L$ consists of pairs $(u, f)$, where $u \in U$ and $f$ is a finite one-to-one function from $B$ to $\mathcal{O}_u$. Let $\widehat{\ell} = (u, \varnothing)$, where $u$ consists of only the first element. If $\ell = (u, f)$, we denote by $E(\ell)$ the set of atomic sentences and the negations of atomic sentences $\varphi(\bar{b})$ involving constants $\bar{b}$ from $\text{dom}\,(f)$ such that $f$ makes $\varphi(\bar{b})$ true in $\mathcal{O}_u$. If $\ell = (u, f)$ and $\ell' = (v, g)$, we assume that $\ell \leqslant_0 \ell'$ if $g \circ f^{-1}$ preserves order. Suppose that $\ell \leqslant_\beta \ell'$ if it preserves order and sends elements of a single copy of $\omega^\beta$ to elements at the corresponding positions, also in the single copy of $\omega^\beta$. Let $\ell \subseteq \ell'$ if $u \subseteq v$ and $f \subseteq g$.

Denote by $P$ the set of finite alternating sequences $\widehat{\ell}u_1\ell_1u_2\ell_2\ldots$, where

(1) $u_n \in U$ is an ordering on the first $n + 1$ constants in $A$,

(2) $u_n \subseteq u_{n+1}$,

(3) $\ell_n \subseteq \ell_{n+1}$,

(4) if $\ell_n = (u, f)$, then $u = u_n$, $\mathrm{dom}\,(f)$ includes the first $n$ elements of $B$, and $\mathrm{ran}\,(f)$ includes the first $n$ elements of $\mathcal{O}_k$, for all $k \leqslant n$,

Thus, we defined the ingredients of the $\alpha$-system. As usual, conditions (1)–(3) are trivially satisfied. Relative to condition (4), we suppose that $\sigma \ell^0 u \in P$, where $\ell^0 \leqslant_{\beta_0} \ell^1 \leqslant_{\beta_1} \ldots \leqslant_{\beta_{k-1}} \ell^k$ and $\alpha > \beta_0 > \beta_1 > \ldots > \beta_{k-1} > \beta_k$. We set $\ell_m = (u_m, f_m)$. Thus, we find $\ell'_m \supseteq \ell_m$ such that $\ell'_k = \ell_k$ and $\ell'_{m+1} \leqslant_{\beta_{m+1}} \ell'_m$. On the top, we have $\ell'_0 \supseteq \ell_0$. We set $\ell'_0 = (u_0, f)$. We have $u \supseteq u_0$. Let $\ell = (u, g)$, where $g \supseteq f$ includes suitable elements in the domain and range so that $\sigma \ell^0 u \ell \in P$. This $\ell$ is what we need to verify condition (4).

Thus, we have an $\alpha$-system. Define a $\Delta^0_\alpha$ instruction function $q$ such that if $\sigma = \widehat{\ell} u_1 \ell_1 \ldots \ell_n$ is an element of $P$ of length $2n + 1$. Then $q(\sigma)$ is the substructure of $\mathcal{A}$ whose basic set consists of the first $n + 1$ constants. Now, we can use the Ash metatheorem. We find a $\Delta^0_\alpha$ run $\pi = \widehat{\ell} u_1 \ell_1 u_2 \ell_2 \ldots$ of $(P, q)$ ($\pi$ is a path through the tree $P$ with $u_n$ chosen by the instruction function $q$) such that $E(\pi) = \cup_n E(\ell_n)$ is c.e.

We set $\ell_n = (u_n, f_n)$. Then $\cup_n f_n$ is a one-to-one function from $B$ onto $\mathcal{A}' = \cup_n \mathcal{O}_{u_n}$, where $\mathcal{A}'$ is a copy of $\omega^\alpha \mathcal{A}$. Let $F$ be the inverse, and let $\mathcal{B}$ be the copy of $\mathcal{A}'$ induced on $B$ by $F$. Then $D(\mathcal{B}) = E(\pi)$, so that $\mathcal{B}$ is computable. Now, $F$ and $\mathcal{A}'$ are $\Delta^0_\alpha$. For a given $a \in \mathcal{A}$ we can use $\Delta^0_\alpha$ to find the first element of the corresponding copy of $\omega^\alpha$ in $\mathcal{A}'$. Similarly, for a given $b \in \mathcal{B}$ we can find $F^{-1}(b)$. Since we know the position of $F^{-1}(b)$ in its copy of $\omega^\alpha$, we also know the position of $a$. This completes the proof of Lemma 5.12. □

PROOF OF THEOREM 5.11. Relativizing the above lemma to $\Delta^0_\alpha$, we obtain a fixed computable ordering $\mathcal{B}^*$ of type $\omega^2$ such that for any $\Sigma^0_{\alpha+2}$ set $S$ there is a computable sequence of indices for $\Delta^0_\alpha$ orderings $(\mathcal{C}^*_n)_{n \in \omega}$ such that $\mathcal{C}^*_n \cong_{\Delta^0_\alpha} \mathcal{B}^*$ for $n \in S$ and $\mathcal{C}^*_n \not\cong \mathcal{B}^*$ in the opposite case.

Let $\mathcal{B}$ and $\mathcal{C}_n$ be obtained from $\mathcal{B}^*$ and $\mathcal{C}_n^*$ as in Lemma 5.12. The sequence $(\mathcal{C}_n^*)_{n \in \omega}$ is uniformly computable. It is easy to see that if $n \in S$, then $\mathcal{C}_n^*$ has order type $\omega^{\alpha+2}$, whereas if $n \notin S$, then $\mathcal{C}_n$ has type $\omega^{\alpha+1}$. We are interested in $\Delta_\alpha^0$ isomorphisms.

**Claim 5.13.** *If $n \in S$, then $\mathcal{C}_n^* \cong_{\Delta_\alpha^0} \mathcal{B}^*$ (with no uniformity).*

PROOF. There is a $\Delta_\alpha^0$ isomorphism from $\mathcal{C}_n$ onto $\mathcal{B}$. There is a $\Delta_\alpha^0$ procedure that can be applied to $\mathcal{C}_n$ and $\mathcal{B}$ for determining the first element of every copy of $\omega$ and the successor relation on these elements. Consequently, there is a $\Delta_\alpha^0$ procedure that can be applied to $\mathcal{C}_n^*$ and $\mathcal{B}^*$ for determining the Cantor normal form for the interval preceding every element. Therefore, there is a $\Delta_\alpha^0$ isomorphism from $\mathcal{C}_n^*$ onto $\mathcal{B}^*$. This proves the claim. $\qquad\square$

The proof of Theorem 5.11 is complete. $\qquad\square$

Let $E_{\Delta_1^1}(K)$ be the set of pairs $(a, b)$ such that $a, b \in I(K)$ and there is a hyperarithmetical isomorphism between $\mathcal{A}_a$ and $\mathcal{A}_b$. We might think of the statement that $E_{\Delta_1^1}(K)$ is hyperarithmetical as an alternative classification statement. If $I(K)$ is hyperarithmetical, the sets $E_{\Delta_\alpha^0}(K)$ are hyperarithmetical for all computable ordinals $\alpha$. In the cases where we can show that $E(K)$ is hyperarithmetical, it is because there is a bound on ranks and $E(K)$ is equal to one of these sets.

**Proposition 5.14** ([66]). *If $E_{\Delta_1^1}(K)$ is hyperarithmetical, then it is equal to $E_{\Delta_\alpha^0}(K)$ for some computable ordinal $\alpha$.*

To prove this assertion, we can use the Barwise–Kreisel compactness. By assumptions, $I(K)$ is hyperarithmetical. We form a hyperarithmetical structure including all of the structures from $K^c$, their indices, and the relation $E_{\Delta_1^1}(K)$. Then we produce a hyperarithmetical set of computable infinitary sentences, say new constants $a$, $b$ a pair of indices in the relation $E_{\Delta_1^1}(K)$, and there is no $\Delta_\alpha^0$ isomorphism between the corresponding structures. Since we cannot satisfy the whole set, we try to get a suitable bound on the complexity of isomorphisms.

## 5.2.1. Third approach.

Here, we discuss characterization and classification statements involving lists (enumerations). In particular, a list is often taken for classification. Consider the classification of finite simple groups. A good list means that isomorphism types (other equivalence classes) are represented. It is natural to require that no isomorphism type (or equivalence class) appears twice.

**Definition 5.15.** An *enumeration of $K^c/\cong$* is a sequence $(\mathcal{A}_n)_{n\in\omega}$ representing every isomorphism type in $K^c$. An *enumeration of $K^c/\cong_{\Delta^0_\alpha}$* is a sequence representing every equivalence class in $K^c$ under a $\Delta^0_\alpha$ isomorphism.

**Definition 5.16.** A *Friedberg enumeration* of $K^c/\cong$ or $K^c/\cong_{\Delta^0_\alpha}$ is an enumeration such that every isomorphism type or every equivalence class under $\Delta^0_\alpha$ isomorphism is represented only once.

**Definition 5.17.** An enumeration is *computable* (a $\Delta^0_\alpha$ *enumeration*) if there is a computable ($\Delta^0_\alpha$-) sequence of computable indices for the structures.

*Computable characterization.* $K$ has a computable characterization if there is a hyperarithmetical enumeration of $K^c/\cong$ (other equivalence can be substituted for an isomorphism).

*Computable classification.* $K$ has a computable classification if there is a hyperarithmetical Friedberg enumeration of $K^c/\cong$ (other equivalence can be substituted for an isomorphism).

A computable enumeration $(\mathcal{A}_n)_{n\in\omega}$ of $K^c$ is *universal* up to an isomorphism if for a given computable index for $\mathcal{B} \in K^c$ there exists $n$ such that $\mathcal{B} \cong \mathcal{A}_n$. An enumeration is *principal* if for any other enumeration $(\mathcal{B}_n)_{n\in\omega}$ up to an isomorphism there is a computable function $f$ such that $\mathcal{B}_n \cong \mathcal{A}_{f(n)}$. It is clear that a universal enumeration is principal.

### 5.2.2. Computable enumerations.

The following result of Nurtazin [130] yields the existence condition for computable enumerations of $K^c/{\cong}$.

**Theorem 5.18** ([130]). *Suppose that $K$ is a class of structures such that for some $\mathcal{U} \in K^c$ and every $\mathcal{A} \in K^c$ there is a computable embedding of $\mathcal{A}$ into $\mathcal{U}$ and every c.e. subset $W$ of $\mathcal{U}$ generates a unique structure $\mathcal{B} \subseteq \mathcal{U}$ in $K$. Then there is a computable enumeration of $K^c/{\cong}$ determined up to an isomorphism. If for a given index for $\mathcal{A}$ there is an index for a computable embedding of $\mathcal{A}$ into $\mathcal{U}$, then there exists a computable universal enumeration of $K^c/{\cong}$.*

**Corollary 5.19** ([66]). *A computable universal enumeration of $K^c/{\cong}$ exists for each of the following classes $K$ :*

  (a) *linear orderings,*

  (b) *Boolean algebras,*

  (c) *equivalence structures,*

  (d) *Abelian p-groups (not necessarily reduced),*

  (e) *algebraic fields of characteristic p,*

  (f) *structures for a fixed computable relational language.*

In case (f), a universal model $\mathcal{U}$ can be obtained as the union of a chain of finite structures, where, at every stage, new elements are added in order to satisfy all possible open types over the set of "old" elements. The structure $\mathcal{U}$ is computably categorical, and its theory is $\aleph_0$ categorical.

Further, we can obtain the conclusion of Nurtazin's theorem without the assumption of computable embeddings.

**Proposition 5.20.** *If $K$ is the class of vector spaces over $Q$, then there is a computable enumeration of $K^c/{\cong}$. In fact, there is a principal enumeration.*

## 5.2.3. Existence of Friedberg enumerations.

For some classes with simple invariant it is easy to produce computable Friedberg enumerations.

**Proposition 5.21** ([66]). *There is a computable Friedberg enumeration of $K^c/_\cong$ for the following classes $K$ :*

(a) *vector spaces over $Q$,*

(b) *algebraically closed fields of a given characteristic,*

(c) *well orderings of type less than a fixed computable ordinal $\alpha$.*

For computable equivalence structures there are natural invariants, but they are not so simple as the above examples. We suspect that there is no computable Friedberg enumeration up to an isomorphism. We have the following result.

**Theorem 5.22** ([66]). *If $K$ is the class of equivalence structures with infinitely many infinite classes, then $K^c/_\cong$ has a computable Friedberg enumeration.*

A direct proof of the nonexistence of a computable Friedberg enumeration is apparently a rather difficult question. Suppose that there exists a computable bound on the ranks of elements of $K^c$ and there exists a computable Friedberg enumeration $(\mathcal{C}_n)_{n\in\omega}$ of $K/_\cong$. To obtain a contradiction, we try to find a computable $\mathcal{A} \in K$ satisfying the condition $\mathcal{A} \not\cong \mathcal{C}_n$ for all $n$. It is difficult to work out a suitable strategy even if we restrict ourselves to only one of these conditions, for some $n$. The following assertions clarify the difficulties we meet in this way. The assumptions of the first assertion are the same as in Nurtazin's theorem.

**Theorem 5.23** ([66]). *Suppose that there is a $\mathcal{U} \in K^c$ such that*

(1) *for every index $e$ for a structure $\mathcal{A} \in K^c$ it is possible to find an index for a computable embedding of $\mathcal{A}$ into $\mathcal{U}$,*

(2) *every c.e. set $W \subseteq \mathcal{U}$ generates a unique substructure $\mathcal{B} \subseteq \mathcal{U}$ in $K$.*

*Then there is no partial computable function $f$ such that for any index $e$ of $\mathcal{A} \in K^c$, $f(e)$ is an index of some $\mathcal{B} \in K^c$ such that $\mathcal{B} \not\cong \mathcal{A}$.*

**Corollary 5.24 ([66]).** *For the following classes $K$ there is no effective procedure such that for a given index for a computable $\mathcal{A}$ in $K$ it yields an index for a computable $\mathcal{B}$ in $K$ such that $\mathcal{A} \not\cong \mathcal{B}$:*

(a) *linear orderings,*

(b) *Boolean algebras,*

(c) *equivalence structures,*

(d) *arbitrary structures for a computable relational language.*

Thus, we clarified some of the difficulties arising in the attempts to obtain a direct proof of the nonexistence of Friedberg enumerations. Therefore, we use some results on the complexity of the isomorphism problems.

**Proposition 5.25 ([66]).** *Suppose that $I(K)$ is hyperarithmetical and $E(K)$ is properly $\Sigma_1^1$. Then there is no hyperarithmetical Friedberg enumeration of $K^c/{\cong}$.*

**Corollary 5.26 ([66]).** *There is no hyperarithmetical Friedberg enumeration of $K^c/{\cong}$ for the following classes $K$:*

(a) *linear orderings,*

(b) *Boolean algebras,*

(c) *Abelian p-groups,*

(d) *structures for a computable language with at least one binary relation symbol.*

Some of these results are known for isomorphisms of a fixed complexity. In particular, one of the results concerns the class $K$ of vector spaces over $Q$. Note that the computable members of $K$ are isomorphic only if they are $\Delta_2^0$ isomorphic. As we seen, $K^c/{\cong}$ has a computable principal enumeration and a computable Friedberg enumeration.

**Proposition 5.27** ([66]). *If $K$ is a class of vector spaces over $Q$, then there is no computable enumeration of $K^c/_{\Delta^0_1}$.*

Like Proposition 5.25, the following assertion concerns the nonexistence of Friedberg enumerations of various classes, up to a $\Delta^0_\alpha$ isomorphism.

**Proposition 5.28** ([66]). *Suppose that $I(K)$ is $\Delta^0_{\alpha+2}$ and the $\Delta^0_\alpha$ isomorphism problem for $K$ is properly $\Sigma^0_{\alpha+2}$. Then there is no $\Delta^0_{\alpha+2}$ Friedberg enumeration of $K^c/_{\cong_{\Delta^0_\alpha}}$.*

**Corollary 5.29** ([66]). *Let $K$ be one of the following classes:*

(a) *linear orderings,*

(b) *structures for a computable relational language with at least one binary relation symbol,*

(c) *Boolean algebras,*

(d) *Abelian p-groups.*

*Then $K^c$ has no $\Delta^0_3$ Friedberg enumeration up to a computable isomorphism and for a computable ordinal $\alpha$, $K^c$ has no $\Delta^0_{\alpha+2}$ Friedberg enumeration up to a $\Delta^0_\alpha$ isomorphism.*

We write $K^c/_{\cong_{\Delta^1_1}}$ for the set of equivalence classes of elements of $K^c$ under a hyperarithmetical isomorphism. The assertion that $K^c/_{\cong_{\Delta^1_1}}$ has a hyperarithmetical Friedberg enumeration is an alternative classification statement.

**Proposition 5.30** ([66]). *If $(A_n)_{n \in \omega}$ is a hyperarithmetical enumeration of $K^c$ determined up to a $\Delta^1_1$-isomorphism then it is an enumeration of $K^c$ determined up to a $\Delta^0_\alpha$ isomorphism for some computable ordinal $\alpha$.*

Using the Barwise–Kreisel compactness, we obtain a $\Pi^1_1$ set of computable infinitary sentences describing a structure $B$, an index $e$, and a function $F$ such that $B$ is computable, $F$ is an isomorphism from $A_e$ onto $B$, and there is no $\Delta^0_\alpha$ isomorphism from $\mathcal{A}_e$ onto $\mathcal{B}$ for any computable ordinal $\alpha$. If such an $\alpha$ does not exist, then every $\Delta^1_1$ subset is satisfied. Hence we obtain a model of the whole set, which leads to a contradiction.

### 5.2.4. Relationship between three approaches.

We present the relationship between the basic characterization statements in the following form:

I. $K^c$ is the set of computable models of a computable infinitary sentence

⇓ ⇑

II. $I(K)$ is hyperarithmetical

⇓ ⇑̸

III. $K^c/_{\cong}$ has a hyperarithmetical enumeration

It is easy to see that I $\Rightarrow$ II $\Rightarrow$ III. The result below asserts that II $\Rightarrow$ I.

**Theorem 5.31** ([66]). *Suppose that $K$ is a class of structures closed under an isomorphism and $I(K)$ is hyperarithmetical. Then there is a computable infinitary sentence for which $K^c$ is the class of computable models.*

The following result asserts that III $\not\Rightarrow$ II.

**Proposition 5.32** ([66]). *Let $K$ consist of copies of $\omega_1^{CK}(1 + \eta)$ and the linear orderings of rank at most $\omega$. Then $K^c/_{\cong}$ has a hyperarithmetical Friedberg enumeration. However, $I(K)$ is not hyperarithmetical.*

In the first classification statement, we added the corresponding characterization statement. A computable bound on the ranks of elements of $K^c$, without a sentence whose computable models are these elements, does not tell us much. The remaining classification statements imply the corresponding classification statements. We summarize the relations among the basic classification statements as follows.

I. There is a computable bound on the ranks of elements of $K^c$, in addition to a computable infinitary sentence whose computable models are these structures

⇓ ⇑ ?

II. $E(K)$ is hyperarithmetical

⇓ ⇑̸

III. $K^c/_\cong$ has a hyperarithmetical Friedberg enumeration

It is easy to see that I ⇒ II ⇒ III. By Proposition 5.32, III ⇏ II. For a class $K$ in this proposition, $K^c/_\cong$ has a hyperarithmetical Friedberg enumeration, but $I(K)$ is not hyperarithmetical. Hence $E(K)$ cannot be hyperarithmetical. Under the assumption that there is a computable infinitary sentence $\psi$ whose computable models are elements of $K^c$, it is not known whether III ⇒ II.

We formulate a partial result concerning the implication II ⇒ I or III ⇒ I.

**Theorem 5.33** ([66]). *Suppose that $K^c/_\cong$ has a hyperarithmetical Friedberg enumeration. Then there is computable ordinal $\alpha$ such that for $\mathcal{A}$, $\mathcal{B}$ in $K^c$, if every computable $\Pi_\alpha$ sentence true in $\mathcal{A}$ is true in $\mathcal{B}$, then $\mathcal{A} \cong \mathcal{B}$.*

The first approach to the characterization problem is natural from the mathematical point of view. Known classes of structures (for example, groups and fields) are described by using axioms. The second approach, involving index sets, seems to be far from common practice in mathematics. Nevertheless, the characterization statements for the first and second approaches are equivalent. The third approach to the classification problems which yields a list without repetition of invariants is very natural from the mathematical point of view. As we seen, the classification statements obtained by the second and third approaches are not equivalent, although there are relations between them. For some classes with nice invariants (for example, vector spaces over the rational numbers) we can give a computable Friedberg enumeration. However, in the majority of cases, where we established the nonexistence of computable Friedberg enumerations, the proof was indirect and used the complexity of the isomorphism problem.

It is very important to determine precisely the complexity of the isomorphism problem for various classes. Having a classification, it is reasonable to look for the least computable ordinal $\alpha$ such that $E(K) = E_{\Delta_\alpha^0}(K)$.

**Problem 31.** Whether there is an example of a class $K$ for which the isomorphism problem is properly at some level, $\Sigma_1^1$, $\Pi_3^0$, etc., but is not complete at this level?

In all cases where we located $E(K)$ properly at some level of complexity (by proving that it is $\Sigma_1^1$, but not $\Delta_1^1$ or by proving that it is $\Pi_3^0$, but not $\Delta_3^0$), it turned out to be complete at that level. Problem 31 is related to the long-standing challenge of finding a "natural" example of a c.e. set such that it is neither computable nor complete.

The following problem concerns a special case of the missing implication II $\Rightarrow$ I for the classification problems, where $K$ consists of copies of a single computable structure $\mathcal{A}$. In this case, $E(K)$ is essentially the same as $I(\mathcal{A})$.

**Problem 32.** Whether $R^c(\mathcal{A})$ is computable provided that $I(\mathcal{A})$ is hyperarithmetical?

**Definition 5.34.** Let $\mathcal{A}$ be a computable structure such that $R^c(\mathcal{A}) = \omega_1^{CK}$. We say that $\mathcal{A}$ is *computably approximable* if every computable infinitary sentence true in $\mathcal{A}$ is also true in some computable $\mathcal{B} \not\cong \mathcal{A}$.

The known examples of computable structures of noncomputable rank (for example, the Harrison ordering) are computably approximable. This can be explained by the fact that they were obtained from a family of computable approximations by using the Barwise–Kreisel compactness or by some other similar methods.

**Problem 33.** Let $\mathcal{A}$ be computable, and let $R^c(\mathcal{A}) = \omega_1^{CK}$.

(a) Whether $\mathcal{A}$ is computably approximable?

(b) Whether any true computable infinitary sentence in $\mathcal{A}$ is also true in some computable structure $\mathcal{B}$ of computable rank?

By Proposition 3.11, if $\mathcal{A}$ is computable and $R^c(\mathcal{A}) = \omega_1^{CK}$, then any computable infinitary sentence true in $\mathcal{A}$ is also true in some *hyperarithmetical* $\mathcal{B} \not\equiv \mathcal{A}$.

REMARK. Problems 32 and 33 (a) are equivalent. If Problem 33 (a) has a negative answer confirmed by a computable structure $\mathcal{A}$, then Problem 32 has a negative answer confirmed by the same structure $\mathcal{A}$. If Problem 33 (a) has a positive answer, then we can use the Barwise–Kreisel compactness to show that Problem 32 has a positive answer.

**Problem 34.** Let $K$ be a class of equivalence structures. Whether a computable Friedberg enumeration of $K^c$ exists up to an isomorphism?

## 5.3. Definability and index sets of natural classes of computable models

As was already mentioned, the study of the complexity of index sets for computable models is important for understanding structural properties, classifications of models, and complexity level of classifications. On the other hand, if there exists a universal enumeration of computable classes of models of a given structure numbering, we can compare classes by the complexity of their description and choose the most adequate description corresponding to their real algorithmic complexity.

One of the goals of the theory of computable models is to characterize the complexity of classes of autostable models of finite or infinite algorithmic dimension with the Scott family of $\exists$ formulas in a finite enrichment by constants. This question is also of interest for a certain level of arithmetic hierarchy and its extension by notations of constructive ordinals, and the interaction of the complexity of the definition of these classes of models. Related topics are the complexity of index sets of computable models of a given Scott rank; in particular, the case of nonconstructive Scott ranks models of Scott rank $\omega_1^{CK}$ and $\omega_1^{CK} + 1$.

Another cycle of problems is connected with the complexity of finding computable models with theories of a given type and, in particular, the case of a theory categorical in uncountable power, a finitely axiomatizable theory, an Ehrenfeucht theory, a theory without prime model, a theory with countably many countable models, an $\omega$-stable theory, a stable theory, a theory with countably many types, a decidable theory, an elementary theory of a given complexity, a theory with a given complexity of decidability of computable models with respect to Turing degrees, a theory with one computable model, a strongly minimal theory, etc. It is of interest to clarify whether the Turing degree of a theory from the above list is universal in the corresponding hierarchy class of complexity.

**Acknowledgements.** I thank Yu. Ershov, J. Knight, S. Lempp, B. Khoussainov, V. Dobrica, A. Sorbi, V. Harizanov, R. Downey, T. Millar, D. Hirshfeldt, V. Dzgoev, B. Drobotun, R. Solomon, Ch. McCoy, W. Calvert for fruitful discussions of problems and approaches presented in this paper.

# References

1. C. J. Ash, *Categoricity in hyperarithmetical degrees*, Ann. Pure Appl. Logic **34** (1987), 1–14.

2. C. J. Ash and J. F. Knight, *Pairs of computable structures*, Ann. Pure Appl. Logic **46** (1990), 211–234.

3. C. J. Ash and J. F. Knight, *Possible degrees in computable copies*, Ann. Pure Appl. Logic **75** (1995), no. 3, 215–221.

4. C. J. Ash and J. F. Knight, *Possible degrees in computable copies II*, Ann. Pure Appl. Logic **87** (1997), no. 2, 151–165.

5. C. J. Ash and J. F. Knight, *Computable Structures and the Hyper-arithmetical Hierarchy*, Elsevier, 2000.

6. C. J. Ash, J. F. Knight, M. Manasse, and T. Slaman, *Generic copies of countable structures*, Ann. Pure Appl. Logic **42** (1989), no. 3, 195–205.

7. C. J. Ash and T. S. Millar, *Persistently finite, persistently arithmetic theories*, Proc. Am. Math. Soc. **89** (1983), no. 3, 487–492.

8. C. J. Ash and A. Nerode, *Intrinsically computable relations*, In: Aspects of Effective Algebra, J. N. Crossley (Ed.), Upside Down A Book Co., Steel's Creek, Australia, pp. 26-41.

9. E. Baldwin and A. Lachlan, *The quantity of non-autoequivalent constructivizations*, J. Symb. Log. **36** (1971), 79–96.

10. E. Barker, *Intrinsically $\Sigma_\alpha^0$ relations*, Ann. Pure Appl. Logic **39** (1988), 105–130.

11. K. J. Barwise, *Infinitary logic and admissible sets*, J. Symb. Log. **34** (1969), 226–252.

12. I. Bucur and A. Deleanu, *Introduction to the Theory of Categories and Functors*, John Wiley & Sons, London–New York–Sydney, 1968.

13. W. Calvert, S. S. Goncharov, and J. F. Knight, *Computable Structures of Scott Rank $\omega_1^{CK}$ in Familiar Classes*, Preprint.

14. W. Calvert, J. F. Knight, and J. M. Young [Millar], *Computable Trees of Scott Rank $\omega_1^{CK}$ and Computable Approximation*, Preprint.

15. R. Camerlo and S. Gao, *The completeness of the isomorphism relation for countable Boolean algebras*, Trans. Am. Math. Soc. **353** (2000), 491–518.

16. C. C. Chang and H.-J. Keisler, *Model Theory*, North-Holland, Amsterdam, 1990.

17. J. Chisholm, *Effective model theory versus computable model theory*, J. Symb. Log. **55** (1990), 1168–1191.

18. P. Cholak, S. S. Goncharov, B. Khoussainov, and R. A. Shore, *Computably categorical structures and expansions by constants*, J. Symb. Log. **64** (1999), 13–37.

19. R. Downey and C. Jockusch, *Array noncomputable degrees and genericity*, In: Computability, Enumerability, Unsolvability, London Math. Soc. Lect. Notes Ser. **224**, Cambridge Univ. Press, 1996, pp. 93–104.

20. D. Deutsch, *The Fabric of Reality*, The Penguin Press, New York, 2000.

21. V. P. Dobritsa, *Recursively numbered classes of constructive extensions and autostability of algebras*, Sib. Math. J. **16** (1975), no. 6, 879–883.

22. V. P. Dobritsa, *Computability of certain classes of constructive algebras*, Sib. Math. J. **18** (1977), no. 3, 406–413.

23. V. D. Dzgoev and S. S. Goncharov, *Autostable models*, Algebra Log. **19** (1980), 28–37.

24. Yu. L. Ershov and E. A. Palutin, *Mathematical Logic* [in Russian], Mir Publisher, Moscow, 1984.

25. Yu. L. Ershov, *On a hierarchy of sets. I*, Algebra Log. **7** (1968), no. 1, 47–74.

26. Yu. L. Ershov, *On a hierarchy of sets. II*, Algebra Log. **7** (1968), no. 4, 15–47.

27. Yu. L. Ershov, *Computable enumerations*, Algebra Log. **7** (1968) no. 5, 71–99.

28. Yu. L. Ershov, *Numbered fields* In: Logic, Methodology Philos. Sci. III (Proc. Third Internat. Congr., Amsterdam, 1967), North-Holland, Amsterdam, 1968, pp. 31–34,

29. Yu. L. Ershov, *On a hierarchy of sets. III*, Algebra Log. **9** (1970), no. 1, 20–31.

30. Yu. L. Ershov, *On index sets*, Sib. Math. J. **11** (1970), no. 2, 246–258.

31. Yu. L. Ershov, *Existence of constructivizations*, Soviet Math. Dokl. **13** (1972), no. 5, 779–783.

32. Yu. L. Ershov, *Constructive models* [in Russian], In: Selected Questions of Algebra and Logic, Novosibirsk, 1973, pp. 111–130.

33. Yu. L. Ershov, *Skolem functions and constructive models*, Algebra Log. **12** (1973), no. 6, 368–373.

34. Yu. L. Ershov, *Numbering Theory. 3* [in Russian], In: Constructive Models, Novosibirsk State Univ., Novosibirsk, 1974.

35. Yu. L. Ershov, *Theorie der Numerierungen. III*, Z. Logik Grundlag. Math. **23** (1977), 289–371.

36. Yu. L. Ershov, *Theory of Numberings* [in Russian], Nauka, Moscow, 1977.

37. Yu. L. Ershov, *The Decidability Problems and Constructive Models*, Nauka, Moscow, 1980.

38. Yu. L. Ershov, *Definability and Computability*, Consultants Bureau, New York, 1996.

39. Yu. L. Ershov, S. S. Goncharov, A. Nerode, J. B. Remmel, and V. W. Marek, *Handbook of Recursive Mathematics*, Elsevier, Amsterdam, 1998.

40. Yu. L. Ershov and S. S. Goncharov, *Constructive Models*, Kluwer Academic/Plenum Press, New York, 2000.

41. Yu. L. Ershov and I. A. Lavrov, *Upper semilattice $L(S)$*. Algebra Log. **12** (1973), no. 2, 167–189.

42. H. Friedman, *Computableness in $\Pi_1^1$ paths through $\mathcal{O}$*, Proc. Am. Math. Soc. **54** (1976), 311–315.

43. E. Fokina, *About complexity of categorical theories with computable models* [in Russian], Vestnik NGU **5** (2005), no. 2, 78–86.

44. A. Fröhlich and J. Shepherdson, *Effective procedures in field theory*, Philos. Trans. Roy. Soc. London, Ser. A **248** (1956), 407–432.

45. S. S. Goncharov, *Autostability and computable families of constructivizations*, Algebra Log. **14** (1975), 392–408.

46. S. S. Goncharov, *The quantity of non-autoequivalent constructivizations*, Algebra Log. **16** (1977), 257–282.

47. S. S. Goncharov, *Constructive models of $\aleph_1$-categorical theories*, Math. Notes **23** (1978), no. 6, 486–487.

48. S. S. Goncharov, *Problem of the number of non-self-equivalent constructivizations*, Algebra Log. **19** (1980), 401-414.

49. S. S. Goncharov, *Problem of the number of non-self-equivalent constructivizations*, Soviet Math. Dokl. **21** (1980), 411-414.

50. S. S. Goncharov, *Computable single-valued numerations*, Algebra Log. **19** (1980), 325–356.

51. S. S. Goncharov, *The quantity of non-autoequivalent constructivizations*, Algebra Log. **16** (1977) 169–185.

52. S. S. Goncharov, *Strong constructivizability of homogeneous models*, Algebra Log. **17** (1978), no. 4, 247–262.

53. S. S. Goncharov, *A totally transcendental decidable theory without constructivizable homogeneous models*, Algebra Log. **19** (1980), no. 2, 85–93.

54. S. S. Goncharov, *Countable Boolean Algebras* [in Russian], Nauka, Novosibirsk, 1988.

55. S. S. Goncharov, *Computable classes of constructivizations for models of finite computability type*, Sib. Math. J. **34** (1993), no. 5, 812–824.

56. S. S. Goncharov, *Effectively infinite classes of weak constructivizations of models*, Algebra Log. **32** (1993), no. 6, 342–360.

57. S. S. Goncharov, *Countable Boolean Algebras and Decidability*, Consultants Bureau, New York, 1997.

58. S. S. Goncharov, *Computable models and computable numberings* In: Proc. Logic Colloquium 2005, Athens, 2006. [To appear]

59. S. S. Goncharov and B. N. Drobotun, *Numerations of saturated and homogeneous models*, Sib. Math. J. **21** (1980), no. 2, 164–175.

60. S. S. Goncharov and B. N. Drobotun, *Algorithmic dimension of nilpotent groups*, Sib. Math. J. **30** (1989), no. 2, 210–216.

61. S. S. Goncharov, V. S. Harizanov, M. Laskowski, S. Lempp, and Ch. Mccoy, *Trivial, strongly minimal theories are model complete after naming constants*, Proc. Am. Math. Soc. **131** (2003), no. 12, 3901–3912.

62. S. S. Goncharov and B. Khoussainov, *On the spectrum of the degrees of decidable relations*, Dokl. Akad. Nauk **352** (1997), no. 3, 301–303.

63. S. S. Goncharov, A. V. Molokov, and N. S. Romanovskii, *Nilpotent groups of finite algorithmic dimension*, Sib. Math. J. **30** (1989), no. 1, 63–67.

64. S. S. Goncharov and A. A. Novikov, *Examples of nonautostable systems*, Sib. Math. J. **25** (1984), no. 4, 538–544.

65. S. S. Goncharov and A. T. Nurtazin, *Constructive models of complete solvable theories*, Algebra Log. **12** (1973), no. 2, 67–77.

66. S. S. Goncharov and J. F. Knight, *Computable structure/nonstructure theorems*, Algebra Log. **41** (2002), 351–373.

67. S. S. Goncharov, and B. Khoussainov, *Open problems in the theory of constructive algebraic systems*, Contemp. Math. (AMS) **257** (2000), 1 45–170.

68. S. S. Goncharov, and B. Khoussainov, *Complexity of categorical theories with computable models*, Algebra Log. **43** (2004), no. 6, 365–373.

69. S. S. Goncharov, and B. N. Drobotun, *Numerations of saturated and homogeneous models*, Sib. Math. J. **30** (1989), 164–175.

70. S. S. Goncharov, V. S. Harizanov, J. F. Knight, and R. Shore, $\Pi_1^1$-*relations and paths through*, J. Symb. Log. **69** (2004), no. 2, 585–611.

71. S. S. Goncharov, V. S. Harizanov, J. F. Knight, A. Morozov, and A. Romina, *On automorphic tuples of elements in computable models*, Sib. Math. J. **46** (2005), no. 3, 405–412.

72. S. S. Goncharov, V. S. Harizanov, J. F. Knight, C. McCoy, R. G. Miller, and R. Solomon, *Enumerations in computable structure theory*, Ann. Pure Appl. Logic **136** (2005), no. 3, 219–246.

73. S. S. Goncharov and A. Sorbi, *Generalized computable numerations and nontrivial Rogers semilattices*, Algebra Log. **36** (1997), no. 4, 359–369.

74. L. Harrington, *Recursively presentable prime models*, J. Symb. Log. **39** (1974), no. 2, 305–309.

75. V. S. Harizanov, *Some effects of Ash–Nerode and other decidability conditions on degree spectra*, Ann. Pure Appl. Logic **55** (1991), 51–65.

76. J. Harrison, *Computable pseudo well-orderings*, Trans. Am. Math. Soc. **131** (1968), 526–543.

77. D. R. Hirschfeldt, B. Khoussainov, R. A. Shore, and A. M. Slinko, *Degree spectra and computable dimensions in algebraic structures*, Ann. Pure Appl. Logic **115** (2002), 71-113.

78. W. Hodges, *What is a structure theory?* Bull. London Math. Soc. **19** (1987), 209–237.

79. C. G. Jockusch, *Computableness of initial segments of Kleene's $\mathcal{O}$*, Fund. Math. **87** (1975), 161–167.

80. C. G. Jockusch and T. G. McLaughlin, *Countable retracing functions and $\Pi_2^0$ predicates*, Pac. J. Math. **30** (1969), 67–93.

81. C. G. Jockusch and R. I. Soare, *Degrees of orderings not isomorphic to computable linear orderings*, Ann. Pure Appl. Logic **52** (1991), 39–64.

82. I. Kaplansky, *Infinite Abelian Groups*, Univ. Michigan Press, 1954.

83. N. G. Khisamiev, *Strongly constructive models*, Izv. Akad. Nauk Kaz.SSR, Ser. Fiz-Mat. (1971), no. 3, 59–63.

84. N. G. Khisamiev, *Strongly constructive models of a decidable theory*, Izv. Akad. Nauk Kaz.SSR, Ser. Fiz-Mat. (1974), no. 1, 83–84.

85. B. Khoussainov and R. A. Shore, *Computable isomorphisms, degree spectra of relations and Scott families*, Ann. Pure Appl. Logic **93** (1998), 153–193.

86. B. Khoussainov, S. Lempp, and R. Solomon. Preprint

87. A. B. Khoutoretskiĭ, *On the cardinality of the upper semilattice of computable enumerations*, Algebra Log. **10** (1971), no. 5, 348–352.

88. H. J. Keisler, *Model Theory for Infinitary Logic*, North-Holland, 1971.

89. J. F. Knight, *Degrees coded in jumps of orderings*, J. Symb. Log. **51** (1986), 1034–1042.

90. J. F. Knight, *Nonarithmetical $\aleph_0$-categorical theories with recursive models*, J. Symb. Log. **59** (1994), no. 1, 106–112.

91. J. F. Knight, *Effective transfer of invariants*, J. Symb. Log. [To appear]

92. J. F. Knight and J. M. Young [Millar], *Computable structures of rank $\omega_1^{CK}$*, J. Math. Logic [To appear]

93. G. Kreisel, *Set-theoretic problems* In: Infinitistic Methods: Proc. of Symp. on Found. of Math. pergamon, Warsaw, 1961, pp. 103–140.

94. C. Karp, *Languages with Expressions of Infinite Length*, PhD Thesis, Univ. South. California, 1959.

95. A. H. Lachlan, *Solution to a problem of Spector*, Canad. J. Math. **23** (1971), 247–256.

96. M. Lerman and J. H. Schmerl, *Theories with recursive models*, J. Symb. Log. **44** (1979), no. 1, 59–76.

97. *Logic Notebook. Unsolved Problems in Mathematical Logic*, Novosibirsk, 1986.

98. E. Lopez-Escobar, *On definable well-orderings*, Fund. Math. **57** (1966), 299–300.

99. M. Makkai, *An example concerning Scott heights*, J. Symb. Log. **46** (1981), 301–318.

100. A. I. Mal'tsev, *Constructive algebras. I* [in Russian], Usp. Mat. Nauk **16** (1961), no. 3, 3–60.

101. A. I. Mal'tsev, *On recursive Abelian groups*, Dokl. Akad. Nauk SSSR **146** (1962), no. 5, 1009–1012.

102. A. I. Mal'tsev, *On the theory of computable families of objects*, Algebra Log. **3** (1964), no. 4, 5–31.

103. A. I. Mal'tsev, *Algorithms and Recursive Functions*, Wolters-Noordhoff, Groningen, 1970.

104. A. I. Mal'tsev, *Algebraic Systems*, Springer, New York–Heidelberg, 1973.

105. A. I. Mal'tsev, *Selected Works. Mathematical Logic and the General Theory of Algebraic Systems. Vol. 2* [in Russian], Nauka, Moscow, 1976.

106. M. S. Manasse, *Techniques and Counterexamples in Almost Categorical computable Model Theory*, PhD Thesis, Univ. Wisconsin-Madison, 1982.

107. T. S. Millar, *Persistantly finite theories with hyperarithmetic models*, Trans. Am. Math. Soc. **278** (1983), no. 1, 91–99.

108. T. S. Millar, *Foundations of recursive model theory*, Ann. Math. Logic **13** (1978), no. 1, 45–72.

109. T. S. Millar, *A complete, decidable theory with two decidable models*, J. Symb. Log. **44** (1979), no. 3, 307–312.

110. T. S. Millar, *Homogeneous models and decidability*, Pac. J. Math. **91** (1980), no. 2, 407–418.

111. T. S. Millar, *Vaught's theorem recursively revisited*, J. Symb. Log. **46** (1981), no. 2, 397–411.

112. T. S. Millar, *Counterexamples via model completions*, Lect. Notes Math. **859** Springer, 1981, pp. 215–229.

113. T. S. Millar, *Type structure complexity and decidability*, Trans. Am. Math. Soc. **271** (1982), no. 1, 73–81.

114. T. S. Millar, *Omitting types, type spectrums, and decidability*, J. Symb. Log. **48** (1983), no. 1, 171–181.

115. T. S. Millar, *Decidability and the number of countable models*, Ann. Pure Appl. Logic **27** (1984), no. 2, 137–153.

116. T. S. Millar, *Decidable Ehrenfeucht theories*, Proc. Sympos. Pure Math. **42** (1985), 311–321.

117. T. S. Millar, *Prime models and almost decidability*, J. Symb. Log. **51** (1986), no. 2, 412–420.

118. T. S. Millar, *Recursive categoricity and persistence*, J. Symb. Log. **51** (1986), no. 2, 430–434.

119. T. S. Millar, *Bad models in nice neighborhoods*, J. Symb. Log. **51** (1986), no. 4, 1043–1055.

120. T. S. Millar, *Finite extensions and the number of countable models*, J. Symb. Log. **54** (1989), no. 1, 264–270.

121. T. S. Millar, *Homogeneous models and almost decidability*, J. Aust. Math. Soc. Ser. A **46** (1989), no. 3, 343–355.

122. T. S. Millar, *Tame theories with hyperarithmetic homogeneous models*, Proc. Am. Math. Soc. **105** (1989), no. 3, 712–726.

123. T. S. Millar, *Model completions and omitting types*, J. Symb. Log. **60** (1995), no. 2, 654–672.

124. F. Montagna and A. Sorbi, *Universal recursion-theoretic properties of r.e. preordered structures*, J. Symb. Log. **50** (1985), no. 2, 397–406.

125. M. Morley, *Decidable models*, Israel J. Math. **25** (1976), 233–240.

126. M. Morley, *Omitting classes of elements*, In: The Theory of Models, M. Addison, L. Henkin, and A. Tarski (Eds.), North-Holland, 1965, pp. 265–273.

127. M. E. Nadel, *Scott sentences for admissible sets*, Ann. Math. Log. **7** (1974), 267-0294.

128. M. E. Nadel, $L_{\omega_1\omega}$ *and admissible fragments*, In: Model-Theoretic Logics, K. J. Barwise and S. Feferman (Eds.), Springer, 1985, pp. 271–316.

129. A. T. Nurtazin, *Strong and weak constructivization and computable families*, Algebra Log. **13** (1974), no. 3, 177–184.

130. A. T. Nurtazin, *Computable Models and Computable Classes*, PhD Thesis, Novosibirsk, 1975.

131. R. Parikh, *A note on paths through* $\mathcal{O}$, Proc. Am. Math. Soc. **39** (1973), 178–180.

132. M. G. Peretyat'kin, *Strongly constructive models and numerations of the Boolean algebra of recursive sets*, Algebra Log. **10** (1971), no. 5, 332–345.

133. M. G. Peretyat'kin, *Every recursively enumerable extension of a theory of linear order has a constructive model*, Algebra Log. **12** (1973), no. 2, 120–124.

134. M. G. Peretyat'kin, *Strongly constructive model without elementary submodels and extensions*, Algebra Log. **12** (1973), no. 3, 178–183.

135. M. G. Peretyat'kin, *On complete theories with a finite number of denumerable models*, Algebra Log. **12** (1973), no. 5, 310–326.

136. M. G. Peretyat'kin, *Criterion for strong constructivizability of a homogeneous model*, Algebra Log. **17** (1978), no. 4, 290–301.

137. M. O. Rabin, *Effective computability of winning strategies*, In: Contributions to the Theory of Games, Vol. 3, Princeton Univ. Press, Princeton, 1957, 147–157.

138. R. Reed, *A decidable Ehrenfeucht theory with exactly two hyperarithmetic models*, Ann. Pure Appl. Logic **53** (1991), no. 2, 135–168.

139. J. P. Ressayre, *Models with compactness properties relative to an admissible language*, Ann. Math. Log. **11** (1977), 31–55.

140. H. Rogers, Jr., *Theory of Computable Functions and Effective Computability*, McGraw-Hill, New York, 1967.

141. G. E. Sacks, *On the number of countable models*, In: Southeast Asian Conference on Logic (Singapore, 1981), C. T. Chong and M. J. Wicks (Eds.), North-Holland, 1983, pp. 185–195.

142. G. E. Sacks, *Higher Recursion Theory*, Springer, 1990.

143. D. Scott, *Logic with denumerably long formulas and finite strings of quantifiers*, In: The Theory of Models, J. Addison, L. Henkin, and A. Tarski (Eds.), North-Holland, 1965, 329–341.

144. V. L. Selivanov, *Enumerations of families of general computable functions*, Algebra Log. **15** (1976), 205–226.

145. T. Slaman, *Relative to any non-computable set*, Proc. Am. Math. Soc. **126** (1998), 2117–2122.

146. R. I. Soare, *Recursively Enumerable Sets and Degrees*, Springer, 1987.

147. I. N. Soskov, *Intrinsically* $\Pi_1^1$ *relations*, Math. Log Quart. **42** (1996), 109–126.

148. I. N. Soskov, *Intrinsically* $\Delta_1^1$ *relations*, Math. Log Quart. **42** (1996), 469–480.

149. C. Spector, *Computable well-orderings*, J. Symb. Log. **20** (1955), 151–163.

150. S. Tennenbaum, *Non-archimedean models for arithmetic*, Notices Am. Math. Soc. **6** (1959), 270.

151. R. L. Vaught, *Non-recursive-enumerability of the set of sentences true in all constructive models*, Bull. Am. Math. Soc. **63** (1957), no. 4, 230.

152. R. L. Vaught, *Sentences true in all constructive models*, J. Symb. Log. **25**, (1960), no. 1, 39–58.

153. M. C. Venning, *Type Structures of $\omega_0$-Categorical Theories*, Thesis, Cornell Univ., Ithaca, 1976.

154. S. Wehner, *Enumerations, countable structures, and Turing degrees*, Proc. Am. Math. Soc. **126** (1998), 2131–2139.

# First-Order Logic Foundation
# of Relativity Theories

## Judit X. Madarász
*Alfréd Rényi Institute of Mathematics HAS*
*Budapest, Hungary*

## István Németi
*Alfréd Rényi Institute of Mathematics HAS*
*Budapest, Hungary*

## Gergely Székely
*Eötvös Loránd University Budapest*
*Alfréd Rényi Institute of Mathematics HAS*
*Budapest, Hungary*

Motivation and perspective for an exciting new research direction interconnecting logic, spacetime theory, relativity—including such revolutionary areas as black hole physics, relativistic computers, new cosmology—are presented in this paper. We would like to invite the logician reader to take part in this grand enterprise of the new century. Besides general perspective and motivation, we present initial results in this direction.

Mathematical Problems from Applied Logic. Logics for the XXIst Century. II. Edited by Dov M. Gabbay *et al.* / International Mathematical Series, 5, Springer, 2007

# 1. Introduction
## (Logic and Spacetime Geometry)

Throughout their intimately intertwined histories, logic and geometry immensely profited from their interactions. In particular, logic greatly profited from its applications to geometry. Indeed, the very birth of logic was brought about by the needs of geometry in the times of Socrates, Euclid and their predecessors. Ever since, their interactions had rejuvenating, invigorating effects on logic. For brevity, here we mention only Hilbert's axiomatization of geometry, Tarski's improvements on this in the framework of first-order logic (FOL) [62], Tarski's school of FOL approaches to geometry as a small sample. It is no coincidence that Tarskian algebraic logic is geometrical in spirit.

In this paper, we try to show that this fruitful cooperation promises new blessings for logic. This is so because there are breathtaking revolutions in our understanding of space and time, i.e., in relativity, cosmology, and black hole physics.

What is the subject matter of geometry? Traditionally, geometry was created as a mathematical theory of a physical entity called space. But recent developments in spacetime theory/general relativity show that there is no such thing as physical space. Space is only an illusion and as such is subjective. Space is a "slice" of a larger entity called spacetime. Spacetime, on the other hand, is objective, it exists. What is subjective about space is the, necessarily ad hoc, way we decide to "slice" spacetime up into spacelike slices. Actually, it was logician Kurt Gödel who first discovered and emphasized that in certain non-negligible cases such slicing is impossible (non-foliazibility, in the technical terminology) [25].

So, a great challenge for logic and logicians is to continue the tradition sketched above of providing foundation and conceptual analysis for geometry by doing the same to spacetime theory, hence to relativity.

A further motivation for geometry-friendly logicians is the following. Relativity theory can be conceived of geometrizing parts of physics in a sense (cf. [46]). Special relativity (SR) geometrizes some basic aspects of motion (kinematics) including light propagation; general relativity (GR) geometrizes gravitation + SR; the Kaluza–Klein style extension of GR geometrizes electromagnetic phenomena + GR; and currently intensively researched extensions of GR (for example, string theory) search for extending the scope of this aim for geometrizing more and more aspects of our understanding of the world.

Why is this interesting for logicians? Well, because history tells us that logic is applicable to geometry in an essential way. Hence if relativity (and its extensions) is the act of geometrizing more and more of physics, then it also can be regarded as a potential act of "logicizing" these areas, inviting logicians to take part in this grandiose adventure of mankind.

## 2. More Concrete Introduction (Foundation of Spacetime)

The idea of elaborating the foundational analysis of the logical structure of spacetime theory and relativity theories (foundation of relativity) in a spirit analogous with the rather successful foundation of mathematics was initiated by several authors including David Hilbert [34] (cf. also Hilbert's 6th problem [33], Patrick Suppes [59], Alfred Tarski [32] and leading contemporary logician Harvey Friedman [22, 23]).

There are several reasons for seeking an axiomatic foundation of a physical theory [60]. One is that the theory may be better understood by providing a basis of explicit postulates for the theory. Another reason is that if we have an axiom system we can ask ourselves what axioms are responsible for which theorems. For more on this kind of foundational thinking called reverse mathematics, see, for example, Friedman [22] and Simpson [56]. Furthermore, if we have an axiom system for special relativity or general relativity, we can ask what happens with the theory if we change one or more of the axioms. This could lead us to a new physically interesting theory. This is what happened with Euclid's axiom system for geometry when Bolyai and Lobachevsky altered the axiom of parallelism and discovered hyperbolic geometry.

Seeking a logical foundation for spacetime theory (i.e., roughly, relativity) is a worthwhile attempt for several reasons. One of these is that spacetime can be regarded as a foundation of physics since spacetime is the arena in which physical phenomena take place. Another reason for seeking a logical foundation for spacetime is that throughout its history, logic benefited the most from those applications of logic which were aiming at branches of learning going through a turmoil or a revolutionary phase, and at the same time being important for our understanding of the world [35]. As a quick glance to recent issues of, for example, Scientific American can convince the reader, spacetime theory and relativity/cosmology certainly qualify. So we believe that it serves the best interest of logic community to apply logic to spacetime theory, relativity, cosmology, and black hole physics. Indeed, logic can benefit from such studies in many ways. As a bonus, as indicated in [14] or [38], spacetime theory can give a feedback to the foundation of mathematics itself.

For certain reasons, the foundation of mathematics has been carried through strictly within the framework of first-order logic (FOL). One of these reasons is that staying inside FOL helps us to avoid tacit assumptions. Another reason is that FOL has a complete inference system while higher-order logic cannot have one by Gödel's incompleteness theorem (cf., for example, Väänänen [65, p.505]). (For more motivation for staying inside FOL as opposed to higher-order logic cf., for example, [1], [2, Appendix 1], and [6, 21, 48, 67].) The same reasons motivate the effort of keeping the foundation of spacetime and relativity theory inside FOL.

The interplay between logic and relativity theory goes back to around 1920 and has been playing a non-negligible role in works of researchers like Reichenbach, Carnap, Suppes, Ax, Szekeres, Malament, Walker, and of many other contemporaries. For more details cf., for example, [1]. Also, it is no coincidence that relativity was the main motivating example for the logical positivists of the Vienna Circle.

Axiomatizations of SR have been quite extensively studied in the literature (cf., for example, the references of [1]). However, these works usually stop with a kind of representation theorem for their axiomatizations. As a contrast, what we call the foundation of relativity begins with the axiomatization (and representation theorems), but the real work and the real fun (the conceptual analysis) comes afterwards when we investigate, for example, what axioms are responsible for which statements, what happens if we change the axioms etc.

While some FOL axiomatizations of the theory of inertial observers and for SR can be found in the literature (cf. [**6, 26,** **1**]), axiom systems—let alone FOL axiom systems—for accelerated observers and for GR are not too many in the literature (but cf. [**44**] for an exception).

In Section 4, we recall a streamlined FOL axiomatization AccRel of SR extended with accelerated observers. In Section 5, we take one step toward GR and investigate an aspect of time warp, that is the effect of gravitation on clocks, in our FOL setting. There we use Einstein's equivalence principle to talk about gravitation and prove the gravitational time dilation effect, that is that "gravity causes time to run slow," from AccRel in more than one sense (cf. Theorems 5.1–5.3). We will also see that gravity can slow time down arbitrarily (cf. Theorems 5.4–5.6). Furthermore, we investigate the role of the "direction" and the "magnitude" of gravitation in gravitational time dilation (cf. Theorems 5.7 and 5.8). We note that the most exotic features of black holes, wormholes and the like (mentioned in Section 3 below) can be traced back to this effect of time warp (to be analyzed in Section 5).

## 3. Intriguing Features of GR Spacetimes (Challenges for the Logician)

Both SR and GR have many interesting consequences. Most of them show that we have to refine our common sense concepts of space and time. They are full of surprising predictions and paradoxes which seriously challenge our common sense picture of the world. But it is exactly this negation of common sense which makes this area an attractive field to apply logic.

Gravitation has many surprising effects on time. The common name for these effects is *time warp*.

For example, in the Schwarzschild spacetime, which is associated with a non-rotating black hole (or star), we face one of the simplest aspect of time warp called *gravitational time dilation*. There we see that if we suspend an observer closer to the black hole and another observer farther away from it, then the clock of the closer one will run slower than the clock of the one which is farther away. So in some sense we see that "gravity causes time run slow." There are places where this time warp effect becomes infinite, i.e., some clocks entirely stop ticking, i.e., freeze from the point of view of some other observers. Moreover, time and space may get interchanged. These effects are part of the reason why we said in Section 1 that space does not exist while spacetime does.

The above-mentioned time warp effect leads to even stronger effects. We meet new interesting aspects of time warp in the Reissner–Nordström, Kerr and Kerr–Newman spacetimes that are associated with charged, rotating and charged-rotating black holes, respectively. For astronomical evidence for the existence of rotating black holes cf., for example, [49, 58]. In these spacetimes, there is an event whose causal past contains timelike curves which are infinitely long in the future direction. Such a curve can be the life-line of an observer (or computer) who has infinite time for working and sending light-signals that can be received before the distinguished event. The spacetimes in which these kinds of events occur are called Malament–Hogarth spacetimes (cf., for example, Earman [15, § 4] and [38]). In Malament–Hogarth spacetimes, we can design a computer that decides non-Turing-computable sets (cf., for example, [38, 16, 14, 19]). Thus inside these spacetimes, we can decide whether an axiom system of set theory (for example ZFC) is consistent or not. Therefore, in contrast with the consequence of Gödel's second incompleteness theorem, we can find out whether mathematics is consistent or not. For more detail on these kinds of computers in the physically reasonable Kerr spacetime we refer, for example, to [14, 19]). Recently, the acceleration of the expansion of the universe made anti-de-Sitter spacetimes very popular with cosmologists. These also have the Malament–Hogarth property, hence are also suitable for harboring computers breaking the Turing barrier.

There are several models of GR in which there are so-called Closed Timelike Curves (CTC). Such are Gödel's rotating universe [25], Kerr and Kerr–Newman spacetimes [47], Gott's spacetime [27], Tipler's rotating cylinder [64], van Stockum's spacetime [57], Taub-NUT spacetime [31], to mention only a few. Since timelike curves correspond to possible life-lines of observers, in these spacetimes an observer can go through the same event more than once. This situation can be interpreted as *time travel*. This leads to nontrivial philosophical problems, in analysing/understanding which the methods of logic can considerably help. We believe, currently logic is the discipline best positioned for clarifying the apparent problems with CTC's, i.e., with time travel. Namely, the only problem with time travel is that it represents a kind of circularity, because of the following: a time traveler goes back into his past, changes his past so as to prevent his own existence, but then who went back into the past? etc. This circularity is not more vicious than the Liar paradox or self-reference implemented, for example, in Gödel's second incompleteness proof. Logic has been extremely successful in understanding and "de-mystifying" self-referential situations and the Liar paradox. Examples are provided by literature of Gödel's incompleteness method [30], the book on "The Liar" by Barwise and Etchemendy [7] which used non-well-founded set theory for providing an explicit semantic analysis for self-referential situations, [55]. So logic seems to be best suited for providing rational understanding of situations like the circularity represented by CTC's or time travel. (For more on CTC's cf., for example, [15, § 6] and [17, 28].)

These are only a few of the many examples that show that turning Relativity Theory into a real FOL theory, axiomatizing it and analyzing its logical structure seem to be a promising, worth-while undertaking.

What could science gain from such a logical analysis of relativity theory? Turning GR into a FOL theory will make it more flexible. By flexibility we mean that we can change some of the axioms whenever we would like to change the theory, without having to re-build the whole theory from scratch. By changing the axioms, we can control the changes of theory better than by changing Einstein's field equations. This might be useful when we would like to understand the connection of GR to other theories of gravitation like the Brans–Dicke theory (cf. [8, 9, 20]). This flexibility can also be useful when we would like to extend GR. We indeed would like to extend GR since we do not have a good theory of Quantum Gravity (QG) which is a common extension of the quantum theory and GR. Some eminent researchers of relativity formulated an even more optimistic goal of searching for the geometrization of all physical phenomena known today into a so-called theory of everything (TOE). Of course, one wants both QG and TOE to be some kinds of extensions of GR.

Recent astronomical observations provided strong evidence that the expansion of our universe is accelerating. This discovery leads to many questions and to the idea that the cosmological constant might be replaced with a dynamical parameter, i.e., with a scalar field, under the name of Quintessence or "dark energy" (cf., for example, [10, 13]). But this leads to a new need for modifying or at least fine-tuning GR. This also shows the merit of making GR more flexible by providing a FOL axiom system for it.

So far we have talked mainly about the significance of the logical foundation of GR, but the logical analysis of SR is also important since GR is built on SR. Moreover, there are other different relativity theories such as the Reichenbach–Grünbaum version (cf. [50, 51, 29] or the Lorentz-Poincaré version of special relativity (cf. [41]). Their logical structures and connection with Einstein's relativity are also worth analyzing in order to get a more refined understanding of relativity theory. Our research group has done some work in this direction (cf. [2, § 4.5]).

In the following sections we try to give a sample of the work done by our research group in Budapest in the direction of a FOL investigation of relativity theories (including GR).

## 4. A FOL Axiom System of SR Extended with Accelerated Observers

We recall one of our axiom systems for SR extended with accelerated observers (hence extended with a handle on gravity). We try to be as self contained as possible. First occurrences of concepts used in this work are set in italic to make them easier to find.

The motivation for our choice of vocabulary is summarized as follows. Here, we deal with the kinematics of relativity only, that is we deal with motion of *bodies* (or *test-particles*). We will represent motion as changing spatial location in time. To do so, we will have reference-frames for coordinatizing events and, for simplicity, we will associate reference-frames with special bodies which we will call *observers*. We visualize an observer-as-a-body as "sitting" in the origin of the space part of its reference-frame, or equivalently, "living" on the time-axis of the reference-frame. We will distinguish *inertial* observers from non-inertial (accelerated) ones. There will be another special kind of bodies which we will call *photons*. For coordinatizing events, we will use an arbitrary *ordered field* in place of the field of the real numbers. Thus the elements of this field will be the *"quantities"* which we will use for marking time and space. Allowing arbitrary ordered fields in place of the field of the reals increases flexibility of our theory and minimizes the amount of our mathematical presuppositions (cf., for example, Ax [6] for further motivation in this direction). Similar remarks apply to our flexibility oriented decisions below, for example, keeping the dimension of spacetime a variable. Using observers in place of coordinate systems or reference frames is only a matter of didactic convenience and visualization. Using observers (or coordinate systems, or reference-frames) instead of a single observer-independent spacetime structure has many reasons. One of them is that it helps us in weeding out unnecessary axioms from our theories; but we state and emphasize the equivalence/duality between observer-oriented and observer-independent approaches to relativity theory (cf. [**42**, § 4.5]). Motivated by the above, we now turn to fixing the first-order language of our axiom systems.

We fix a natural number $d \geqslant 2$ for the dimension of spacetime. Our language contains the following non-logical symbols:

- unary relation symbols B (for *Bodies*), Ob (for *Observers*), IOb (for *Inertial Observers*), Ph (for *Photons*) and Q (for *Quantities*),
- binary function symbols $+$, $\cdot$ and a binary relation symbol $\leqslant$ (for the field operations and the ordering on Q), and
- a $2 + d$-ary relation symbol W (for *World-view relation*).

The bodies will play the role of the "main characters" of our spacetime models and they will be "observed" (coordinatized using the quantities) by the observers. This observation will be coded by the world-view relation W. Our bodies and observers are basically the same as the "test particles" and the "reference-frames," respectively, in some of the literature.

We read $B(x)$, $Ob(x)$, $IOb(x)$, $Ph(x)$, and $Q(x)$ as "$x$ is a body," "$x$ is an observer," "$x$ is an inertial observer," "$x$ is a photon," and "$x$ is a quantity." We use the world-view relation W to talk about coordinatization, by reading $W(x, y, z_1, \ldots, z_d)$ as "observer $x$ observes (or sees) body $y$ at coordinate point $\langle z_1, \ldots, z_d \rangle$." This kind of observation has no connection with seeing via photons, it simply means coordinatization.

$B(x)$, $Ob(x)$, $IOb(x)$, $Ph(x)$, $Q(x)$, $W(x, y, z_1, \ldots, z_d)$, $x = y$ and $x \leqslant y$ are the so-called atomic formulas of our first-order language, where $x, y, z_1, \ldots, z_d$ can be arbitrary variables or terms built up from variables by using the field-operations "$+$" and "$\cdot$". The *formulas* of our first-order language are built up from these atomic formulas by using the logical connectives *not* ($\neg$), *and* ($\wedge$), *or* ($\vee$), *implies* ($\Longrightarrow$), *if-and-only-if* ($\Longleftrightarrow$) and the quantifiers *exists* $x$ ($\exists x$) and *for all* $x$ ($\forall x$) for every variable $x$.

The *models* of this language are of the form

$$\mathfrak{M} = \langle U; B, Ob, IOb, Ph, Q, +, \cdot, \leqslant, W \rangle,$$

where $U$ is a nonempty set and B, Ob, IOb, Ph and Q are unary relations on $U$, etc. A unary relation on $U$ is just a subset of $U$. Thus we use B, Ob etc. as sets as well, for example, we write $m \in Ob$ in place of $Ob(m)$.

$Q^d := Q \times \ldots \times Q$ ($d$-times) is the set of all $d$-tuples of elements of Q. If $\vec{p} \in Q^d$, then we assume that $\vec{p} = \langle p_1, \ldots, p_d \rangle$, i.e., $p_i \in Q$ denotes the $i$-th component of the $d$-tuple $\vec{p}$. We write $W(m, b, \vec{p})$ in place of $W(m, b, p_1, \ldots, p_d)$, and we write $\forall \vec{p}$ in place of $\forall p_1, \ldots, p_d$ etc.

Let us begin formulating our axioms. We formulate each axiom at two levels. First we give an intuitive formulation, then we give a precise formalization using our logical notation (which easily can be translated into first-order formulas by substituting the definitions into the formalizations). We aspire to formulate easily understandable axioms in FOL.

The first axiom expresses our very basic assumptions like: both photons and observers are bodies, inertial observers are also observers, etc.

**AxFrame:** $Ob \cup Ph \subseteq B$, $IOb \subseteq Ob$, $U = B \cup Q$, $B \cap Q = \emptyset$, $W \subseteq Ob \times B \times Q^d$, $+$ and $\cdot$ are binary operations on Q, $\leqslant$ is a binary relation on Q.

To be able to add, multiply and compare measurements of observers, we put some algebraic structure on the set of quantities Q by the next axiom.

**AxEOF:** A FOL axiom stating that the *quantity part* $\langle Q; +, \cdot, \leqslant \rangle$ is a Euclidean [1] ordered field.

For the first-order definition of linearly ordered field see, for example, Chang–Keisler [**11**].

---

[1] That is a linearly ordered field in which positive elements have square roots.

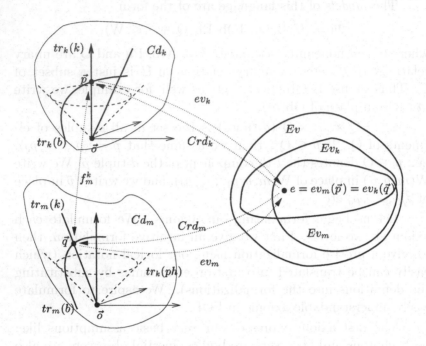

FIGURE 1.   Illustration for the basic definitions

We need some definitions to formulate our other axioms. Let $0, 1, -, /, \sqrt{\ }$ be the usual field operations which are definable from "+" and ".". We use the vector-space structure of $Q^d$, i.e., if $\vec{p}, \vec{q} \in Q^d$ and $\lambda \in Q$, then $\vec{p} + \vec{q}, -\vec{p}, \lambda\vec{p} \in Q^d$; and $\vec{o} := \langle 0, \dots, 0 \rangle$ denotes the *origin*. $Q^d$ is called the *coordinate system* and its elements are referred to as *coordinate points*. We use the notation $\vec{p}_s := \langle p_2, \dots, p_d \rangle$ for the *space component* of $\vec{p}$ and $p_t := p_1$ for the *time component* of $\vec{p} \in Q^d$. The *event* (the set of bodies) observed by observer $m$ at coordinate point $\vec{p}$ is:

$$ev_m(\vec{p}) := \{ b \in \mathrm{B} : \mathrm{W}(m, b, \vec{p}) \}.$$

The *coordinate-domain* of observer $m$ is the set of coordinate points where $m$ observes something:

$$Cd_m := \{ \vec{p} \in Q^d : ev_m(\vec{p}) \neq \emptyset \}.$$

Now we formulate our first axiom on observers. This natural axiom goes back to Galileo Galilei and even to d'Oresme of around 1350 (cf., for example, [1, p.23, § 5]). It simply states that each observer thinks that he rests in the origin of the space part of his coordinate system.

**AxSelf⁻:** An observer sees himself in an event if and only if the space component of the coordinate of this event is the origin:

$$\forall m \in \mathrm{Ob} \ \forall \vec{p} \in Cd_m \ \ ( m \in ev_m(\vec{p}) \iff \vec{p}_s = \vec{o} ).$$

To formulate our axiom about the constancy of this speed, we choose 1 for the speed of photons. Below, the *Euclidean-length* of $\vec{p} \in Q^n$ is defined as $|\vec{p}| := \sqrt{p_1^2 + \ldots + p_n^2}$, for any $n \geqslant 1$.

**AxPh$_0$:** For every *inertial* observer, there is a photon through two coordinate points $\vec{p}$ and $\vec{q}$ if and only if the slope of $\vec{p} - \vec{q}$ is 1:

$$\forall m \in \mathrm{IOb} \ \forall \vec{p}, \vec{q} \in Q^d \ ( |\vec{p}_s - \vec{q}_s| = |p_t - q_t| \iff$$
$$\mathrm{Ph} \cap ev_m(\vec{p}) \cap ev_m(\vec{q}) \neq \emptyset ).$$

Motivations for this axiom can be found, for example, in [3], or in d'Inverno [12, § 2.6].

The set of events seen by observer $m$ is:

$$Ev_m := \{ ev_m(\vec{p}) \ : \ \vec{p} \in Cd_m \},$$

and the set of all events is

$$Ev := \{ e \in Ev_m \ : \ m \in \mathrm{Ob} \}.$$

With the next axiom, we assume that every inertial observer sees the same set of events.

**AxEv:** Every *inertial* observer sees the same events:

$$\forall m, k \in \mathrm{IOb} \ \ Ev_m = Ev_k.$$

One can prove from $AxPh_0$ and AxEOF that if $m$ is an inertial observer and $e \in Ev_m$, then there is a unique coordinate point $\vec{p} \in Q^d$ such that $e = ev_m(\vec{p})$. This unique coordinate point $\vec{p} \in Q^d$ is denoted by $Crd_m(e)$.

**Convention 4.1.** Whenever we write "$Crd_m(e)$," we mean that there is a unique $\vec{q} \in Cd_m$ such that $ev_m(\vec{q}) = e$, and $Crd_m(e)$ denotes this unique $\vec{q}$. That is, if we talk about the value $Crd_m(e)$, we postulate that it exists and is unique (by the present convention).

We say that events $e_1$ and $e_2$ are *simultaneous* for observer $m$, in symbols $e_1 \sim_m e_2$, if and only if $e_1$ and $e_2$ have the same time-coordinate in $m$'s coordinate-domain, i.e., if $Crd_m(e_1)_t = Crd_m(e_2)_t$. To talk about time differences measured by observers, we use $\mathsf{time}_m(e_1, e_2)$ as an abbreviation for $|Crd_m(e_1)_t - Crd_m(e_2)_t|$ and we call it the *elapsed time* between events $e_1$ and $e_2$ measured by observer $m$. We note that, if $m \in e_1 \cap e_2$, then $\mathsf{time}_m(e_1, e_2)$ is called the *proper time* measured by $m$ between $e_1$ and $e_2$, and $e_1 \sim_m e_2$ if and only if $\mathsf{time}_m(e_1, e_2) = 0$. We use $\mathsf{dist}_m(e_1, e_2)$ as an abbreviation for $|Crd_m(e_1)_s - Crd_m(e_2)_s|$ and we call it the *spatial distance* of events $e_1$ and $e_2$ according to an observer $m$. We note that when we write $\mathsf{dist}_m(e_1, e_2)$ or $\mathsf{time}_m(e_1, e_2)$, we assume that $e_1$ and $e_2$ have unique coordinates by Convention 4.1.

**AxSimDist:** If events $e_1$ and $e_2$ are simultaneous for both *inertial* observers $m$ and $k$, then $m$ and $k$ agree on the spatial distance between $e_1$ and $e_2$:

$$\forall m, k \in \text{IOb } \forall e_1, e_2 \in Ev_m \left( e_1 \sim_m e_2 \wedge e_1 \sim_k e_2 \implies \mathsf{dist}_m(e_1, e_2) = \mathsf{dist}_k(e_1, e_2) \right).$$

We collect these axioms in an axiom system, called $SpecRel_d$:

$$SpecRel_d := \{ \text{AxFrame}, \text{AxEOF}, \text{AxSelf}^-, \text{AxPh}_0, \text{AxEv}, \text{AxSimDist} \}.$$

Now for each natural number $d \geqslant 2$, we have a FOL theory of SR. Usually we omit the dimension parameter $d$. From the few axioms introduced so far, we can deduce the most frequently quoted predictions, called paradigmatic effects, of SR:

(i) "moving clocks slow down,"

(ii) "moving meter-rods shrink,"

(iii) "moving pairs of clocks get out of synchronism."

For more detail see, for example, [**1, 2, 3**]. Here, we concentrate on the behavior of clocks and indicate a connection with Minkowski geometry.

**Theorem 4.2.** *Assume* $\text{SpecRel}_d$, $d \geqslant 3$. *Then*

$$\text{time}_m(e_1, e_2)^2 - \text{dist}_m(e_1, e_2)^2 = \text{time}_k(e_1, e_2)^2 - \text{dist}_k(e_1, e_2)^2$$

*for any* $m, k \in \text{IOb}$ *and* $e_1, e_2 \in Ev_m$.

The above theorem is the starting point for building Minkowski geometry, which is the "geometrization" of SR. It also indicates that time and space are intertwined in SR. Here, we only concentrate on its corollary usually stated as "moving clocks slow down." Theorem 4.2 shows that SpecRel is a good axiom system for SR if we restrict our interest to textitinertial motion.

**Corollary 4.3** (moving clocks slow down). *Assume* $\text{SpecRel}_d$, $d \geqslant 3$. *Let* $m, k \in \text{IOb}$, $e_1, e_2 \in Ev_k$. *Assume that* $k \in e_1 \cap e_2$ *and* $\text{dist}_m(e_1, e_2) \neq 0$. *Then*

$$\text{time}_m(e_1, e_2) > \text{time}_k(e_1, e_2).$$

In Corollary 4.3, a "moving clock" is represented by observer $k$, that he is moving relative to $m$ is expressed by $\text{dist}_m(e_1, e_2) \neq 0$, $k \in e_1 \cap e_2$, and that $k$'s time is slowing down relative to $m$'s is expressed by $\text{time}_m(e_1, e_2) > \text{time}_k(e_1, e_2)$. This "clock slowing down" effect is only relative, i.e., "clocks moving relative to $m$ slow down relative to $m$." But this relative effect leads to a new kind of gravitation-oriented "absolute slowing time down" effect, as our next theorem as well as the whole of Section 5 will show.

To extend SpecRel, we now formulate axioms about non-inertial observers. The non-inertial observers are called *accelerated observers*. Note that AxSelf⁻ is the only axiom introduced so far that talks about non-inertial observers, too. We assume the following very natural axiom for all observers.

**AxEv⁺:** Whenever an observer participates in an event, he also sees this event:

$$\forall m \in \mathrm{Ob} \ \forall e \in Ev \ \ (m \in e \implies e \in Ev_m).$$

The set of positive elements of Q is denoted by $Q^+ := \{x \in Q : x > 0\}$. The *interval* between $x, y \in Q$ is defined as $(x, y) := \{z \in Q : x < z < y\}$. Let $H \subseteq Q$. We say that $H$ is *connected* if and only if $\forall x, y \in H \ (x, y) \subseteq H$, and we say that $H$ is *open* if and only if $\forall x \in H \ \exists \varepsilon \in Q^+ \ (x - \varepsilon, x + \varepsilon) \subseteq H$.

We assume the following technical axiom.

**AxSelf⁺:** The set of time-instances in which an observer is present in its own world-view is connected and open:

$$\forall m \in \mathrm{Ob} \ \{p_t : m \in ev_m(\vec{p})\} \quad \text{is connected and open.}$$

To connect the coordinate-domains of the accelerated and the inertial observers, we are going to formulate the statement that at each moment of his life, each accelerated observer sees the nearby world for a short while as an inertial observer does. To formalize this, first we introduce the relation of being a co-moving observer. To do so, we define the (coordinate) *neighborhood* of event $e$ with radius $r \in Q^+$ according to observer $k$ as:

$$B_k^r(e) := \{\vec{p} \in Cd_k : \exists \vec{q} \in Cd_k \ \ ev_k(\vec{q}) = e \land |\vec{p} - \vec{q}| < r\}.$$

We note that $B_k^r(e) = \emptyset$ if $e \notin Ev_k$ by this definition. Observer $m$ is a *co-moving observer* of observer $k$ at event $e$, in symbols $m \succ_e k$, if and only if the following holds:

$$\forall \varepsilon \in Q^+ \ \exists \delta \in Q^+ \ \forall \vec{p} \in B_k^\delta(e) \ |\vec{p} - Crd_m(ev_k(\vec{p}))| \leqslant \varepsilon |\vec{p} - Crd_k(e)|.$$

Note that $Crd_m(e) = Crd_k(e)$ and thus also $e \in Ev_m$ if $m \succ_e k$ and $e \in Ev_k$. Note also that $m \succ_e k$ for every observer $m$ if $e \notin Ev_k$, by definition. Behind the definition of the co-moving observers is the following intuitive image: as we zoom into smaller and smaller neighborhoods of the coordinate point of the given event, the coordinate-domains of the two observers are more and more similar. This intuitive picture is symmetric while the co-moving relation $\succ_e$ is not. Thus we introduce a symmetric version. We say that observers $m$ and $k$ are *strong co-moving observers* at event $e$, in symbols $m \,\maltese_e\, k$, if and only if both $m \succ_e k$ and $k \succ_e m$ hold. The following axiom gives the promised connection between the coordinate-domains of the inertial and the accelerated observers.

**AxAcc$^+$:** At any event in which an observer sees himself, there is a strong co-moving *inertial* observer:

$$\forall k \in \mathrm{Ob} \;\; \forall e \in Ev_k \quad ( k \in e \implies \exists m \in \mathrm{IOb} \;\; m \,\maltese_e\, k ).$$

The axioms introduced so far are not strong enough to prove properties of accelerated clocks like the Twin Paradox (cf. [**44**, Theorems 3.5 and 3.7 and Corollary 3.6]). The additional property we need is that every bounded non-empty subset of the quantity part has a supremum. This is a second-order logic property (because it concerns all subsets) which we cannot use in a FOL axiom system. Instead, we will use a kind of "induction" axiom schema. It will state that every non-empty, bounded subset of the quantity part which can be defined by a FOL-formula using possibly the extra part of the model, for example, using the world-view relation, has a supremum. To formulate this FOL induction axiom schema, we need some more definitions.

If $\varphi$ is a formula and $x$ is a variable, then we say that $x$ is a *free variable* of $\varphi$ if and only if $x$ does not occur under the scope of either $\exists x$ or $\forall x$. Sometimes we introduce a formula $\varphi$ as $\varphi(\vec{x})$, this means that all the free variables of $\varphi$ lie in $\vec{x}$.

If $\varphi(x, y)$ is a formula and $\mathfrak{M} = \langle U; \ldots \rangle$ is a model, then whether $\varphi$ is true or false in $\mathfrak{M}$ depends on how we associate elements of $U$ to the free variables $x, y$. When we associate $a, b \in U$ to $x, y$, respectively, then $\varphi(a, b)$ denotes this truth-value, thus $\varphi(a, b)$ is either true or false in $\mathfrak{M}$. For example, if $\varphi$ is $x \leqslant y$, then $\varphi(0, 1)$ is true while $\varphi(1, 0)$ is false in any ordered field. A formula $\varphi$ is said to be *true* in $\mathfrak{M}$ if $\varphi$ is true in $\mathfrak{M}$ no matter how

we associate elements to the free variables. We say that a *subset*
$H$ *of* Q *is* (parametrically) *definable by* $\varphi(y, \vec{x})$ if and only if there
is $\vec{a} \in U^n$ such that $H = \{b \in Q : \varphi(b, \vec{a}) \text{ is true in } \mathfrak{M}\}$. We say
that a subset of Q is *definable* if and only if it is definable by a
FOL-formula.

Let $\varphi(x, \vec{y})$ be a FOL-formula of our language.

**AxSup$_\varphi$:** Every subset of Q definable by $\varphi(x, \vec{y})$ has a supremum
if it is non-empty and *bounded*.

A FOL formula expressing AxSup$_\varphi$ can be found in [**44**].
Our axiom scheme IND below says that every non-empty bounded
subset of Q that is definable in our language has a supremum:

IND := { AxSup $_\varphi$ : $\varphi$ is a FOL-formula of our language } .

Note that IND is true in any model whose quantity part is the
field of real numbers. For more detail about IND we refer to [**44**].

We call the collection of the axioms introduced so far AccRel$_d$:

$$\text{AccRel}_d := \text{SpecRel}_d \cup \{\text{AxEv}^+, \text{AxSelf}^+, \text{AxAcc}^+\} \cup \text{IND}.$$

The so-called Twin Paradox is provable in AccRel (cf. [**44,
61**]). We formulate the Twin Paradox with our logical notation.

The *set of events encountered by* $m \in \text{Ob}$ *between* $e_1, e_2 \in Ev$
is denoted by

$$Ev_m(e_1, e_2) := \{ e \in Ev_m : m \in e \wedge$$
$$Crd_m(e_1)_t < Crd_m(e)_t < Crd_m(e_2)_t \} .$$

Now we can formulate the Twin Paradox in our FOL setting.

**TwP:** Every *inertial* observer $m$ measures more time than or
equal time as any other observer $k$ between any two meeting events $e_1$ and $e_2$; and they measure the same time if

and only if they have encountered the same events between $e_1$ and $e_2$:

$$\forall e_1, e_2 \in Ev \;\; \forall m \in \text{IOb} \;\; \forall k \in \text{Ob}\Big( k, m \in e_1 \cap e_2 \implies$$

$$\Big( \text{time}_m(e_1, e_2) = \text{time}_k(e_1, e_2) \iff Ev_m(e_1, e_2) = Ev_k(e_1, e_2) \Big)$$

$$\wedge \; \text{time}_m(e_1, e_2) \geqslant \text{time}_k(e_1, e_2) \Big).$$

The following theorem states that the Twin Paradox is provable in AccRel$_d$ if $d \geqslant 3$.

**Theorem 4.4.** AccRel$_d \models$ TwP   *if* $d \geqslant 3$.

For the proof of this theorem cf. [**44, 61**].

We note that there are non-trivial models of AccRel. For
example, the construction in Misner–Thorne–Wheeler [**46**, § 6,
especially pp. 172–173 and § 13.6, pp. 327–332] can be used for
constructing models for AccRel.

# 5. One Step toward GR
# (Effect of Gravitation on Clocks)

We would like to investigate the effect of gravitation on clocks in
our FOL setting. As a first step we prove theorems about the
Gravitational Time Dilation that roughly says that "gravitation
makes time flow slower," i.e., the clocks in the bottom of a tower
run slower than the clocks in the top of the tower. We will use Einstein's equivalence principle to treat gravitation in AccRel. This
principle says that a uniformly accelerated frame of reference is in-
distinguishable from a rest frame in a uniform gravitational field
(cf., for example, d'Inverno [**12**, § 9.4]). So, instead of gravitation
we will talk about acceleration and instead of towers we will talk
about spaceships. This way the Gravitational Time Dilation will
become the following statement: "the time in the aft of an accelerated spaceship flows slower than in the front of the spaceship."
We begin to formulate this statement in our FOL language.

To talk about spaceships, we will need a concept of distance between events and observers. We have the following two natural candidates for this:

- Event $e$ is at *radar-distance* $\lambda \in Q^+$ from observer $k$ if and only if there are events $e_1$ and $e_2$ and photons $ph_1$ and $ph_2$ such that $k \in e_1 \cap e_2$, $ph_1 \in e \cap e_1$, $ph_2 \in e \cap e_2$ and $\mathsf{time}_k(e_1, e_2) = 2\lambda$. Event $e$ is at *radar-distance* 0 from observer $k$ if and only if $k \in e$ (cf. Fig. 2, (a)).

- Event $e$ is at *Minkowski–distance* $\lambda \in Q$ from observer $k$ if and only if there is an event $e'$ such that $k \in e'$, $e \sim_m e'$ and $\mathsf{dist}_m(e, e') = \lambda$ for every inertial co-moving observer $m$ of $k$ at $e'$ (cf. Fig. 2, (b)).

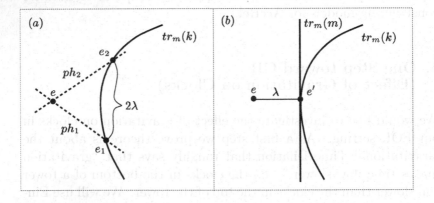

FIGURE 2.    (a) for the radar-distance and (b) for the Minkowski–distance

We say that observer $k$ thinks that body $b$ is at constant radar (Minkowski) distance from him if and only if the radar-distance (Minkowski–distance) of every event which $b$ participates in is the same.

The *life-line*[2] (or *trace*) of body $b$ according to observer $m$ is defined as the set of coordinate points where $b$ was observed by $m$:

$$tr_m(b) := \{ \vec{p} \in Q^d : W(m, b, \vec{p}) \}.$$

Note that $tr_m(b) = \{ \vec{p} \in Q^d : b \in ev_m(\vec{p}) \}$. For stating that the *spaceship does not change its direction* we introduce the following concept. We say that observers $k$ and $b$ are *coplanar* if and only if $tr_m(k) \cup tr_m(b)$ is a subset of a plane containing a line parallel with the time-axis, in the coordinate system of an *inertial* observer $m$.

We now introduce two concepts for spaceships. Observers $b, k$ and $c$ form a *radar-spaceship*, in symbols $\gg\!b, k, c\rangle_{rad}$, if and only if $b$, $k$ and $c$ are coplanar and $k$ thinks that $b$ and $c$ are at constant radar-distances from him. The definition of the *Minkowski–spaceship*, in symbols $\gg\!b, k, c\rangle_{\mu}$, is analogous.

We say that event $e_1$ (causally) *precedes* event $e_2$ according to observer $k$ if and only if $Crd_m(e_1)_t \leqslant Crd_m(e_2)_t$ for all *inertial* co-moving observers $m$ of $k$. In this case, we also say that $e_2$ *succeeds* $e_1$ according to $k$.

We need some concept for deciding which events happened at the same time according to an accelerated observer. The following three natural concepts offer themselves:

- Events $e$ and $e'$ are *radar-simultaneous* for observer $k$, in symbols $e \sim_k^{rad} e'$, if and only if $k \in e$ and there are events $e_1$ and $e_2$ and photons $ph_1$ and $ph_2$ such that $k \in e_1 \cap e_2$, $ph_1 \in e \cap e_1$, $ph_2 \in e \cap e_2$ and $\mathsf{time}_k(e_1, e) = \mathsf{time}_k(e, e_2)$ or there is an event $e_3$ such that $e \sim_k^{rad} e_3$ and $e_3 \sim_k^{rad} e'$ (cf. Fig. 3, (a)).

- Events $e_1$ and $e_2$ are *photon-simultaneous* for observer $k$, in symbols $e_1 \sim_k^{ph} e_2$, if and only if there is an event $e$ and photons $ph_1$ and $ph_2$ such that $k \in e$, $ph_1 \in e \cap e_1$, $ph_2 \in e \cap e_2$ and $e_1$ and $e_2$ precedes $e$ according to $k$ (cf. Fig. 3, (b)).

- Events $e_1$ and $e_2$ are *Minkowski–simultaneous* for observer $k$, in symbols $e_1 \sim_k^{\mu} e_2$, if and only if there is an event $e$ such

---

[2] Life-line is called world-line in some of the literature.

that $k \in e$ and $e_1$ and $e_2$ are simultaneous for any inertial co-moving observer of $k$ at $e$ (cf. Fig. 3, (c)).

We note that, for *inertial* observers, the concepts of radar–simultaneity and Minkowski–simultaneity coincide with the concept of simultaneity introduced on p. 230.

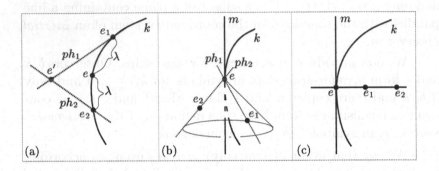

FIGURE 3.   (a) is for $e \sim_k^{rad} e'$, (b) is for $e_1 \sim_k^{ph} e_2$ and (c) is for $e_1 \sim_k^{\mu} e_2$

We will distinguish the front and the aft of the spaceship by the direction of the acceleration. Thus we need a concept for direction. We say that the *directions of $\vec{p} \in Q^d$ and $\vec{q} \in Q^d$ are the same*, in symbols $\vec{p} \uparrow\uparrow \vec{q}$, if and only if there is a $\lambda \in Q^+$ such that $\lambda\vec{p}_s = \vec{q}_s$ (cf. Fig. 4, (a)).

Now let us turn our attention towards the definition of acceleration in our FOL setting.

We define the *life-curve* of observer $k$ according to observer $m$ as the life-line of $k$ according to $m$ *parameterized by the time measured by $k$*, formally:

$$Tr_m^k := \{ \langle t, \vec{p} \rangle \in Q \times Cd_m :$$
$$\exists \vec{q} \in tr_k(k) \quad q_t = t \wedge ev_m(\vec{p}) = ev_k(\vec{q}) \}.$$

The *domain* of a binary relation $R$ is defined as $Dom\, R := \{x : \exists y \ \langle x, y \rangle \in R\}$.

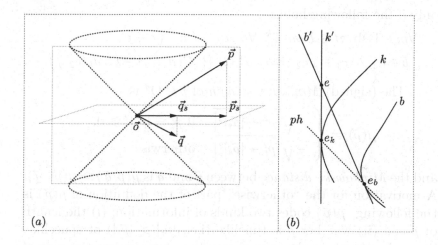

FIGURE 4.    (a) is for illustrating $\vec{p} \uparrow\uparrow \vec{q}$ and (b) is for
illustrating observer $b$ approaching to observer $k$, as seen
by $k$ with photons

Both the life-curves of observers (according to any *inertial*
observer) and the derivative $f'$ of a given function $f$ are first-order
logic definable concepts (cf. [**44**]). Thus the following definitions
are also FOL ones: The *relative-velocity* $\vec{v}_m^k$ of observer $k$ according
to observer $m$ at instant $t \in Q$ is the derivative of the life-curve
of $k$ according to $m$ at $t$, i.e., $\vec{v}_m^k(t) = (Tr_m^k)'(t)$ if $t \in Dom\, Tr_m^k$
and undefined otherwise. The *relative-acceleration* $\vec{a}_m^k$ of observer
$k$ according to observer $m$ at instant $t \in Q$ is the derivative of
the relative-velocity at $t$ if it is differentiable at $t$ and undefined
otherwise.

Events $e_1$ and $e_2$ are called *spacelike separated*, in symbols
$e_1 \equiv_s e_2$, if and only if $Crd_m(e_1)$ and $Crd_m(e_2)$ can be connected
by a line of slope more than 1 for every *inertial* observer $m$, i.e., if
and only if $|(Crd_m(e_1) - Crd_m(e_2))_s| > |(Crd_m(e_1) - Crd_m(e_2))_t|$
for every inertial observer $m$. We say that *the direction of the
spaceship* $\rangle b, k, c\rangle$ *agrees with that of the acceleration of $k$ if and*

only if the following holds:

$$\forall m \in IOb \ \forall t \in Dom\, \vec{a}_m^k \ \forall e_1, e_2 \in Ev \quad (c \in e_1 \wedge$$

$$b \in e_2 \wedge e_1 \equiv_s e_2 \implies \vec{a}_m^k(t) \uparrow\uparrow (Crd_k(e_1) - Crd_k(e_2))).$$

The (signed) *Minkowski–length* of $\vec{p} \in Q^d$ is

$$\mu(\vec{p}) := \begin{cases} \sqrt{|p_t^2 - |\vec{p}_s|^2|} & \text{if } p_t^2 - |\vec{p}_s|^2 \geqslant 0, \\ -\sqrt{|p_t^2 - |\vec{p}_s|^2|} & \text{otherwise} \end{cases}$$

and the *Minkowski–distance* between $\vec{p}$ and $\vec{q}$ is $\mu(\vec{p}, \vec{q}) := \mu(\vec{p} - \vec{q})$. A motivation for the "otherwise" part of the definition of $\mu(\vec{p})$ is the following. $\mu(\vec{p})$ codes two kinds of information, (i) the length of $\vec{p}$ and (ii) whether $\vec{p}$ is timelike (i.e., $|p_t| > |\vec{p}_s|$) or spacelike. Since the length is always non-negative, we can use the sign of $\mu(\vec{p})$ to code (ii).

The *acceleration* of an observer $k$ at instant $t \in Q$ is defined as the Minkowski–length of the relative-acceleration seen by any *inertial* observer $m$ at $t$, that is:

$$a_k(t) := \mu\big(\vec{a}_m^k(t)\big).$$

The acceleration is a well defined concept since it is independent of the choice of the inertial observer $m$. We say that observer $k$ is *positively accelerated* if and only if $a_k(t) \neq 0$ for all $t \in Dom\, Tr_k^k$. Observer $k$ is called *uniformly accelerated* if and only if there is an $a \in Q^+$ such that $a_k(t) = a$ for all $t \in Dom\, Tr_k^k$.

We say that *the clock of $b$ runs slower than the clock of $c$ as seen by $k$ with radar (photons; Minkowski–simultaneity)* if and only if $\mathsf{time}_b(e_b, e_b') < \mathsf{time}_c(e_c, e_c')$ for all events $e_b, e_b', e_c, e_c'$ for which $b \in e_b \cap e_b'$, $c \in e_c \cap e_c'$ and $e_b \sim_k^{rad} e_c$, $e_b' \sim_k^{rad} e_c'$. ($e_b \sim_k^{ph} e_c$, $e_b' \sim_k^{ph} e_c'$; $e_b \sim_k^{\mu} e_c$, $e_b' \sim_k^{\mu} e_c'$).

Now we can state our first theorem about the clock-slowing effect of gravitation:

**Theorem 5.1.** *Assume* $AccRel_d$ *and* $d \geqslant 3$. *Let* $\rangle b, k, c \langle_{rad}$ *be a radar-spaceship such that*

(1) *$k$ is positively accelerated,*

(2) *the direction of the spaceship agrees with that of the acceleration of k.*

*Then*

(i) *the clock of b runs slower than the clock of c as seen by k with radar and*

(ii) *the clock of b runs slower than the clock of c as seen by each of k, b and c with photons.*

To state a similar theorem in Minkowski–spaceships, we need the following concept. We say that observer $b$ *is not too far behind* positively accelerated observer $k$ if and only if the following holds:

$$\forall m \in \text{IOb} \ \forall t \in \text{Dom} \, Tr_m^k \ \forall \vec{p}, \vec{q} \in Cd_m \quad \left( \vec{p} \in tr_m(k) \wedge \right.$$

$$\vec{q} \in tr_m(b) \ \wedge \ ev_m(\vec{p}) \sim_k^{\mu} ev_m(\vec{q}) \ \wedge \ \vec{a}_m^k(t) \uparrow\uparrow (\vec{p} - \vec{q}) \implies$$

$$\left. \forall \tau \in \text{Dom} \, \vec{a}_m^k \quad \mu(\vec{p} - \vec{q}) < 1/a_k(\tau) \right).$$

Now we can state our second theorem about the clock-slowing effect of gravitation:

**Theorem 5.2.** *Assume* AccRel $_d$ *and* $d \geqslant 3$. *Let* $\rangle b, k, c \langle_{\mu}$ *be a Minkowski–spaceship such that*

(1) *k is positively accelerated,*

(2) *the direction of the spaceship agrees with that of the acceleration of k,*

(3) *b is not too far behind k.*

*Then*

(i) *the clock of b runs slower than the clock of c as seen by k with Minkowski–simultaneity or with photons and*

(ii) *the clock of b runs slower than the clock of c as seen by each of k, b and c with photons.*

In the following theorem we will see that the flow of time as seen by photons is strongly connected with the following two concepts. We say that observer $b$ is *approaching to* (*moving away from*) observer $k$ as seen by $k$ with photons if and only if for all events $e_k$ and $e_b$, if $b \in e_b$, $k \in e_k$ and $e_k \sim_k^{ph} e_b$, then there is an event $e$ such that $k', b' \in e$ for all inertial co-moving observers $k'$ and $b'$ of $k$ and $b$ at events $e_k$ and $e_b$, respectively, and $e_k$ precedes (succeeds) $e$ according to $k$ (cf. Fig. 4, (b)).

**Theorem 5.3.** *Assume* AccRel$_d$ *and* $d \geqslant 3$. *Let* $b, k \in$ Ob *be such that* $b$ *and* $k$ *are coplanar.*

(1) *If* $b$ *is approaching to* $k$ *as seen by* $k$ *with photons, then the clock of* $k$ *runs slower than the clock of* $b$ *as seen by* $k$ *with photons.*

(2) *If* $b$ *is moving away from* $k$ *as seen by* $k$ *with photons, then the clock of* $b$ *runs slower than the clock of* $k$ *as seen by* $k$ *with photons.*

None of the axioms introduced so far require the existence of accelerated (non-inertial) observers. Our following axiom scheme says that every definable timelike curve is the life-line of an observer. Since from AxSelf$^-$, AxPh$_0$ and AxEv it follows that the life-lines of inertial observers are straight lines (cf., for example, [1, 39, 40]), this will ensure the existence of many non-inertial observers.

A differentiable function $\gamma$ is called *timelike curve* if and only if the slope of $\gamma'(t)$ is less than 1 (i.e., $|\gamma'(t))_s| < |\gamma'(t))_t|$) for all $t \in Dom\,\gamma$ and $Dom\,\gamma$ is an open and connected subset of Q. It is clear that this is a first-order logic definable concept since every fragment of it is such. We say that a *function $f$ is* (parametrically) *definable by* $\psi(x, \vec{y}, \vec{z})$ if and only if there is $\vec{a} \in U^n$ such that $f(b) = \vec{p} \iff \psi(b, \vec{p}, \vec{a})$ true in $\mathfrak{M}$.

Let $\psi$ be a FOL-formula of our language.

**Ax∃Ob$_\psi$:** If a function parametrically definable by $\psi$ is a timelike curve, then there is an observer whose life-line is the range of this function.

Now we introduce the promised axiom scheme about the existence of observers:

$Ax\exists Ob := \{Ax\exists Ob_\psi : \psi$ is a FOL-formula of our language$\}$

The following three theorems say that the clocks can run arbitrarily slow or fast, as seen with the three different methods.

**Theorem 5.4.** *Assume* AccRel $_d$, Ax $\exists$ Ob, *and* $d \geqslant 3$. *Let* $m \in$ Ob *be positively accelerated such that* $Dom\, Tr^m_m = Q$, *and let* $e, e' \in Ev$ *be such that* $e \neq e'$ *and* $m \in e \cap e'$. *Then for all* $\lambda \in Q^+$, *there is an observer* $b$ *and events* $e_b, e'_b \in Ev$ *such that* $b \in e_b \cap e'_b$, $e \sim^{rad}_m e_b$, $e' \sim^{rad}_m e'_b$ *and* $\mathsf{time}_b(e_b, e'_b) = \lambda\, \mathsf{time}_m(e, e')$.

**Theorem 5.5.** *Assume* AccRel $_d$, Ax $\exists$ Ob, *and* $d \geqslant 3$. *Let* $m \in$ Ob *be uniformly accelerated, and let* $e, e' \in Ev$ *be such that* $e \neq e'$ *and* $m \in e \cap e'$. *Then for all* $\lambda \in Q^+$, *there is an observer* $b$ *and events* $e_b, e'_b \in Ev$ *such that* $b \in e_b \cap e'_b$, $e \sim^{\mu}_m e_b$, $e' \sim^{\mu}_m e'_b$ *and* $\mathsf{time}_b(e_b, e'_b) = \lambda\, \mathsf{time}_m(e, e')$.

**Theorem 5.6.** *Assume* AccRel $_d$, Ax $\exists$ Ob, *and* $d \geqslant 3$. *Let* $m \in$ Ob *be positively accelerated and* $e, e' \in Ev$ *such that* $e \neq e'$ *and* $m \in e \cap e'$. *Then for all* $\lambda \in Q^+$, *there is an observer* $b$ *and events* $e_b, e'_b \in Ev$ *such that* $b \in e_b \cap e'_b$, $e \sim^{ph}_m e_b$, $e' \sim^{ph}_m e'_b$ *and* $\mathsf{time}_b(e_b, e'_b) = \lambda\, \mathsf{time}_m(e, e')$.

We have seen that gravitation (acceleration) makes "time flow slowly." However, we left open the question what role the "magnitude" and the "direction" of the gravitation play in this effect. The following theorem shows that two observers, say $m$ and $k$, can feel the same gravitation while the clock of $k$ runs slower than the clock of $m$. Thus it is not the "magnitude" of the gravitation that makes "time flow more slowly."

**Theorem 5.7.** *Assume* AccRel $_d$, Ax $\exists$ Ob, *and* $d \geqslant 3$. *There are uniformly accelerated observers* $m$ *and* $k$ *such that* $a_k(t) = a_m(t)$ *for all* $t \in Q$, *but the clock of* $k$ *runs slower than the clock of* $m$ *as seen by both* $m$ *and* $k$ *with photons (or with radar or with Minkowski– simultaneity).*

Now let us see what we can say about the role of the "direction" of gravitation. Being "more down in a gravitational well" becomes being "behind" if we translate it from the language of gravitation into the language of acceleration. This can be formulated by our notation as follows. We say that observer $b$ is *behind* observer $k$ if and only if

$$\forall m \in IOb \ \forall t \in Dom \, Tr_m^k \ \forall \vec{p}, \vec{q} \in Cd_m \quad \vec{p} \in tr_m(k) \wedge$$

$$\vec{q} \in tr_m(b) \ \wedge \ ev_m(\vec{p}) \sim_k^\mu ev_m(\vec{q}) \ \wedge \ \vec{a}_m^k(t) \uparrow\uparrow (\vec{p} - \vec{q}).$$

The following theorem shows that if observer $b$ is at a lower level in the tower than observer $k$ is, then his clock runs slower than the clock of $k$, as seen by $k$ with radar.

**Theorem 5.8.** *Assume* AccRel$_d$ *and* $d \geqslant 3$. *Let* $b, k \in$ Ob *be such that*

(1) $k$ *is positively accelerated,*

(2) $b$ *and* $k$ *are coplanar,*

(3) $b$ *is behind* $k$.

*Then the clock of* $k$ *runs slower than the clock of* $b$, *as seen by* $k$ *with radar.*

The proofs, along with more explanation and motivation, of the theorems presented in this section can be found in [45].

# 6. Questions, Suggestions for Future Research

1. We hope that the perspective outlined in Sections 1–3, and the techniques presented in Sections 4–5, [44] already suggest a research proposal. Sections 4–5 cover only a small fragment of the research proposed in Sections 1–3. So the proposal is: elaborate a larger part of the perspective outlined in Sections 1–3 in the style of Sections 4–5 and [44].

2. The Introduction of [2] contains more ideas both on the general perspective (of applying logic to spacetime theory) and also more of the long-distance goals. However, some of the present results were not available when [2] was written, therefore that introduction does not replace completely the present section.

3. In Section 5, we started to elaborate a purely logical theory of the effects of gravitation on clocks. Elaborate this direction in more detail, and investigate more aspects of gravitation on clocks. For example, assume we bore a hole through the Earth from the North pole to the South pole. Now put a clock into the middle of the Earth. It will levitate "weightlessly" in the middle. Put another clock to the surface of the Earth. It will be squeezed by gravity to the surface. Despite this, the clock levitating in the middle will run slower than the one on the surface. A third clock high above in deep space will run even faster (than the one on the surface). Why? Find a logic style formulation of the above (and prove it) in the manner of Section 5.

4. Investigate/formulate further aspects of the effects of gravity on instruments (like clocks, meter-rods). For example, define the so-called gravitational force-field experienced by an accelerated observer (via acceleration, relative to the observer, of test particles dropped by the observer). Study this force-field and connect this study with the investigations in Section 5. Try to make an integrated coherent picture of gravity, time warp (clock behavior in gravitational fields), and gravitational force. (Remark: gravitational force is often suppressed in the literature because it is not "absolute," i.e., is not observer independent. All the same, if we keep in mind that it is observer dependent, then it is a helpful concept.) Imagine a long, accelerated spaceship. The gravitational force experienced in the aft of the ship will be greater than that in the front of the ship. Why?

5. Continuing in the spirit of Sections 4,5, [44], and the above, elaborate a FOL theory of the spacetime of a Schwarzschild black hole [63]. Streamline that theory, make it logically transparent and illuminating. Apply conceptual analysis to the theory similar in spirit as conceptual analysis of special relativity is started in [2, 1, 43]. Using the theory of accelerated observers and Einstein's equivalence principle, create a logically convincing, illuminating theory of such black holes. In this direction it might be helpful that the analogy between the world-view or reference frame of an

accelerated spaceship and skyscrapers (towers) on the event horizon of a black hole is described in detail in Rindler's relativity book [52, § 12.4, pp. 267–272]. Figure 12.6 is especially useful therein. Also note how in Rindler's arrangement of the skyscrapers above the black hole they are prevented from falling by rigid rods separating them (these rods provide the "acceleration" experienced by the inhabitants of the towers/spaceships). These rods are called struts in [52, p. 270].

So, we suggest combining the presently started FOL theory of accelerated observers and of effects of gravity (acceleration) on instruments of observers with the just quoted part of Rindler's work in order to elaborate a FOL theory of the simplest kind of black holes. Of course, the main point is that we are striving for a very special kind of illuminating (etc.) FOL theory (and not just any FOL theory describing a black hole).

When the above is done, we suggest applying re-coordinatization in order to obtain an Eddington–Finkelstein version of this FOL theory of the black hole. This second (EF) version of the theory will also describe what the in-falling observer sees, for example, from inside the event horizon. For the latter question we suggest assuming that the black hole is huge (galactic size) so that enough stuff remains to be observed after falling through the event horizon.

6. After having streamlined, analyzed, simplified FOL theories of simple (but huge) black holes, we propose turning to what we call double black holes or exotic black holes. Double black holes have two event horizons, an outer one and an inner one. In theory and under certain assumptions, a traveler might fall into the black hole, survive this and may come out at some other point of spacetime (in our universe or in some other universe). So, some of these double black holes may be regarded kind of wormholes. Examples are spinning black holes (Kerr spacetime, Kerr–Newman spacetime), and electrically charged black holes (Reissner-Nordström spacetime) [63].

The task here is again to build up, streamline, and conceptually analyse, simplify FOL theories for such double black holes. They offer logically intriguing issues for the logician as indicated in Section 3.

7. Besides the relatively simple kind of acceleration studied in Sections 4,5, [44], rotation provides a kind of acceleration appearing in the form of the centrifugal force. A further research task is to analyse via FOL the world-view represented by a rotating coordinate lattice (relative to the gyroscopes) and generally, the rotational spacetimes. An example for these is the slowly rotating black hole (Kerr spacedtime), other examples are Gödel's rotating universe, Tipler–Stockum spacetime. In these spacetimes rotation leads to CTC's and to many other exotic effects like the so-called dragging of inertial frames or the drag effect. Finding out more about these is the task of NASA's recent "Probe B." Here again a FOL theory of such spacetimes waits for the creation, conceptual analysis and detailed illuminating explanation of what happens and exactly why. A particular question waiting to be answered is to find out and analyse what the common features/mechanisms/principles of these rotating spacetimes (with CTC's) are. For example, many features of the above mentioned three spacetimes coincide. Is this a coincidence or is there a more general "theory of rotating spacetimes" lurking in the background. For more on this question we refer to [5]. In particular, we are looking for a logical answer to the quasi-philosophical question: "Exactly why and how CTC's are generated in rotating black holes and in Gödel's universe. Why do they counter-rotate with matter?" (More on what we call "counter-rotation" can be found in [5].)

**Acknowledgements.** Thanks go to Victor Pambuccian for many valuable conversations on the subject, for reading an earlier version of this paper and for helpful suggestions leading to the present version. Research supported by the Hungarian National Foundation for scientific research grant T43242 and by Bolyai Grant for Judit X. Madarász.

# References

1. H. Andréka, J. X. Madarász, and I. Németi, *Logical axiomatizations of space-time*, In: Non-Euclidean Geometries: János Bolyai Memorial Volume, A. Prékopa and E. Molnár (Eds.), Springer, 2006, pp. 155–185.
   [http://www.math-inst.hu/pub/algebraic-logic/lstsamples.ps]

2. H. Andréka, J. X. Madarász, and I. Németi with contributions from A. Andai, G. Sági, I. Sain, and Cs. Tőke, *On the Logical Structure of Relativity Theories*, Research report, Alfréd Rényi Institute of Mathematics, Budapest, 2002.
   [http:/www.math-inst.hu/pub/algebraic-logic/Contents.html]

3. H. Andréka, J. X. Madarász, and I. Németi, *The logic of space-time.* [To appear]

4. H. Andréka, J. X. Madarász, and I. Németi, *Logical analysis of relativity theories*, In: First-Order Logic Revisited, Logos, Berlin, 2004, pp. 7–36.

5. H. Andréka, I. Németi, and C. Wüthrich, *A twist in the geometry of rotating black holes: seeking the cause of acausality.* [To appear]

6. J. Ax, *The elementary foundations of spacetime*, Found. Phys. **8** (1978), 507–546.

7. J. Barwise and J. Etchemendy, *The Liar: An Essay on Truth and Circularity*m Oxford Univ. Press, 1987.

8. C. H. Brans, *Gravity and the Tenacious Scalar Field*, 1979.
   [arXiv:gr-qc/9705069]

9. C. H. Brans, *The Roots of Scalar-Tensor Theory: an Approximate History*, 2004. [arXiv:gr-qc/0506063]

10. S. M. Carroll, *Why is the Universe Accelerating*, 2003. [arXiv:astro-ph/0310342]

11. C. C. Chang and H. J. Keisler, *Model Theory*, North–Holland, 1973, 1990.

12. R. d'Inverno, *Introducing Einstein's Relativity*, Clarendon, Oxford, 1992.

13. Gy. Dávid, *Modern cosmology - astronomical, physical and logical approaches*, Abstracts of Invited talks in "Logic in Hungary 2005."
    [http://atlas-conferences.com/cgi-bin/abstract/caqb-64]
    [http://www.logicart.hu/events/lh05/index.php]

14. Gy. Dávid and I. Németi, *Relativistic computers and the Turing barrier*, J. Appl. Math. Comput. (2006). [To appear] [http://renyi.hu/pub/algebraic-logic/beyondturing.pdf]

15. J. Earman, *Bangs, Crunches, Whimpers, and Shrieks*, Oxford Univ. Press, 1995.

16. J. Earman and J. D. Norton, *Forever is a day: supertasks in Pitowsky and Malament–Hogarth spacetimes*, Philos. Sci. **60** (1993), 22–42.

17. J. Earman, C. Smeenk, and C. Wüthrich, *Take a Ride on a Time Machine*, In: Reverberations of the Shaky Game: Festschrift for Arthur Fine, R. Jones and P. Ehrlich (Eds.), Oxford Univ. Press. [To appear]

18. A. Einstein, *Über die spezielle und die allgemeine Relativitätstheorie*, von F. Vieweg, Braunschweig, 1921.

19. G. Etesi and I. Németi, *Non-turing computations via Malament–Hogarth space-times*, Int. J. Theor. Phys. **41** (2002), 341–370. [arXiv:gr-qc/0104023]

20. V. Faroni, *Illusions of general relativity in Brans-Dicke gravity*, Phys. Rev. D **59** (1999), 084021–084027.

21. J. Ferreirós, *The road to modern logic – an interpretation*, Bull. Symb. Log. **7** (2001), no. 4, 441-484.

22. H. Friedman, *On foundational thinking 1*, Posting in Foundations of Mathematics Archives, 2004. [www.cs.nyu.edu].

23. H. Friedman, *On foundations of special relativistic kinematics 1*, Posting No 206 in Foundations of Mathematics Archives, 2004. [www.cs.nyu.edu]

24. L. Fuchs, *Partially Ordered Algebraic Systems*, Pergamon, Oxford, 1963.

25. K. Gödel, *An example of a new type of cosmological solution of Einstein's field equations of gravitation*, Rev. Mod. Phys. **21** (1949), 447–450.

26. R. Goldblatt, *Orthogonality and Spacetime Geometry*, Springer, 1987.

27. R. J. Gott, *Closed timelike curves produced by pairs of moving cosmic strings: Exact solutions*, Phys. Rev. Lett. **66** (1991), 1126–1129.

28. R. J. Gott, *Time Travel in Einstein's Universe: The Physical Possibilities of Travel Through Time*, Houghton Mifflin, New York, 2002.

29. A. Günbaum, *Philosophical Problems of Space and Time*, Dover, New York, 1963.

30. P. Hájek and P. Pudlák, *Metamathematics of First-Order Arithmetic*, Springer, 1993.

31. S. W. Hawking and G. F. R. Ellis, *The Large Scale Structure of Space-Time*, Cambridge Univ. Press, 1973.

32. L. Henkin, A. Tarski, and P. Suppes (Eds.) *The Axiomatic Method with Special Reference to Geometry and Physics*, North-Holland, 1959.

33. D. Hilbert, *Mathematische Behandlung der Axiome der Physik*, Akad. Wiss. Göttingen (1990), pp. 272–273.
[www.mathematik.uni-bielefeld.de/~kersten/hilbert/prob6.html]

34. D. Hilbert, *Über den Satz von der Gleichheit der Basiswinkel im gleichschenkligen Dreieck*, Proc. London Math. Soc. **35** (1902/1903), 50–68.

35. R. Hirsch, *Logic and Dialectics*, Cultural Logic, Spring, 2004.

36. W. Hodges, *Model Theory*, Cambridge Univ. Press, 1997.

37. M. L. Hogarth, *Deciding arithmetic using SAD computers*, Br. J. Phil. Sci. **55** (2004), no.4, 681–691.

38. M. L. Hogarth, *Predictability, Computability and Spacetime*, PhD Thesis, Univ. Cambridge, UK, 2002.
[http://www.renyi.hu/pub/algebraic-logic/Hogarththesis.ps.gz]

39. R. Horváth, *An Alexandrov–Zeeman type theorem and relativity theory*, Eötvös Loránd Univ. Budapest, 2005.

40. R. Horváth and G. Székely, *An Alexandrov–Zeeman type theorem and its consequences in special relativity*. [To appear]

41. H. A. Lorentz *Electromagnetic phenomena in a system moving with any velocity less than that of light*, Proc. Royal Acad. Amsterdam **6** (1904), 809–831.

42. J. X. Madarász, *Logic and Relativity (in the Light of Definability Theory)*, PhD Thesis, Eötvös Loránd Univ., Budapest, 2002.
[http://www.math-inst.hu/pub/algebraic-logic/Contents.html]

43. J. X. Madarász, I. Németi, and Cs. Tőke, *On generalizing the logic-approach to space-time towards general relativity: first steps*, In: First-Order Logic Revisited, Logos, Berlin, 2004, pp. 225–268.

44. J. X. Madarász, I. Németi, and G. Székely, *Twin paradox and the logical foundation of relativity theory*, Found. Phys. (2006). [To appear]
[arXiv:gr-qc/0504118]

45. J. X. Madarász, I. Németi, and G. Székely, *A logical analysis of the time-warp effect of general relativity.* [To appear]

46. C. W. Misner, K. S. Thorne, and J. A. Wheeler, *Gravitation*, W.H. Freeman, 1973.

47. B. O'Neill, *The Geometry of Kerr Black Holes* A. K. Peters, 1995.

48. V. Pambuccian, *Axiomatizations of hyperbolic and absolute geometries*, In: Non-Euclidean Geometries: János Bolyai Memorial Volume, A. Prékopa and E. Molnár (Eds). Springer, 2006, pp.119–153.

49. C. S. Reynolds, L. W. Brenneman, and D. Garofalo, *Black hole spin in AGN and GBHCs*, Astrophys. Space Sci. **300** (2005), 71–79.
[arXiv:astro-ph/0410116]

50. H. Reichenbach, *Axiomatik der relativistische Raum-Zeit-Lehre*, 1924; English transl.: Axiomatization of the Theory of Relativity, Univ. California, Berkeley, 1969.

51. H. Reichenbach, *The Philosophy of Space and Time*, Dover, 1958.

52. W. Rindler, *Relativity. Special, General and Cosmological*, Oxford Univ. Press, 2001.

53. W. Rudin, *Principles of Mathematical Analysis*, McGraw-Hill, 1953.

54. J. Q. Schutz, *Foundations of Special Relativity: Kinematic Axioms for Minkowski Space-Time*, Springer, 1973.

55. Gy. Serény, *Boolos-style proofs of limitative theorems*, Math. Log. Q. **50** (2004), no. 2, 211–216.

56. S. G. Simpson (Ed.), Reverse Mathematics 2001,, Lect. Notes Logic **21**, Association for Symbolic Logic (ASL), 2005.

57. W. J. van Stockum, *The gravitational field of a distribution of particles rotating around an axis of symmetry*, Proc. R. Soc. Edinb. **57** (1937), 135–154.

58. T. E. Strohmayer, *Discovery of a 450 HZ quasi-periodic oscillation from the microquasar GRO J1655-40 with the Rossi X-Ray Timing Explorer*, Astrophys. J. Lett. **552** (2001), L49–L53.
[arXiv:astro-ph/0104487]

59. P. Suppes, *Axioms for relativistic kinematics with or without parity*, In: The Axiomatic Method with Special Reference to Geometry and

Physics, L. Henkin, P. Suppes and A. Tarski (Eds.), North-Holland, 1959.

60. P. Suppes, *The desirability of formalization in science*, J. Philos. **65** (1968), 651–664.

61. G. Székely, *A First-Order Logic Investigation of the Twin Paradox and Related Subjects*, Master's Thesis, Eötvös Loránd Univ. Budapest, 2004.

62. A. Tarski, *A Decision Method for Elementary Algebra and Geometry*, Univ. California, Berkeley, 1951.

63. E. F. Taylor and J. A. Wheeler, *Exploring Black Holes: Introduction to General Relativity*, Addison Wesley, 2000.

64. F. J. Tipler, *Rotating cylinders and the possibility of global causality violation*, Phys. Rev. D **9** (1974), 2203–2206.

65. J. Väänänen, *Second-order logic and foundations of mathematics*, Bull. Symb. Log. **7** (2001), 504–520.

66. R. M. Wald, *General Relativity*, Univ. Chicago, 1984.

67. J. Woleński, *First-order logic: (philosophical) pro and contra*, In: First-Order Logic Revisited, Logos, Berlin, 2004, pp. 369–398.

# Beyond Hybrid Systems

## Anil Nerode

*Cornell University*
*Ithaca, USA*

## Preface

What does the future hold for mathematical logic? In the early 1950's I learned all the logic then existing. Until the mid eighties I read everything published in logic and related computer science. I am a "quick study," but the quantity of papers become enormous, and I now limit my reading. I have watched all the well-known logicians and their subjects evolve for fifty-six years. Can I say anything beyond truisms about future trends?

First, what *are* the truisms? Since 1950 the winds of time have swept the grains of logic into four dunes, four research disciplines: computability theory, model theory, set theory, and proof theory. Each of these four disciplines has developed a coterie of respected specialists. In each there have been suites of applications to mathematics and suites of developments from computer

Mathematical Problems from Applied Logic. Logics for the XXIst Century. II. Edited by Dov M. Gabbay *et al.* / International Mathematical Series, 5, Springer, 2007

science. All four disciplines have bright futures. This is the main truism, which I leave it to others to elaborate.

Specialists in each discipline are acquainted with the basics of the other disciplines, but often not with the proofs of any of the latest theorems or the latest applications of the other three disciplines. Whether and how these disciplines will fractionate further into subdisciplines no one can predict. Fractionation is not peculiar to logic. All of mathematics, perhaps all of knowledge, has been fractionating in the last hundred years. There are no longer any universal mathematicians. There will be no more Hilberts. The terrain is too vast. With fractionation goes a witticism. Robert Maynard Hutchins, President of the University of Chicago, said sixty years ago that we learn more and more about less and less, and less and less about more, until we know everything about nothing and nothing about everything.

There are countervailing winds sweeping the dispersed grains into new dunes. This is the tendency toward merging of disciplines. In my opinion the most significant mergings do not arise by forming abstract theories encompassing several pre-existing theories, which was the E.H. Moore dictum and is represented by Bourbaki's old textbooks. Such generalizations are mostly useful as teaching devices intended to communicate a lot briefly and for reducing what looks new to what has already been done. I see the truly significant mergings as arising from new applied areas which cannot be understood or modelled or designed or controlled using previously available concepts and methods. These areas are often important, murky, and confused. They often demand new ideas or new ways of fusing old ideas. One recent example in computer science is the use of varied logical systems: classical, non-monotonic, intuitionistic, modal, and temporal, developed for automatic theorem proving, program verification, and for the logic of knowledge. Previous to the needs of computer science, the modal and temporal logics were of interest only to speculative philosophers, although they had good mathematical semantics due to Kripke. The new applications required answering new questions and developing new semantics and new proof procedures. Another new application area is the development of logical models, tools, and algorithms for creating reliable secure networks of communicating agents. Just writing agent software without having a particular logical model in mind has created a plethora of badly understood systems with substantial non-robust and non-secure behaviors.

Most of my past research areas such as automata, computability theory, recursive algebra, non-monotonic and intuitionistic and modal logics, complexity theoretic algebra, concurrency, etc., have been pursued by many. I leave others to document the future of those lively areas. I confine myself to a very brief discussion of a still emerging area of more recent vintage which lies at the borderline of mathematics, computer science, and engineering.

"Hybrid Systems" is the area I am referring to. Hybrid systems is a concept intended to unify discrete logic and continuous control engineering. As originally conceived it was logic interacting with differential equations. I introduced the term "Hybrid Systems" in this context in 1992 to describe interacting networks of logical (digital) and physical (continuous) devices, motivated by unfilled needs in business and industry and the military. For instance, hybrid system control is needed to optimize performance when controlling inherently unstable high performance systems such as aerospace vehicles and structures, power grids, and automated intelligent manufacturing facilities, as well as for control of decision systems involving man-machine interactions such as automated decision aids for battlefield commanders or air traffic control. I sponsored several conferences and put out several volumes, with the specific purpose of forming a community around the world to pursue developing this as a merged discipline of logic interacting with continuous physical processes. There had been previous sporadic developments in this direction, but these conferences spurred systematic research. They are listed as the first references at the end of this paper, volumes still worth perusing. Since then the study of Hybrid Systems has become a well established part of computer science and engineering. Despite a burgeoning literature, sampled at the end of this paper, the surface has hardly been scratched. In this essay I suggest the logical and analytic study of far more general structures that more closely fit applications. But first, here is a brief description of hybrid systems as Wolf Kohn and I developed them starting in 1992.

## 1. Digital Programs

Digital programs live in a world in which inputs, outputs, and internal states are expressed using finite alphabets. The mathematical language for describing digital programs is the language of logic. The semantics of the language may be state automata, or continuous maps between cpo's, or process algebras, among other possibilities. These are often described by systems of logical formulas or equations. Digital programs are written to realize program specifications about their behavior. Conclusions about behavior of digital programs are drawn by reasoning using logical deductions expressible in Dynamic, Hoare, Temporal, or other program logics, or using automated exploration of state space. Theorems of logic and algorithms of computer science underlie verifying such properties of programs as termination or fairness or correctness.

## 2. Continuous Plants and Controllers

Continuous devices whose state evolution is controlled by continuous physical controllers we refer to as plants with controllers. They live in a physical world in which inputs, outputs, and internal states of controller and plant are points in manifolds. Sensors sense the state of the plant, the controller responds by altering the state of the plant. The controller exercises control as a function of the state of the plant. The behavior of such a continuous feedback system is modelled by systems of coupled ordinary differential equations for plant, plant controller, and sensors, with the controller governed by a control which is a function of plant state or perhaps a function of time. The theory of optimal control deals with how to choose the control function of state (time) so as to minimize, over the space of possible control functions, the integral of a non-negative function along the resulting plant state trajectory. This involves differential geometry, calculus of variations, numeric and symbolic algorithms for differential equations, etc. These ideas are used to establish such features as controllability, observability and reachability. The reasoning used is that of differential equations and differential geometry.

## 3. Hybrid Systems

A simple example is a closed loop system which consists of a continuous plant subject to both external disturbances and to control by a digital program. The digital control program reads at discrete intervals sensor data about plant state and computes a new control law to govern the plant and substitutes it for the previously used control law. The plant will continue to use this control law until the next such intervention. The sequence consisting of "read the sensors" followed by "compute and impose the next control" constitutes the control cycle. The control law is a function of plant state. How, and when, to make these control law changes is the business of the digital control program.

From the point of view of the digital controller, the world of inputs and outputs is finite sequences of strings, input readings and output control orders. From the point of view of the controlled plant, the world of inputs and outputs is real time streams of real numbers, inputs to the controller and sensor outputs to the digital control chip.

So on the surface hybrid systems are governed by coupled sets of equations, logical and differential. That is where Wolf Kohn and I started. We believe that one should study digital control of continuous plants at the same intellectual level of sophistication as has been applied in computer science for the analysis of concurrent and distributed programs and in control theory for the Pontryagin analysis of optimal control.

## 4. Discretization

A computer scientist's first reaction is to discretize the continuous
controller and continuous plant and time and input and output,
and to end up representing the hybrid system as an interacting
network of digital devices. Some workers in discrete event systems
do this. Then one gets a pure automaton problem of extracting
a control function of state which controls a discretized automaton
model of the plant. There are two difficulties with this approach.
First, it only guarantees control of the discretized version of the
plant with the derived digital control program. There is no guarantee that this digital control program will control the original mixed
continuous-discrete system. Second, if we extract a near optimal
control for this discretized model relative to a cost function, it may
well be nowhere near an optimal control for the original system.
Systems in practice that we care about are highly non-linear, and
these things happen frequently. Traditional control theory han-
dles linear systems extremely well, but has not been much help
for highly non-linear systems.

## 5. Continualization

Wolf Kohn and I took the opposite approach. We "continualized" everything, replaced the discrete by the continuous. Physically this is meaningful though complex. After all, the digital devices interacting with continuous ones ARE in the physical world,
they ARE continuous devices (at least in the quantum mechanical
sense), but are built of components with abrupt but not instan-
taneous changes of state from 0 to 1 and conversely. We did not
model the digital control program in this way. Instead we followed
Boole and replaced Boolean 0, 1 logical conditions such as $x \wedge x = x$
by real number equations such as $x^2 = x$. This results in describing the hybrid system by differential equations for the continuous
elements coupled with algebraic equations for the previously discrete components, where the variables in the algebraic equations

are functions of time, equations with a control function as parameter. This gives a set of algebraic and differential equations which represent the hybrid system in purely continuous form. If we have a non-negative cost function integrated on plant state trajectories to determine cost, it now makes sense to speak of an optimal or near optimal control, one which produces a plant state trajectory with minimal or near minimal cost. This falls under the point of view of Pontryagin's optimal control using the calculus of variations. In this approach the digital world has been completely absorbed into the continuous. Even in the simplest cases with logical constraints this gives a highly non-convex optimization problem. Usually there are no smooth control functions which achieve optimal trajectories. In fact, coding the usual Specker phenomenon of computable non-negative continuous functions on compact spaces with no computable point at which minimum value is taken, it is obvious that there is no general algorithm which will yield a computable optimal control function of state from computable data.

But here is where the magic of classical approximate weak solutions enters, and it is the point at which Kohn and I entered the game. The argument is that no engineer wants to pay for optimal behavior, he or she is merely interested in a control function which yields a trajectory within a prespecified epsilon of minimum cost. The epsilon is determined by the use and how much he or she is willing to spend on the project. The 1930's work of Young and of McShane on convexifying calculus of variations problems and proving the existence of relaxed probability measure valued control functions achieving the minimum shows that for a prescribed epsilon, one can compute a piecewise linear epsilon optimal solution which is implementable. These are piecewise linear chattering controls. The control changes abruptly once in a while based on what part of the plant state space you are in. Chattering controls have jump discontinuities but nevertheless yield piecewise smooth plant state trajectories. We extracted our epsilon optimal digital control programs as finite automata which take as input at discrete times a letter representing current plant state, and produce output a letter at a time which tells what linear control to use next. The discrete logical side is the logic of these automata.

## 6. Methodology

### 6.1. Summary

We continualize the discrete elements of the hybrid system and solve by numerical continuous methods of relaxed control to get an epsilon optimal measure valued optimal control. Then we discretize this control to get a finite digital control automaton enforcing approximate optimal control. The technical tools we use are differential geometric. We think of an integrated cost function along a segment of a trajectory in a suitable manifold as a "metric" and think of the geodesics of the corresponding metric ground form as the optimal trajectories and use approximations to the relaxed controls leading to geodesics as our intended piecewise linear controls, which are then implemented as a digital control program which represents a large finite automaton. We use Finsler and higher jet spaces to allow us to incorporate second order effects (curvature) as necessary. When applied to entirely discrete problems, this gives new continuous methods for solving logic problems. This has been used by us, but not investigated in mathematical depth.

Jennifer Davoren and I wrote a review article on the known logics for hybrid systems and their semantics and syntax, referenced at the end of this essay. These logics describe the discrete and the continuous aspects of the same entity and how they interact and how to go from one to the other. One should also look for inspiration at the continuous logics of Keisler and Hoover in which the existential quantifier is mathematical expectation, and the Hopf algebras of Grossman and Sweedler for quantum hybrid systems.

In 1995 Kohn and I established a corporation to prove our technology in industrial and business contexts. Getting real time algorithms took another ten years and produced some commercial applications. But even with real time algorithms, we found that there is a *huge* additional problem which I now introduce as a good direction for future research–analytic, geometric, and logical.

## 6.2. Multiple models

There are often a multiplicity of useful models for the same process, each giving different information. When you sense plant state, you have a choice of sensors and what to sense. When you choose controllers, you have a wide choice of what devices to use. Further, measurements sensed are necessarily averages of time varying quantities. For control, what sensors and controllers should you use and what average measurements should you record? The problem is that one can make a variety of somewhat predictive models based on different sensors and different controllers and get something out of each, with absolutely no coherent underlying scientific model. A lesson of control practice is that there is always unmodelled dynamics left over to perhaps be treated by a different later model.

To repeat, in constructing hybrid models for industrial and business applications we lack complete control over choice and location of sensors and controllers and measurement schemes and often have to deal with multiple models of the same thing. Some are legacy models. Some are the best that science can provide. Some have coarse granularity in space and time, some are fine grained. Often they cannot be naturally or usefully embedded in any overarching model. But often each contributes unique information about plant state that can be used, in conjunction with the rest, to influence our choice of control to discipline the process.

The hybrid systems concept, while an essential abstraction for understanding and controlling the evolution of digital-continuous systems, is too narrow for many problems of practical interest because it is based on a single model for the behavior of the digital and continuous parts of a real system. This is perhaps an unconscious reflection of the heritage of the mind-body distinction. The role of the mind is taken by the digital control program (remember Turing's original description of a Turing head as a finite state mind); the role of the body is taken by the controlled physical plant; the role of the mind-body interaction is taken by the feedback loop between them. Nowadays science presents many different models for the mind and the body, based on physics, chemistry, mechanics, biochemistry of DNA, and so on. These are different models at different levels of granularity for the body and the brain. They do not impart the same information. An all-encompassing single model is a dream, not an actuality. The hope for a unifying model is based on the paradigm that fundamental physics will eventually explain all. Today there are materials scientists who treat materials on a macro scale, there are materials scientists who treat them on a quantum scale, there are economists who study macroeconomic indicators to divine how to influence the future, there are economists who study the microeconomics of isolated systems for the same purpose, there are financial analysts who use a variety of very different conflicting indicators to predict market behavior, etc. If we want to extract control laws which constrain the behavior of complex systems with partially modelled dynamics we need to develop a calculus of languages and models where many models may represent the same underlying process, and there is a congenial systematic way to incorporate new models while keeping old ones.

How have Kohn and I treated this so far? We regard the actual process as input to its several different models, each of which has its own cost function for plant state trajectories as that model sees them. We form a composite cost function which is a function of the cost functions for the individual models and which has adjustable parameters. For example, we have used a weighted sum of the Lagrangian cost functions for the various models, the weights being the parameters. We control so as to approximately follow the trajectories of minimal cost according to this composite cost function. We use feedback based on sensor measurements of plant trajectory performance to adjust the parameters continuously, therefore to adjust the cost function continuously, therefore to adjust the geodesic field being followed. This is a powerful ad hoc device compatible with the differential geometric and calculus of variations point of view. Embedded in this is a combination of continualized versions of the logics of the individual models. Continualization of logics will be a central theme. Can we isolate out how the logics of such compound views of one process arise from the logics of the individual views, both in syntax and semantics, and use them as tools for engineering and objects of mathematical study? Only time will tell.

# References

## Original Hybrid System Series

**1993** *Hybrid systems*, R. L. Grossman, A. Nerode, A. P. Ravn, and H. Rischel (Eds.), Lect. Notes Compt. Sci. **736**, Springer, 1993.

**1995** *Hybrid systems. II*, P. Antsaklis, W. Kohn, A. Nerode, and S. Sastry (Eds.), Lect. Notes Compt. Sci. **999**, Springer, 1995.

**1996** *Hybrid Systems III*, R. Alur, T. A. Henzinger, and E. D. Sontag (Eds.), Lect. Notes Compt. Sci. **1066**, Springer, 1996.

**1997** *Hybrid Systems IV*, P. Antsaklis, W. Kohn, A. Nerode, and S.Sastry (Eds.), Lect. Notes Compt. Sci. **1273**, Springer, 1997.

**1999** *Hybrid Systems V*, P. Antsaklis, W. Kohn, M. Lemmon, A. Nerode, and S. Sastry (Eds.), Lect. Notes Compt. Sci. **1567**, Springer, 1999.

## Successor Series

**1998** *Hybrid Systems: Computation and Control: First International Workshop, HSCC'98 Berkeley, California, USA, April 13-15, 1998 Proceedings*, Th.A. Henzinger and S. Sastry (Eds.), Lect. Notes Compt. Sci. **1386**, Springer, 1998.

**1999** *Hybrid Systems: Computation and Control: Second International Workshop, HSCC'99, Berg en Dal, The Netherlands, March 1999. Proceedings*, F.W. Vaandrager and J.H. van Schuppen (Eds.), Lect. Notes Compt. Sci. **1569**, Springer, 1999.

**2000** *Hybrid Systems: Computation and Control: Third InternationalWorkshop, HSCC 2000 Pittsburgh, PA, USA, March 23-25,*

*2000 Proceedings*, N. Lynch and B. Krogh (Eds.), Lect. Notes Compt. Sci. **1790**, Springer, 2000.

**2001** *Hybrid Systems: Computation and Control: 4th International Workshop, HSCC 2001 Rome, Italy, March 28-30, 2001, Proceedings*, M.D. Di Benedetto and A. Sangiovanni-Vincentelli (Eds.), Lect. Notes Compt. Sci. **2034**, Springer, 2001.

**2002** *Hybrid Systems: Computation and Control: 5th International Workshop, HSCC 2002, Stanford, CA, USA, March 25-27, 2002. Proceedings*, C.J. Tomlin and M.R. Greenstreet (Eds.), Lect. Notes Compt. Sci. **2289**, Springer, 2002.

**2003** *Hybrid Systems: Computation and Control: 6th International Workshop, HSCC 2003, Prague, Czech Republic, April 3-5, 2003. Proceedings*, O. Maler and A. Pnueli (Eds.), Lect. Notes Compt. Sci. **2623**, Springer, 2003.

**2004** *Hybrid Systems: Computation and Control: 7th International Workshop, HSCC 2004, Philadelphia, PA, USA, March 25-27, 2004. Proceedings* R. Alur and G. J. Pappas (Eds.), Lect. Notes Compt. Sci. **2993**, Springer, 2004.

**2005** *Hybrid Systems: Computation and Control: 8th International Workshop, HSCC 2005, Zurich, Switzerland, March 9-11, 2005. Proceedings*, M. Morari and L. Thiele (Eds,), Lect. Notes Compt. Sci. **3414**, Springer, 2005.

**2006** *Hybrid Systems: Computation and Control: 9th International Workshop, HSCC 2006, Santa Barbara, CA, USA, March 29-31, 2006. Proceedings*, J. Hespanha and A. Tiwari (Eds.), Lect. Notes Compt. Sci. **3927**, Springer, 2006.

## Other Recent Work on Hybrid Systems

**1997** *Hybrid and Real-Time Systems: International Workshop, HART'97 Grenoble, France, March 26-28, 1997 Proceedings*, O.Maler (Ed.), Lect. Notes Compt. Sci. **1201**, Springer, 1997.

**1999** *An Introduction to Hybrid Dynamical Systems*, A. J. Van Der Schaft and J. M. Schumacher (Eds.), Lect. Notes Control Inf. Sci. **251**, Springer, 1999.

**2000** A. Matveev and A. Savkin, *Qualitative Theory of Hybrid Dynamical Systems*, Boston, Birkhauser, 2000.

**2002** *Modelling, Analysis, and Design of Hybrid Systems*, S. Engell, G. Frehse, and E. Schnieder (Eds.), Lect. Notes Control Inf. Sci. **279**, Springer, 2002.

**2003** *Proceedings of the 15th IFAC World Congress on the International Federation of Automatic Control: Hybrid Systems*, E. F. Camacho, L. Basenez, and J. A. De la Puenta (Eds.), Elsevier, 2003.

**2005** *Control of Nonlinear and Hybrid Process Systems: Designs for Uncertainty, Constraints and Time-Delays*, P. D. Christofides and N. H. El-Farra (Eds.), Lect. Notes Control Inf. Sci. **324**, Springer, 2005.

**2006** *Impulsive and Hybrid Dynamical Systems: Stability, Dissipativity, and Control*, W. Haddad, V. Chellaboina, and S. Nersesov (Eds.), Princeton Series Appl. Math., Princeton Univ. Press, 2006.

## Review Papers

**1998** *Special issue on hybrid control systems*, P. J. Antsaklis and A. Nerode (Eds.), IEEE Trans. Autom. Control **43** (1998), no. 4, 453–587.

**2000** J. M. Davoren and A. Nerode, *Logics for hybrid systems*, Proceedings of the IEEE **88** (2000), no. 7, 985–1010.

## Kohn-Nerode/Hybrid Systems Theory

**1993** W. Kohn and A. Nerode, *Models for hybrid systems: automata, topologies, controllability and observability*, In: *Hybrid systems*, Lect. Notes Compt. Sci. **736**, Springer, 1993, pp. 317–356.

**1996** W. Kohn, A. Nerode, and J. Remmel, *Continualization: A hybrid systems control technique for computing*, In: Proceedings of CESA'96, IMACS Multiconference, Vol 2, 1996, pp. 507–511.

**1996** W. Kohn, A. Nerode, and J. Remmel, *Feedback derivations: Near optimal controls for hybrid systems*, In: Proceedings of CESA'96, IMACS Multiconference, Vol 2, 1996, pp. 517–521.

**1997** W. Kohn and J. Remmel, *Hybrid dynamic programming*, In: *Hybrid and Real Time Systems Hybrid and Real-Time Systems: International Workshop, HART'97 Grenoble, France, March*

*26-28, 1997 Proceedings*, Lect. Notes Compt. Sci. **1201**, Springer, 1997, pp. 391–396.

**1997**  W. Kohn, A. Nerode, and J. Remmel, *Digital to hybrid program transformations*, In: Proceedings of the 1997 IEEE International Symposium on Intelligent Control, pp. 342–347.

**1997**  W. Kohn, A. Nerode, and J. Remmel, *Automaton comparison procedure for verification of hybrid systems*, In: Proceedings of the 5th IEEE Mediterranean Conference on Control and Systems.

**1997**  W. Kohn, A. Nerode, and J. Remmel, *Agent based velocity control of highway systems*, In: *Hybrid Systems IV*, Lect. Notes Compt. Sci. **1273**, Springer, 1997, pp. 174–214.

**1999**  W. Kohn, A. Nerode, and J. Remmel, *Scalable data and sensor fusion via multiple agent hybrid systems*, In: *Hybrid Systems V*, Lect. Notes Compt. Sci. **1567**, Springer, 1999, pp. 122–140.

**2000**  W. Kohn, V. Brayman, and J. Ritcey, *Enterprise dynamics via non-equilibrium membrane models*, Open Syst. Inf. Dyn. **7** (2000), no. 4, 327–348.

**2002**  W. Kohn and V. Brayman, *Automated sales and supply control of enterprise systems via agent cluster networks*, In: Proceedings of the International Conference on Internet Computing, IC'2002, Las Vegas, Nevada, USA, June 24-27, 2002. CSREA Press, 2001, pp. 713–718.

**2002**  W. Kohn, V. Brayman, and A. Nerode, *Control synthesis in hybrid systems with finsler dynamics*, Houston J. Math. **28** (2002), no. 2, 353–375.

**2003**  W. Kohn, V. Brayman, P. Cholewinski, and A. Nerode, *Control in hybrid systems*, International J. Hybrid Systems **3** (2003), no. 1&2, 109–150.

# Region-Based Theory of Space: Algebras of Regions, Representation Theory, and Logics

## Dimiter Vakarelov

*Sofia University*
*Sofia, Bulgaria*

## Preface

In this paper, we present recent results in the region-based theory of space that concern algebras of regions, the corresponding topological and discrete models, and representation theory. We also discuss applications to Qualitative Spatial Reasoning (QSR), an actively developing branch of AI and Knowledge Representation (KR). In particular, we show how new results in some practically motivated areas of QSR and KR can be obtained by combining methods from such established classical disciplines as Boolean algebras, topology and logic.

Mathematical Problems from Applied Logic. Logics for the XXIst Century. II. Edited by
Dov M. Gabbay *et al.* / International Mathematical Series, 5, Springer, 2007

The paper is organized as follows. Section 1 is a historical excursion into the region-based theory of space. We discuss the "pointless approach" to this theory, whose roots can be found in some philosophical ideas of de Laguna [13] and Whitehead [64]. We show connections of region-based theory of space with mereology (the theory of *part–whole relations*) and applications to QSR. In Section 2, we consider algebras of regions known as contact algebras. We study topological and discrete point-based models of contact algebras, and discuss different definitions of a point depending on the choice of axioms. We also consider representation theorems establishing a correspondence between the chosen axiomatizations and the required point-based models. In Section 3, we deal with a class of spatial logics. Some of them are related to the well-known system of Region Connection Calculus (RCC). In that section, we obtain completeness and decidability results by using representation theorems.

Some of the most important statements and new results are supplied with brief proofs. Standard definitions and facts from Boolean algebra can be found in [52], from topology in [24], from proximity spaces in [42], and from modal logic in [7, 8].

## 1. Historical Excursion into the Region-Based Theory of Space

One of the oldest theories of space is classical Euclidean geometry. It can be regarded as a *point-based* theory in the sense that the notion of a point is basic, whereas all other geometrical figures are defined as sets of points. The same can be said about topology considered as a more abstract kind of geometry. In general, by a point we mean the simplest spatial entity without dimension and internal structure. However, this notion is too abstract to have an adequate analog in reality, in contrast to many geometrical figures for which we can find their images in nature. The following idea then arises: to develop an alternative theory of space where the basic notion is not a point, but some other objects that are more closely related to the real world, for example, solid bodies. As basic relations between solids we could take, for instance, "one solid is part of another solid," "two solids overlap," or "one solid touches another solid," etc. This point of view is close to the ideas of some abstract philosophical disciplines such as *ontology*, the theory of "Existent," and especially *mereology* understood as a theory of "part-whole" relations. One of the founders of mereology was Leśnewski [**38**], who developed it as part of an ambitious and nonorthodox programme of constructing new foundations of mathematics. But due to Tarski [**58**], the mathematical content of mereology can be clearly presented in terms of complete Boolean algebras (cf. also [**33**] for such a presentation, and [**53**] for some other systems of mereology). The only difference between mereology and complete Boolean algebras is that Boolean algebras have an analog of the empty set (zero element), whereas mereology excludes such a zero individual.

The pointless approach does not mean that points are not considered at all. The notion of a point is necessary for a pointless theory of space to be equivalent in some sense to the classical point-based theory. But, in this case, points must be defined in terms of new primitive notions. This idea, as well as the necessity to use mereology for constructing a pointless theory of space, was expressed in the philosophical paper *"Point, line and surface as sets of solids"* [**13**] by de Laguna in 1922 and in the famous book *"Process and Reality"* [**64**] by Whitehead in 1929.

De Laguna considered a ternary relation between solids, *"x connects y with z,"* and defined a point, a line, and a surface via certain collections of solids. Whitehead developed this idea and simplified the ternary connection relation to the following binary relation: *"x is connected with y,"* which he called the *connection relation*. Here we use the term *contact relation*. Whitehead called solids *regions*, which later gave the name *region-based theory of space*.

As a primitive relation Whitehead took the contact relation between regions. He also introduced mereological relations such as *part-of*, *overlap* and some new relations called *external connection*, *tangential inclusion*, and *nontangential inclusion*. From the intuitive point of view, two regions are *in contact* if they have a common point. However, according to Whitehead, this property cannot be taken as a definition because a point is not defined and points must be defined by means of regions and the contact relation.

In [64] Whitehead listed explicitly a large number of assumptions and definitions about regions and the contact relation, and illustrated some of them by pictures. He did not make any attempt to reduce the number of his assumptions to a logical minimum. To define the notion of point, he introduced quite complicated notions of *geometrical element* and the relation of *incidence* between geometrical elements (see Definitions 13 and 15 in [64]). Then the definition of a point (Definition 16) sounds as follows: "A geometrical element is called a *point* when there is no geometrical element incident with it." Whitehead pointed out an analogy of his definition with the first definition of Euclid's *Elements*: "A point is that of which there is no part." This analogy shows that some mereological foundations of "pointless" geometry have their roots even in the old Euclid's Elements. Whitehead's final goal was to approach the Euclidean notions of a *straight line* and of *plane* in a similar way. Note that Whitehead's pointless theory of space is quite vague, and it is still a problem to extract a readable axiomatization and present it in a standard mathematical format (we refer the reader to the nice survey of pointless geometry by Gerla [31]). However, the idea to define points via regions is quite remarkable. Something similar can be found in Boolean algebras which can be considered as pointless analogs of sets. In Stone's representation theory of Boolean algebras [55] (1937) points in a given Boolean algebra are identified with ultrafilters, sets of elements of the algebra. So de Laguna–Whitehead's ideas of pointless approach to the theory of space could be regarded as early predecessors of the representation theory of Boolean algebras.

For further references we summarize here some formal properties of the contact relation and some other Whitehead's spatial relations between regions. We write $aCb$ for "region $a$ is in a contact with region $b$."

(W1) $(\forall a)(aCa)$,

(W2) $(\forall a, b)(aCb \to bCa)$,

(W3) $a = b$ if, and only if, $(\forall c)(aCc \leftrightarrow bCc)$,

(W4) $a$ is included in $b$ $(a \leqslant b)$ if, and only if, $(\forall c)(aCc \to bCc)$,

(W5) $a$ and $b$ overlaps $(aOb)$ if, and only if, $(\exists c)(c \leqslant a$ and $c \leqslant b)$,

(W6) $a$ is externally connected with $b$ $(aC^{\text{ext}}b)$ if, and only if, $aCb$ and not $aOb$,

(W7) $a$ is tangentially included in $b$ $(a \leqslant^\circ b)$ if, and only if, $a \leqslant b$ and $(\exists c)(cC^{\text{ext}}a$ and $cC^{\text{ext}}b)$,

(W8) $a$ is non-tangentially included in $b$ $(a \ll b)$ if, and only if, $a \leqslant b$ and not $a \leqslant^\circ b$.

Axiom (W3), known as the *axiom of extensionality* of contact, is very important. It can be proved that it is equivalent to axiom (W4), which says that part-of relation in Whitehead's system is definable by means of contact.

Another, much simpler, pointless reconstruction of Euclidean geometry was given by Tarski [57] in 1927. He called his system *Geometry of solids*. Geometry of solids is an extension of Leśnewski's mereology with the primitive notion of sphere. To define points, Tarski first introduced the relation of two spheres being concentric, and then points were identified with certain sets of concentric spheres. A simplified version of Tarski's system can be found in [4], where similar approaches are also discussed.

Another attempt to build a pointless theory of space was made by Grzegorczyk [34] in 1960. Independently from de Laguna [13] and Whitehead [64], Grzegorczyk developed a system that was close to Whitehead's system.

As primitives he took the relations of part-of and separation, which, in fact, is the complement of the Whitehead contact relation. Grzegorczyk's results were presented in [6], where the notion of contact was used instead of separation. According to [6], Grze-gorczyk's pointless geometry $(R, \leqslant, C)$ is given by the following axioms:

(G0) $(R, \leqslant)$ is a *mereological field*, i.e., a complete Boolean al-gebra with deleted zero element.

(G1) $C$ is a reflexive relation in $R$,

(G2) $C$ is a symmetric relation in $R$,

(G3) $C$ is monotone with respect to $\leqslant$ in the sense that we have: $a \leqslant b \rightarrow (\forall c \in R)(aCc \rightarrow bCc)$.

Then the relation of non-tangential inclusion $\ll$ is defined in the same way as by Whitehead (see axiom (W8) above). A set $p$ of re-gions is called a *representative of a point* if the following conditions are satisfied:

(1) $p$ has no minimum and is totally ordered by the relation $\ll$,

(2) given two regions $u$ and $v$, if we have $uOc$ and $vOc$, for every $c \in p$, then $uCv$.

A filter $P$ in $R$ is called a *point* if it is generated by a repre-sentative of a point. We say that $P$ *belongs to a region* $a$ if $a$ is a member of $P$.

Then two additional axioms are introduced:

(G4) every region has at least one point,

(G5) if $aCb$ then there is a point $P$ such that $a$ and $b$ overlap with every member of $P$.

Denote by $\mathbb{P}$ the set of all points of $(R, \leqslant, C)$ and by $\pi(r)$ the set of all points of a region $r$.

Grzegorczyk proved the following two important theorems.

**Theorem 1.** *Let $(X, \tau)$ be a Hausdorff topological space, and let $R$ be a family of nonempty regular open sets of $(X, \tau)$. For any $a, b \in R$, we set $aCb$ if, and only if, $\mathrm{Cl}(a) \cap \mathrm{Cl}(b) \neq \varnothing$. Then $(R, \subseteq, C)$ satisfies (G0)–(G3). If every point of $X$ is the* *intersection of a decreasing (with respect to $\ll$) family of open sets, then axioms* (G4) *and* (G5) *are also satisfied.*

**Theorem 2.** *Suppose that* $(R, \leqslant, C)$ *satisfies* (G0)–(G5). *Let* $\tau$ *be a topology in* $\mathbb{P}$ *generated by the set* $\{\pi(r) : r \in R\}$. *Then* $\{\pi(r) : r \in R\}$ *coincides with the set of all nonempty regular open sets of* $(\mathbb{P}, \tau)$, *and* $\pi$ *is an isomorphism.*

As was noted in [6], the implication in axiom (G3) can be replaced with equivalence, which eliminates the part-of relation from the primitives. This means that, as in the case of Whitehead, the system can be based on the unique primitive $C$.

Theorems 1 and 2 show that there is an equivalence between the point-based and pointless theories of space. Theorem 1 also shows the importance of regular (open or closed) sets in topological spaces as models of regions. In fact, Theorem 2 is the first representation theorem of a special system of region-based theory of space which is an extension of mereology with the primitive of Whitehead's contact relation. Since the models of such extended mereologies are topological, some authors prefer to call them *mereotopologies* or *region-based topologies*.

An interesting comparison between the notions of a point used by Whitehead [64] and Grzegorcyk [34] was given by Biacino and Gerla in [6]. They proved that these definitions are equivalent in some sense if the relation of non-tangential inclusion $\ll$ satisfies the following additional axiom:

(G6)    if $a \ll b$, then $a \ll c \ll b$ for some region $c \in R$.

Using the complement $a^*$, we can equivalently express axiom (G6) in terms of $C$:

(G6′)    if $a\overline{C}b$, then $a\overline{C}c$ and $c^*\overline{C}b$ for some $c \in R$.

This axiom is referred to as the *normality axiom*, since it is satisfied by regular open (closed) sets in a Hausdorff space provided that the space is normal.

One unpleasant feature of Grzegorczyk's system is that it includes axioms containing the second-order definition of a point and, consequently, it is not a first-order system.

It is of interest to note that by accepting the normality axiom (G6) one can obtain first-order axiomatizations of pointless theory of space. This was done independently by several authors: [**63, 62, 61, 15**]. The first to do this was de Vries [**63**] (1962) in his thesis "*Compact Spaces and Compactifications*." This work, independent from Whitehead [**64**] and Grzegorczyk [**34**], was completely unknown to the community of authors interested in the region-based theory of space. Thus, de Vries is mentioned neither in Gerla's survey of pointless geometry [**31**], nor in later papers on region-based theory of space.

Note that axiom (G6) is well known among specialists in the theory of Proximity spaces. Proximity spaces are abstract spaces [**42**] with the proximity relation $A\delta B$ between subsets satisfying almost all axioms for the contact relation $C$. They can also be axiomatized using the relation $A \ll B$ definable by $\delta$ in the same way as $\ll$ is definable by $C$. By analogy with the axioms of proximity spaces based on the relation $\ll$, de Vries considered Boolean algebras $(B, 0, 1, ., +, *, \ll)$ with the additional relation $\ll$, called *compingent algebras*, which satisfy the following first-order axioms:

(P0)  $(B, 0, 1, ., +, *)$ is a Boolean algebra with $*$ as the Boolean complement,

(P1)  $0 \ll 0$,

(P2)  $a \ll b$ implies $a \leqslant b$,

(P3)  $a \leqslant a' \ll b$ implies $a \ll b$,

(P4)  $a \ll b$ and $c \ll d$ imply $a.c \ll b.d$,

(P5)  $a \ll b$ implies $b^* \ll a^*$,

(P6)  $a \ll b \neq 0$ implies $\exists c \neq 0$ with $a \ll c \ll b$.

Note that axiom (P6) can be replaced by two axioms:

(P6′)  $a \ll b$ implies $\exists c$ with $a \ll c \ll b$ (which is just the normality axiom), and

(P7)  if $b \neq 0$ then $\exists a \neq 0$ with $a \ll b$.

Observe that axioms (P1)–(P5), (P6′) are algebraic analogs of the axioms of Efremovič's proximity spaces [25] (cf. also [42]). We will see in Section 2 that axiom (P7) is equivalent to Whitehead's extensionality axiom for the contact relation.

Using the well-known techniques from the proximity spaces and Smirnov's theory of compactifications, de Vries proved that each compingent algebra is isomorphic to a subalgebra of the algebra of regular open sets of a compact Hausdorff space with the compingent relation on regular open sets defined as follows: $a \ll b$ if, and only if, $\mathrm{Cl}\,(a) \subseteq b$. The points defined by de Vries, called *compingent filters*, are just lattice analogs of the *ends*, special filters used in proximity theory. In fact, de Vries established a one-to-one correspondence between complete compingent algebras and compact Hausdorff spaces. Similar results were obtained also by Fedorčuk [26].

Another, more general than de Vries–Fedorčuk's, first-order axiomatization of a region-based theory of space was given by Roeper [49] in 1997. His theory corresponds to the point-based theory of locally compact Hausdorff spaces, and his approach is a skillful combination of de Vries–Fedorčuk's methods and Leader's compactification theory of local proximity (cf. [37], [42]). Roeper's axiomatization is based, like Leader's notion of local proximity, on two primitive spatial relations: the contact and the unary relation of *limitedness*. An attempt to give a different formulation of the same theory using only one primitive relation, called *interior parthood*, was made by Mormann [41] (see also [61]).

We continue our historical excursion into the region-based theory of space by mentioning the contribution made by Clarke [10, 11]. Clarke noted that his system should be understood as a formalization of the ideas of Whitehead [64]. Clarke's system $(R, C)$ is based on a unique primitive relation $C$ of contact satisfying Whitehead's axioms and definitions (W1)–(W8). Clarke assumed also the so-called *fusion axiom*:

If $A$ is a nonempty subset of $R$, then there exists $a \in R$ (called a *fusion* of $A$) such that $C(a) = \bigcup\{C(x) : x \in A\}$, where $C(x) = \{y \in R : xCy\}$.

Points in Clarke's system are identified with certain subsets $P$ of $R$ satisfying some closure conditions. He needed also the following axiom, containing a definable notion of point:

If $aCb$, then there exists a point $P$ such that $a, b \in P$.

Biacino and Gerla [5] studied this system in detail and proved that $(R, C)$ is equivalent to a complete Boolean algebra with zero element removed (mereological field). It follows from this fact that the contact $C$ coincides with the overlap $O$, which is not satisfactory. Another unsatisfactory feature is that the system has an axiom containing the second-order notion of a point and, consequently it is not a first-order one. Nevertheless, Clarke's system had a remarkable impact on some research areas in AI for which the pointless approach to the theory of space was important. One such area is the so-called *Qualitative Spatial Reasoning* (QSR). It is related to a new generation of information systems dealing with geographical information and known as Geographical Information Systems (GIS). It has been recognized that reasoning techniques in GIS using *quantitative* methods of classical theory of space are not efficient and tractable. This motivated researchers in these areas to look for new, *qualitative* models of space. Similar problems have appeared in robotics, computer vision, natural language semantics related to a commonsense spatial vocabulary, etc. Models of space based on mereology proved to fit well into the problems of QSR, and this made region-based theory of space important for AI and computer science (see [48]). Several attempts to build systems similar to that of Clarke have been made within the QSR community. One of the most important and popular systems is *Region Connection Calculus* (RCC), proposed by Randel, Cui and Cohn [48] in 1992. Now RCC is in the center of an intensive research in the realm of QSR, and one of the most active is Cohn's group at the University of Leeds. A comprehensive overview of the QSR research and related work was given by Cohn and Hazarika [12] (2001). Recent collections of papers on QSR are the special issues of Fundamenta Informaticae (2001) edited by I. Düntsch [17] and the Journal of Applied Non-Classical Logics (2002) edited by Balbiani [1].

Stell [54] and Düntsch et al. [20] presented an equivalent version of RCC based on Boolean algebras satisfying all axioms for contact given by Whitehead plus an additional axiom of *connectedness* forcing topological models to be connected spaces. So connected regular spaces form a correct semantics for RCC. A representation theorem for RCC in a class of more general spaces, called *weakly regular*, was proved by Dünch and Winter [21] in 2005. A representation theorem for a variety of related systems was proved in [15] (2006).

The main point-based models of the region-based theory of space considered in QSR are the contact algebras of regular open or regular closed sets in certain topological spaces. Since topology aims to formalize some *continuous, indiscrete* features of space we may call this kind of models continuous or indiscrete. More special models of regions generated by *polygonal regions* were considered by Pratt and Schoop [46, 47]. It has been pointed out by several authors that continuous models are not so convenient in computer modelling of space, and a modified and generalized region-based theory of space, admitting discrete models, is required. One solution was proposed by Galton [29, 30]. Instead of topological spaces, Galton proposes to consider the so-called *adjacency spaces*. An adjacency space is a relational system of the form $(W, R)$, where $W$ is a nonempty universe whose elements are called *cells* and $R$ is a binary relation between cells, called an adjacency relation. Galton defines regions to be arbitrary sets of cells, and the contact relation between regions is defined by taking $aCb$ if, and only if, $\exists x \in a, \exists y \in b$ with $xRy$. This definition relates Galton's adjacency spaces to the Kripke semantics of modal logic [7] which makes it possible to use methods from modal logic for studying discrete region-based theories of space [3]. Pointless formulations of Galton's theory of discrete spaces and the corresponding representation theory was given in [19]. It was shown in [14] that the algebras corresponding to discrete spaces have also standard topological representations in which regions are represented by regular closed or open sets. In this way both kinds of models of region-based theory of space—discrete and indiscrete—can be considered in a unified way. This unified approach is presented in more detail in Section 2.

To conclude this historical excursion, we mention that, in the realm of QSR, different kinds of logical systems for reasoning about space have been developed and their computational properties have been studied. Some authors advocated logical systems based on first-order languages (cf., for example, [43, 45]). One of the practical motivations for dealing with first-order systems of region-based theories of space is that this makes it possible to employ first-order provers for some applications. Using some results of Grzegorczyk [32], one can show, however, that most of these systems are undecidable. That is why weaker, quantifier-free systems with better computational properties have been designed. Examples are the system RCC-8 introduced by Egenhofer and Franzosa [23] and its extension with Boolean terms introduced by Wolter and Zakharyaschev [65]. Completeness theorems and decidability results for these and other related to RCC quantifier-free systems with respect to their topological and discrete semantics are given in [3]. For more information on these logics see Section 3 below. A decidable system with predicates of component-counting was presented by Pratt-Hartmann [44]. Dynamic Logics for discrete region-based theory of space have been studied in [2]. Modal logics with Kripke frames based on the RCC-8 relations have been introduced by Lutz and Wolter [40]. For various combinations of spatial and temporal logics see Gabelaia et al. [28] and Konchakov et al. [36].

This section does not cover all aspects of the region-based theory of space. We have only concentrated on pointless approaches similar to those of de Laguna and Whitehead. Of course, this is not the only way to look at the region-based theory of space: an alternative one is described, for example, by Pratt-Hartmann [43, 45]. Another alternative is given by Schoop [50] who motivates the idea of taking both regions and points as primitives. We hope that the survey above presents the region-based theory of space as an active and developing area. Started from some very abstract philosophical ideas of de Laguna and Whitehead, it has reached its flourishing stage, with a clear mathematical theory and multiple applications in practically oriented areas of QSR, GIS and KR.

## 2. Algebras of Regions, Models, and Representation Theory

### 2.1. Contact algebras

Following [15], by a *contact algebra* we mean any system $\underline{B} = (B, C) = (B, 0, 1, ., +, *, C)$, where $(B, 0, 1, ., +, *)$ is a nondegenerate Boolean algebra, $*$ denotes the complement, and $C$ is a binary relation in $B$, called a *contact*, such that

(C1)   if $xCy$, then $x, y \neq 0$,

(C2)   $xC(y + z)$ if and only if $xCy$ or $xCz$,

(C3)   if $xCy$, then $yCx$,

(C4)   if $x.y \neq 0$, then $xCy$.

Elements of $B$ are called *regions*. The negation of $C$ is denoted by $\overline{C}$. The relation $\ll$ of *nontangential inclusion* is defined as follows: $x \ll y$ if and only if $x\overline{C}y^*$. We say that $\underline{B}$ is *complete* if $B$ is complete.

Axiom (C2) implies the monotonicity of $C$ with respect to $\leqslant$:

(Mono)   if $aCb$ and $a \leqslant a'$ and $b \leqslant b'$, then $a'Cb'$.

A contact algebra can be equivalently defined in terms of $\ll$ (cf. the axioms of de Vries in Section 1):

($\ll 1$)   $1 \ll 1$,

($\ll 2$)   if $x \ll y$, then $x \leqslant y$,

($\ll$ 3)    if $x \leqslant y \ll z \leqslant t$, then $x \ll t$,

($\ll$ 4)    if $x \ll y$, then $y^* \ll x^*$,

($\ll$ 5)    if $x \ll y$ and $x \ll z$, then $x \ll y.z$.

Axioms (C1)–(C4) are Boolean versions of the axioms of basic proximity spaces (known as Čech proximity spaces, cf. [**9, 56**]). Note that the main intended models of contact algebras are not basic proximity spaces, but some other models of topological nature that can be constructed in the following way.

**Example 2.1.1.** (1) *Contact algebra of regular closed sets.* Let $(X, \tau)$ be a topological space with closure $\text{Cl}(a)$ and interior operations $\text{Int}(a)$. A subset $a$ of $X$ is *regular closed* if $a = \text{Cl}(\text{Int}(a))$. The set of all regular closed subsets of $(X, \tau)$ is denoted by $\text{RC}(X, \tau)$ or $\text{RC}(X)$. As is known, the regular closed sets with operations $a + b = a \cup b$, $a.b = \text{Cl}(\text{Int}(a \cap b))$, $a^* = \text{Cl}(X \setminus a) = \text{Cl}(-a)$, $0 = \varnothing$, and $1 = X$ form a Boolean algebra. Moreover, if we consider the infinite join operation $\sum_{i \in I} a_i = \text{Cl}(\bigcup_{i \in I} a_i)$, then the Boolean algebra $\text{RC}(X)$ is complete. The contact is defined as follows: $a \, C_X \, b$ if and only if $a \cap b \neq \varnothing$. It satisfies axioms (C1)–(C4). This contact is called the *standard contact for regular closed sets* and the corresponding contact algebra is called the *standard contact algebra of regular closed sets*. The nontangential inclusion is defined as follows: $a \ll b$ if and only if $a \subseteq \text{Int}(b)$.

(2) *Contact algebra of regular open sets.* A subset $a$ of $(X, \tau)$ such that $a = \text{Int}(\text{Cl}(a))$ is called a *regular open set*. The set of all regular open subsets of $(X, \tau)$ is denoted by $\text{RO}(X, \tau)$ or $\text{RO}(X)$. The Boolean operations and contact in $\text{RO}(X)$ are defined as follows: $a + b = \text{Int}(\text{Cl}(a \cup b))$, $a.b = a \cap b$, $a^* = \text{Int}(X \setminus a) = \text{Int}(-a)$, $0 = \varnothing$, $1 = X$, and $a C_X b$ if and only if $\text{Cl}(a) \cap \text{Cl}(b) \neq \varnothing$ (consequently, $a \ll b$ if and only if $\text{Cl}(a) \subseteq b$). Then $(\text{RO}(X), C_X)$ is a contact algebra and it is complete relative to the infinite meet $\prod_{i \in I} a_i = \text{Int}(\bigcap_{i \in I} a_i)$. In this case, $C_X$ is called the *standard contact for regular open sets* and the corresponding contact algebra is called the *standard contact algebra of regular open sets*.

Note that $(\mathrm{RO}(X), C_X)$ and $(\mathrm{RC}(X), C_X)$ are isomorphic contact algebras. The corresponding isomorphism $f$ is defined as $f(a) = \mathrm{Cl}(a)$ for every $a \in \mathrm{RO}(X)$. This fact explains, why we will consider only models with regular closed sets.

In Section 2.5, we will establish the existence of topological models of contact algebras related to proximity spaces, where elements of the algebra are regular closed (open) sets, but the contact is not standard, unlike these examples.

Note that the Boolean part in the definition of a contact algebra incorporates the mereological component of the notion. Although the zero element is not traditionally accepted in mereology, we consider the zero element, which makes the definition more suitable for our considerations.

For a Boolean algebra we introduce the following basic mereological relations between regions:

*part-of relation*     $a \leqslant b$ is the lattice ordering of $B$,

*overlap*        $aOb$ if and only if $a.b \neq 0$.

This definition of an overlap agrees with that introduced by Whitehead: $\exists c \in B \smallsetminus \{0\} : c \leqslant a$ and $c \leqslant b$. Indeed, it suffices to take $c = a.b \neq 0$.

Another mereological relation is the following:

*dual overlap*    $a\check{O}b$ if and only if $a^*Ob^*$

or, equivalently:

   $a\check{O}b$ if and only if $a + b \neq 1$.

It is natural to find a general definition of a "mereological relation" and one possibility to do this is to identify them with all Boolean relations definable by open formulas in the first-order theory of Boolean algebras. Having such a definition, we can obtain finitely many mereological relations of given arity, so that for $n = 2$ there are exactly 30 such relations and each of them can be defined by an open first-order formula in terms of $\leqslant$, $O$, and $\check{O}$.

Using the notions of contact, overlap, and nontangential inclusion, it is possible to introduce the so-called RCC-8 basic mereotopological relations between two nonzero regions:

RCC $-8$ *relations*

- disconnected DC$(a, b)$: $a\overline{C}b$,
- external contact EC$(a, b)$: $aCb$ and $a\overline{O}b$,
- partial overlap PO$(a, b)$: $aOb$ and $a \not\leqslant b$ and $b \not\leqslant a$,
- tangential proper part TPP$(a, b)$: $a \leqslant b$ and $a \not\ll b$ and $b \not\leqslant a$,
- tangential proper part$^{-1}$ TPP$^{-1}(a, b)$: $b \leqslant a$ and $b \not\ll a$ and $a \not\leqslant b$,
- nontangential proper part NTPP$(a, b)$: $a \ll b$ and $a \neq b$,
- nontangential proper part$^{-1}$ NTPP$^{-1}(a, b)$: $b \ll a$ and $a \neq b$,
- equal EQ$(a, b)$: $a = b$.

It is easy to see that these relations are pairwise disjoint and exhaustive. Pure topological definitions, introduced by Egenhofer and Franzosa [23] and sometimes referred to as *Egenhofer–Franzosa relations*, were studied by many authors (cf. Wolter and Zakharyaschev [65] for complexity and Lutz and Wolter [40] for more references).

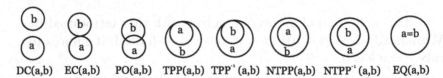

DC(a,b)    EC(a,b)    PO(a,b)    TPP(a,b)   TPP⁻¹(a,b)   NTPP(a,b)   NTPP⁻¹(a,b)   EQ(a,b)

FIGURE 1

In the language of contact algebras, we can define some other mereotopological relations, for example, the one-place predicate Con$(a)$: "the region $a$ is *connected* or $a$ is a *one-piece region*" which is formally expressed as follows:

Con$(a)$ if and only if $(\forall b, c)(b \neq 0$ and $c \neq 0$ and $b + c = a \rightarrow bCc)$.

In the case of Con$(1)$, the contact algebra is said to be *connected*. The negation of Con$(a)$ is denoted by $\overline{\mathrm{Con}}(a)$. From an intuitive point of view, $\overline{Con}(a)$ says that the region $a$ is the sum of at least two disconnected nonzero regions. We can consider a more general predicate by assuming that $c^{\geqslant n}(a)$ is the sum of $n$ pairwise disconnected nonzero regions $b_1, \ldots, b_n$ or, formally:

$c^{\geqslant n}(a)$ if and only if $(\exists b_1 \ldots b_n)(a = b_1 + \ldots + b_n$ and $(\forall i = 1 \ldots n)(b_i \neq 0)$ and $(\forall i \neq j, i, j = 1, \ldots, n)(b_i \overline{C} b_j))$.

It is obvious that $\overline{Con}(a)$ is equivalent to $c^{\geqslant 2}(a)$. The computational complexity of $c^{\geqslant n}$, called the *component counting*, and part-of relation is studied by Pratt-Hartmann [43].

Another interesting mereotopological relation considered by Gabelaia et al [28] is the following $n$-ary contact $C_n(a_1, \ldots, a_n)$ with the standard meaning in the contact algebra of regular closed sets:

$C_n(a_1, \ldots, a_n)$ if and only if $a_1 \cap \ldots \cap a_n \neq \varnothing$.

We do not know whether this relation is definable in the language of contact algebras by a first-order formula. In Section 2.3, we will give a definition using a second-order formula.

## 2.2. Extensions of contact algebras by adding new axioms

Consider contact algebras satisfying some of the following axioms:

(Con)  if $a \neq 0$ and $a \neq 1$, then $aCa^*$        *connectedness*

(Ext)  if $a \neq 1$, then $\exists b \neq 0$ such that $a\overline{C}b$      *extensionality*

(Nor)  if $a \ll b$, then $\exists c$ such that $a \ll c \ll b$      *normality*

A contact algebra satisfying axiom (Con) ((Ext) or (Nor)) is said to be *connected* (*extensional* or *normal*).

Contact algebras satisfying axioms (Con) and (Ext) were introduced by Stell in [54] under the name *Boolean contact algebras* and were considered as an equivalent formulation of the system

RCC [48]. Stell proved that (Ext) is equivalent (under axioms (C1)–(C4)) to each of the following axioms:

(Ext′)   $a \leqslant b$ if and only if $(\forall c \in B)(aCc \rightarrow bCc)$,

(Ext″)   $a = b$ if and only if $(\forall c \in B)(aCc \rightarrow bCc)$,

(Ext‴)   $(\forall b \neq 0)(\exists a \neq 0)(a \ll b)$.

Note that (Ext′) is just Whitehead's definition of the part-of relation and (Ext″) is Whitehead's axiom of extensionality.

Contact algebras satisfying (Nor) and (Ext) were first studied by de Vries [63] and Fedorčuk [26]. Independently, such algebras were introduced in [62, 61], where the authors noted the connection with proximity theory and the possibility to use proximity theory for proving topological and proximity representation theorems for contact algebras.

We recall some topological notions.

A topological space $X$ is said to be
- *semiregular* if it has a base $\mathbb{B}$ of regular closed sets; namely, every closed set is the intersection of elements of $\mathbb{B}$,
- *normal* if every pair of closed disjoint sets can be separated by a pair of open sets,
- *κ-normal* (cf. [51]) if every pair of regular closed disjoint sets can be separated by a pair of open sets,
- *extensional* if $RC(X)$ satisfies axiom (Ext),
- *weakly regular* (cf. [21]) if it is semiregular and for every nonempty open set $a$ there exits a nonempty open set $b$ such that $Cl(a) \subseteq b$,
- *connected* if it cannot be represented as the sum of two disjoint nonempty open sets,
- a $T_0$-*space* if for every two different points $x \neq y$ there exists an open set that contains one of them and does not contain the other,
- a $T_1$-*space* if every one-point set $\{x\}$ is a closed set,
- a *Hausdorff space* (or a $T_2$-*space*) if every two different points can be separated by a pair of disjoint open sets,

- a *compact space* if it satisfies the following condition: if $\{A_i : i \in I\}$ is a nonempty family of closed sets of $X$ such that for every finite subset $J \subseteq I$ we have $\bigcap\{A_i : i \in J\} \neq \varnothing$, then $\bigcap\{A_i : i \in I\} \neq \varnothing$.

**Lemma 2.2.1.** *The following assertions hold.*

(1) *Let $X$ be semiregular. Then $X$ is weakly regular if and only if $\mathrm{RC}(X)$ satisfies* (Ext) [21].

(2) *$X$ is $\varkappa$-normal if and only if $\mathrm{RC}(X)$ satisfies* (Nor) [21].

(3) *$X$ is connected if and only if $\mathrm{RC}(X)$ satisfies axiom* (Con) [5, 21].

(4) *If $X$ is a compact Hausdorff space, then $\mathrm{RO}(X)$ (consequently, $\mathrm{RC}(X)$) satisfies* (Ext) *and* (Nor) [63].

(5) *If $X$ is a normal Hausdorff space, then $\mathrm{RO}(X)$ satisfies* (Nor) [6].

Note that axiom (Con) is equivalent to the axiom

(Con′)   if $a \neq 0$, $b \neq 0$, and $a + b = 1$, then $aCb$.

Similarly, (Nor) is equivalent to the axiom

(Nor′)   if $a\overline{C}b$, then $(\exists a'b')(a\overline{C}a'$ and $b\overline{C}b'$ and $a' + b' = 1)$.

Below, we consider embedding theorems for contact algebras regarded as contact subalgebras of the contact algebras of regular closed sets in some topological spaces. It is important to know the conditions under which an algebra satisfies some of axioms (Con), (Ext), and (Nor) if and only if its subalgebra satisfies the same axioms.

A contact subalgebra $B_1$ of $B_2$ is said to be *dense* if

(Dense)   $(\forall a_2 \in B_2)(a_2 \neq 0 \rightarrow (\exists a_1 \in B_1)(a_1 \neq 0$ and $a_1 \leqslant a_2))$

and *co-dense* if

(Co-dense)   $(\forall a_2 \in B_2)(a_2 \neq 1 \rightarrow (\exists a_1 \in B_1)(a_1 \neq 1$ and $a_2 \leqslant a_1))$.

It is easy to see that (Dense) is equivalent to (Co-dense).

We say that $B_1$ is a $C$-*separable* subalgebra of $B_2$ if

($C$-separation)    $(\forall a_2 b_2 \in B_2)(a_2 \overline{C} b_2 \rightarrow (\exists a_1 b_1 \in B_1)(a_2 \leqslant a_1$
$$\text{and } b_2 \leqslant b_1 \text{ and } a_1 \overline{C} b_1)).$$

If $h$ is an embedding of $B_1$ regarded as a contact subalgebra of $B_2$, then $h$ is a *dense embedding* provided that $h(B_1)$ is a dense subalgebra of $B_2$. We say that $h$ is a $C$-*separable* if $h(B_1)$ is a $C$-separable subalgebra of $B_2$.

The following assertion is important.

**Theorem 2.2.2.** *Let $B_1$ be a $C$-separable contact subalgebra of $B_2$. Then the following assertions hold.*

(1) $B_1$ *satisfies* (Con) *if and only if $B_2$ satisfies* (Con).

(2) *Let $B_1$ be a dense subalgebra of $B_2$. Then $B_1$ satisfies* (Ext) *if and only if $B_2$ satisfies* (Ext).

(3) $B_1$ *satisfies* (Nor) *if and only if $B_2$ satisfies* (Nor).

PROOF. We prove assertion (3) taking (Nor′) instead of (Nor).

($\rightarrow$) Let $B_1$ satisfies (Nor′), and let $a_2 \overline{C} b_2$ for $a_2, b_2 \in B_2$. By ($C$-separation), there exist $a_1$ and $b_1$ in $B_1$ (consequently, in $B_2$) such that $a_2 \leqslant a_1$, $b_2 \leqslant b_1$, and $a_1 \overline{C} b_1$. By (Nor), there exist $a_1'$ and $b_1'$ in $B_1$ (consequently, in $B_2$) such that $a_1' + b_1' = 1$, $a_1 \overline{C} a_1'$, and $b_1 \overline{C} b_1'$. Since $C$ is monotone and symmetric, we have $a_2 \overline{C} a_1'$ and $b_2 \overline{C} b_1'$, which shows that $B_2$ satisfies (Nor′).

($\leftarrow$) Let $B_2$ satisfy (Nor′), and let $a_1 \overline{C} b_1$. for $a_1$ and $b_1$ in $B_1$ (consequently, in $B_2$). By (Nor′), there exist $a_2', b_2' \in B_2$ such that $a_2' + b_2' = 1$, $a_1 \overline{C} a_2'$, and $b_1 \overline{C} b_2'$. By ($C$-separation), if $a_1 \overline{C} a_2'$, then there exist $c_1, d_1 \in B_1$ such that $a_1 \leqslant c_1$, $a_2' \leqslant d_1$ and $c_1 \overline{C} d_1$. Similarly, by ($C$-separation), $b_1 \overline{C} b_2'$ implies that there exist $e_1, f_1 \in B_1$ such that $b_1 \leqslant e_1$, $b_2' \leqslant f_1,$, and $e_1 \overline{C} f_1$. Therefore, $d_1 + f_1 = 1$, $a_1 \overline{C} d_1$, and $b_1 \overline{C} f_1$, which shows that $B_1$ satisfies (Nor′).    $\square$

The following assertion is well known.

**Proposition 2.2.3** ([52]). *If h is a dense embedding of a Boolean algebra $B_1$ in a Boolean algebra $B_2$ and $B_1$ is complete, then h is a complete isomorphism of $B_1$ onto $B_2$.*

## 2.3. Points in contact algebras and topological representation theorems. A simple case

We begin by discussing how to define canonically points in contact algebras. Then we discuss how to introduce canonically a topology in the set of points. Finally, we show that regions in the algebra can be identified with regular closed sets in the topological space by an appropriate canonical isomorphism. This procedure is not unique. Choosing different axioms of contact algebra, we obtain different kinds of points and thereby different canonical constructions implying different kinds of topological spaces. This shows that the notion of a point is not unique and points of a more complicated structure can provide better topological spaces. We illustrate this fact by considering the simplest notion of a point. A more complicated notion of a point and the corresponding canonical constructions will be considered in Section 2.4. We mainly follow [15], However, the presented construction is new and leads to stronger results. Therefore, we give proofs.

Let $X$ be a topological space, and let $x \in X$ be a point. The set $P_x = \{a \in \mathrm{RC}(X) : x \in a\}$ satisfies the following conditions:

(1) $X \in P_x$,

(2) $a \cup b \in P_x$ if and only if $a \in P_x$ or $b \in P_x$.

(3) If $a, b \in P_x$, then $aCb$.

The set $P_x$ is a collection of regions. If the space is at least $\mathrm{T}_0$, then $x \neq y$ implies $P_x \neq P_y$. Another interesting property of $P_x$ is that if regions $a$ and $b$ are in a contact, then there exists $P_x$ such that $a, b \in P_x$. Thus, the sets $P_x$ react like points. This fact can be used to identify points with sets $P_x$. There are no points in contact algebras, but, instead of points, we can consider collections of regions satisfying (1)–(3). The situation is similar to that in the representation theory of Boolean algebras (cf. [55]), where abstract points in a Boolean algebra are associated with ultrafilters, collections of elements of the algebra. Sets satisfying (1)–(3) are similar to ultrafilters and were considered in the theory of proximity spaces, where they were called *clans* (cf. [56]). For contact algebras clans were used in [61, 21, 15]. A clan is defined as follows.

Let $\underline{B} = (B, C)$ be a contact algebra. A set $\Gamma \subseteq B$ of regions is called a *clan* (in $\underline{B}$) if it satisfies the following conditions:

(Clan 1)    $1 \in \Gamma$,

(Clan 2)    $a + b \in \Gamma$ if and only if $a \in \Gamma$ or $b \in \Gamma$,

(Clan 3)    If $a, b \in \Gamma$, then $aCb$.

Clans in $\mathrm{RC}(X)$ in the form $P_x$ are called *point clans*. A clan is said to be *maximal* if it is maximal with respect to inclusion. By the Zorn lemma, every clan is contained in a maximal clan. Denote by $\mathrm{CLANS}(\underline{B})$ ($\mathrm{MaxCLANS}(\underline{B})$) the set of all clans (maximal clans) in $\underline{B}$. For brevity, we write CLANS and MaxCLANS if a contact algebra $\underline{B}$ is fixed. Thus, we have two candidates for points: CLANS and MaxCLANS. In this section, we consider only CLANS.

We show how to construct a clan. First of all, note that every ultrafilter in $B$ satisfies (Clan 1) and (Clan 2) and also (Clan3) by (C4), which means that it is a clan. Another construction is as follows. For two filters $F$ and $G$ in $B$ we define: $F\rho G$ if and only if $F \times G \subseteq C$. It is easy to see that the relation $\rho$ is reflexive and symmetric. Let $\Sigma$ be a nonempty set of maximal filters of $B$ such that for any $F, G \in \Sigma$ we have $F\rho G$. Then the union of all elements of $\Sigma$ is a clan and every clan can be obtained by such a construction (cf. [15]).

The following assertion is a simple consequence of the Zorn lemma.

**Lemma 2.3.1** ([19, 15]). *The following assertions hold.*

(1) *If $F$ and $G$ are filters and $F\rho G$, then there exist maximal filters $F' \supseteq F$ and $G' \supseteq G$ such that $F'\rho G'$.*

(2) *$aCb$ if and only if there exist maximal filters $F$ and $G$ such that $F\rho G$, $a \in F$, and $b \in G$.*

The following assertion characterizes contacts and part-of in terms of clans.

**Lemma 2.3.2** ([15]). *The following assertions hold.*

(1) *$aCb$ if and only if $(\exists \Gamma \in \text{CLANS}(B))(a, b \in \Gamma)$.*

(2) *$a \leqslant b$ if and only if $(\forall \Gamma \in \text{CLANS}(B))(a \in \Gamma \to b \in \Gamma)$.*

(3) *$a = 1$ if and only if $(\forall \Gamma \in \text{CLANS}(B))(a \in \Gamma)$.*

We explain the idea of the proof of (1). If $aCb$, then for the filters $F' = \{a' : a \leqslant a'\}$ and $G' = \{b' : b \leqslant b\}$ we have $F'\rho G'$. By Lemma 2.3.1, $F'$ and $G'$. can be extended to maximal filters $F$ and $G$ such that $F\rho G$. Then the clan $\Gamma = F \cup G$ contains both $a$ and $b$. The converse implication follows from the properties of clans. Assertions (2) and (3) are proved in a standard Boolean way because ultrafilters are clans.

For $a \in B$ we introduce the Stone-like mapping $h(a) = \{\Gamma \in \text{CLANS}(B) : a \in \Gamma\}$.

From Lemma 2.3.2 and the properties of clans we obtain the following assertion.

**Lemma 2.3.3** ([15]). *The following assertions hold.*

(1) *$h(a + b) = h(a) \cup h(b)$, $h(0) = \varnothing$, and $h(1) = \text{CLANS}(B)$.*

(2) *$a \leqslant b$ if and only if $h(a) \subseteq h(b)$.*

(3) *$a = 1$ if and only if $h(a) = \text{CLANS}(B)$.*

(4) *$aCb$ if and only if $h(a) \cap h(b) \neq \varnothing$.*

Our next goal is to turn the set $X = \text{CLANS}$ into a topological space and to establish a representation theorem. For this purpose, as in the Stone representation theory for Boolean algebras, we define a topology $\tau$ taking $\{h(a) : a \in B\}$ for the base of closed sets and considering $h$ as the required embedding. We expect that $h$ will embed the contact algebra $\underline{B}$ into the contact algebra $RC(X)$. Proposition 2.3.4 shows that in a sense regular closed sets cannot be excluded. Recall that the reduct $(B, 0, 1, +)$ of a Boolean algebra $(B, 0, 1, +, ., *)$ is a Boolean algebra, called the *upper semi-lattice* of $B$, and it generates the same ordering relation $\leqslant$ as in $B$.

**Proposition 2.3.4.** *Suppose that $X$ is a topological space, $\underline{B} = (B, 0, 1, +, ., *)$ is a Boolean algebra, and $h$ is an embedding of the upper semi-lattice $(B, 1, +)$ in the upper semi-lattice of closed sets of $X$ such that the set $\{h(a) : a \in B\}$ is a base of closed sets of $X$. Then the following assertions hold:*

(1) $h(a^*) = \mathrm{Cl}(-h(a))$,

(2) *for every $a \in B$, $h(a)$ is a regular closed set in $X$ and, consequently, $X$ is a semiregular space,*

(3) *$h$ is an embedding in $RC(X)$.*

PROOF. (1) Consider an arbitrary point $x \in X$. Assertion (1) follows from the sequence of equivalences

$$x \in \mathrm{Cl}(-h(a)) \Leftrightarrow (\forall b \in B)(-h(a) \subseteq h(b) \to x \in h(b))$$
$$\Leftrightarrow (\forall b \in B)(h(a) \cup h(b) = X \to x \in h(b)),$$
$$\Leftrightarrow (\forall b \in B)(a + b = 1 \to x \in h(b)),$$
$$\Leftrightarrow (\forall b \in B)(a^* \leqslant b \to x \in h(b)),$$
$$\Leftrightarrow (\forall b \in B)(h(a^*) \subseteq h(b) \to x \in h(b)) \Leftrightarrow x \in h(a^*)$$

since $\mathrm{Cl}(-h(a))$ is the intersection of all elements in the base containing $-h(a)$. Here, we repeatedly used the assumption that $h$ is an embedding preserving $1$, $+$, and $\leqslant$.

(2) Applying (1) twice, we find

$$x \in h(a) \Leftrightarrow x \in h(a^{**})$$
$$\Leftrightarrow x \in \mathrm{Cl}(-\mathrm{Cl}(-h(a)))$$
$$\Leftrightarrow x \in \mathrm{Cl}(\mathrm{Int}(h(a))),$$

which shows that for every $a \in B$, $h(a)$ is a regular closed set and, consequently, $X$ is a semiregular space.

(3) This assertion follows from (2), (1), and the assumption that $h$ preserves $+$ and 1. □

Combining Lemmas 2.3.2, 2.3.3, and 2.3.4, we obtain the following assertion.

**Lemma 2.3.5.** *$h$ is an embedding of $\underline{B}$ in* $\mathrm{RC}(X)$ *with* $X = $ CLANS $(\underline{B})$.

Properties of $X = $ CLANS are presented by the following assertion.

**Lemma 2.3.6.** *The space $X = $ CLANS $(\underline{B})$ is semiregular, possesses the $T_0$ property, and is compact.*

PROOF. The space $X$ is semiregular since it has the base of regular closed sets.

To prove the $T_0$ property, we suppose that $\Gamma$ and $\Delta$ are two different points of $X$. Since $\Gamma$ and $\Delta$ are clans, one of them, say $\Gamma$, is not included in the other, $\Delta$. Then there is $a \in \Gamma$ such that $a \notin \Delta$. Hence the open set $-h(a)$ contains $\Delta$ and not $\Gamma$.

To prove the compactness of $X$, it suffices to prove the following. Let $I$ be a nonempty set of indices, and let $A = \bigcap\{h(a) : a \in I\}$. If for every finite set $I_0 \subseteq I$ we have $\bigcap\{h(a) : a \in I_0\} \neq \varnothing$, then $A \neq \varnothing$. Indeed, the condition that $\bigcap\{h(a) : a \in I_0\} \neq \varnothing$ for all finite subsets $I_0$ of $I$ guarantees the existence of an ultrafilter $U$ such that $\{h(a) : a \in I\} \subseteq U$. It is easy to see that the set $\Gamma = \{a : h(a) \in U\}$ is a clan. Hence for every $a \in I$

$$a \in I \rightarrow h(a) \in U \rightarrow a \in \Gamma \rightarrow \Gamma \in h(a).$$

Thus, $\Gamma \in A$ and, consequently, $A \neq \varnothing$. □

We show how the additional axioms (Con), (Ext), and (Nor) affect the properties of the canonical space $X = $ CLANS $(\underline{B})$.

Let $A$ be a regular closed set in the canonical space $X$. The set $F_A = \{a \in B : A \subseteq h(a)\}$ is called the *canonical filter* of $A$.

**Lemma 2.3.7.** *The canonical filter $F_A$ possesses the following properties:*

(1) $F_A$ is a filter,

(2) $(\forall \Gamma \in X)(\Gamma \in A$ if and only if $F_A \subseteq \Gamma)$,

(2) If $A \neq X$, then there is $a \in B$ such that $a \neq 1$ and $A \subseteq h(a)$,

(4) $F_A \times F_B \subseteq C$ if and only if $A \cap B \neq \varnothing$,

(5) $A \cap B = \varnothing$ if and only if $(\exists a, b \in B)(A \subseteq h(a)$ and $B \subseteq h(b)$ and $a\overline{C}b)$.

PROOF. (1) This assertion is a direct consequence of the definition of $F_A$ and Lemma 2.3.3.

(2) Since $A$ is a closed set and the set of all $h(a)$ is a closed base for the topology of $X$, for any clan $\Gamma$

$$\Gamma \in A \Leftrightarrow (\forall a \in B)(A \subseteq h(a) \to a \in \Gamma)$$
$$\Leftrightarrow (\forall a \in B)(a \in F_A \to a \in \Gamma)$$
$$\Leftrightarrow F_A \subseteq \Gamma.$$

(3) Let $A \neq X$. Then there is a clan $\Gamma$ such that $\Gamma \notin A$. By (2), $F_A \not\subseteq \Gamma$ and, consequently, there is $a \in B$ such that $a \in F_A$ and $a \notin \Gamma$. Hence $A \subseteq h(a)$, and $a \neq 1$ by Lemma 2.3.3.

(4) ($\leftarrow$) Assume that there is a clan $\Gamma \in A$ such that $\Gamma \in B$. Then $F_A \subseteq \Gamma$ and $F_B \subseteq \Gamma$. Consequently, $(\forall a, b \in B)(A \subseteq h(a)$ and $B \subseteq h(b) \to a, b \in \Gamma)$. Hence $(\forall a, b \in B)(A \subseteq h(a)$ and $B \subseteq h(b) \to aCb)$, which yields $F_A \times F_B \subseteq C$.

($\to$) Let $F_A \times F_B \subseteq C$. By Lemma 2.3.1, there exist maximal filters $F_1$ and $F_2$ such that $F_A \subseteq F_1$, $F_B \subseteq F_2$, and $F_1 \rho F_2$ , i.e., $F_1 \times F_2 \subseteq C$. Then $\Gamma = F_1 \cup F_2$ is a clan and $F_A \subseteq \Gamma$, $F_B \subseteq \Gamma$. By (2), $\Gamma \in A$, $\Gamma \in B$ and, consequently, $A \cap B \neq \varnothing$.

(5) This assertion is equivalent to (4). $\qquad\qquad\square$

**Corollary 2.3.8.** $h$ is a dense $C$-separable embedding of $\underline{B}$ in $\mathrm{RC}(X)$ with $X = \mathrm{CLANS}(\underline{B})$.

PROOF. The assertion immediately follows from Lemma 2.3.7, (3), (4). $\qquad\qquad\square$

The above results yield the following

**Theorem 2.3.9** (representation of contact algebras). *Let $\underline{B} = (BC)$ be a contact algebra. Then there exists a compact semi-regular $T_0$-space $(X, \tau)$ and a dense C-separable embedding $h$ of $\underline{B}$ in the contact algebra of regular closed sets $\mathrm{RC}(X)$. Moreover,*

(1) *$\underline{B}$ satisfies (Con) if and only if $X$ is connected,*
(2) *$\underline{B}$ satisfies (Ext) if and only if $X$ is weakly regular,*
(3) *$\underline{B}$ satisfies (Nor) if and only if $X$ is $\varkappa$-normal,*
(4) *if $\underline{B}$ is a complete algebra, then $h$ is an isomorphism between $\underline{B}$ and the complete contact algebra $\mathrm{RC}(X)$.*

PROOF. Assertions (1)–(4) follow from Lemmas 2.2.2, 2.2.1, and 2.2.3. □

A similar assertion was proved in [15] with the compactness of $X$ replaced with a stronger notion of *C-semiregularity* (a semiregular $T_0$-space is C-*semiregular* if every clan in $\mathrm{RC}(X)$ is a point clan). Note that any C-semiregular space is compact, but there are compact semiregular spaces that are not C-semiregular.

Based on the definition of a point in a contact algebra, we can give a second-order definition of the $n$-ary contact:

$C_n(a_1, \ldots, a_n)$ if and only if there exists a clan $\Gamma$
such that $\{a_1, \ldots, a_n\} \subseteq \Gamma$.

Using this definition and Theorem 2.3.9, we find

$C_n(a_1, \ldots, a_n)$ if and only if $h(a_1) \cap \ldots \cap h(a_n) \neq \varnothing$,

which shows that the above definition agrees with the notion of the standard topological $n$-ary contact.

## 2.4. Another topological representation of contact algebras

Under additional assumptions, contact algebras can be represented in better topological spaces, $T_1$ or $T_2$. If contact algebras satisfy axiom (Ext), we can prove a representation theorem for compact weakly regular $T_1$-spaces with maximal clans instead of points. By axiom (Ext), it is possible to repeat all the arguments of Section 2.3 to obtain a representation result similar to Theorem 2.3.9, but $X$ should be replaced with $T_1$ in view of the maximality of clans.

**Theorem 2.4.1** (representation of extensional contact algebras). *Let $\underline{B} = (BC)$ be a contact algebra satisfying axiom (Ext). Then there exists a compact weakly regular $T_1$-space $(X, \tau)$ and a dense $C$-separable embedding $h$ of $\underline{B}$ in the contact algebra of regular closed sets* RC$(X)$. *Moreover,*

(1) *$\underline{B}$ satisfies (Con) if and only if $X$ is connected,*

(2) *$\underline{B}$ satisfies (Nor) if and only if $X$ is $\varkappa$-normal,*

(3) *if $\underline{B}$ is a complete algebra, then $h$ is an isomorphism between $\underline{B}$ and* RC$(X)$.

This theorem covers the case of the RCC system. Similar assertions were proved by Düntsch and Winter in [**21**] (without compactness) for RCC system and by Dimov and Vakarelov in [**15**], where the compactness was replaced with the stronger condition of *CM-semiregularity*

For contact algebras satisfying both axioms (Ext) and (Nor) the representation theorem can be improved.

**Theorem 2.4.2** (representation of extensional normal contact algebras, [**61, 15**]). *Let $\underline{B} = (BC)$ be a contact algebra satisfying both axioms (Ext) and (Nor). Then there exists a compact Hausdorff space $(X, \tau)$ and a dense embedding $h$ of $\underline{B}$ in the contact algebra of regular closed sets* RC$(X)$. *Moreover,*

(1) *$\underline{B}$ satisfies (Con) if and only if $X$ is connected,*

(2) *if $\underline{B}$ is a complete algebra, then $h$ is an isomorphism between $\underline{B}$ and* RC$(X)$.

An assertion similar to Theorem 2.4.2 was first proved by de Vries [**63**] for RO$(X)$ instead of RC$(X)$.

To prove Theorem 2.4.2, we introduce another kind of points. A subset $\Gamma$ of $\underline{B}$ is called a *cluster* if it is a clan such that

(Cluster)     if $aCb$ for every $b \in \Gamma$, then $a \in \Gamma$.

Any cluster is a maximal clan. However, to prove the existence of clusters in $\underline{B}$, we need axioms (Ext) and (Nor). Clusters were used in proximity theory for obtaining compactification theorems for topological spaces (cf. [**42**]).

For representing contact algebras in some special topological spaces (for example, *regular* spaces), other (not necessarily first-order) axioms can be required. The role of points in such algebras is played by clusters of special kind, called *co-ends*. Formally, a *co-end* $\Gamma$ is a cluster such that for every $a \notin \Gamma$ there exists $b \notin \Gamma$ such that $a \ll b$.

Contact algebras representable in $RC(X)$ with a regular space $X$ satisfy the following regularity axiom:

(Reg)  if $aCb$, then there exists a co-end $\Gamma$ containing $a$ and $b$.

We refer to [**15**] for details.

Note that (Reg) is not a first-order axiom because it contains the second-order notion of a co-end. It is not known if there is a first-order axiom equivalent to (Reg). The following general question can be posed: For a given class $\Sigma$ of topological spaces find axioms providing representation of algebras in $RC(X)$ with $X \in \Sigma$.

The above representation theorems are of embedding type, i.e., they state that a contact algebra $\underline{B}$ can be embedded in the contact algebra of regular closed sets $RC(X)$ of some topological space $X$. Such representations do not exclude the case where non-isomorphic contact algebras are embedded in the contact algebra of the same space $X$. Moreover, $X_1$ and $X_2$ can be nonhomeomorphic, whereas $RC(X_1)$ and $RC(X_2)$ are isomorphic. To establish a one-to-one correspondence between contact algebras (up to an isomorphism) and topological spaces (up to a homeomorphism), we require the completeness of contact algebras. Then for $X$ we take the so-called C-*semiregular space*, i.e., a semiregular space $X$ such that every clan in $RC(X)$ is a point clan. Representations theorems for complete contact algebras satisfying some axioms like (Con), (Ext), and (Nor) can be found in [**15**].

## 2.5. Models of contact algebras in proximity spaces

Proposition 2.3.4 motivates the following observation: In order for a topological representation $h$ of contact algebras to generate a topology, $h$ must be an embedding of the Boolean part of the contact algebra in the Boolean algebra $\mathrm{RC}(X)$ with some semi-regular space $X$. However, Proposition 2.3.4 does not guarantee that the contact relation $C$ in $\mathrm{RC}(X)$ is defined in the standard way, i.e., $aCb \Leftrightarrow a \cap b \neq \varnothing$.

In this section, we demonstrate topological models for contact algebras, where elements of the Boolean algebra are regular closed sets of some topological space, whereas the relation $aCb$ is not the standard topological contact. To construct such examples, we use *proximity spaces* introduced by Efremovič in [25] (cf. also [42]) and known as *Efremovič proximity spaces* or simply E-proximity spaces.

An *Efremovič proximity space* is a system $(X, \delta)$, where $X$ is a nonempty set and $\delta$ is a binary relation, called *proximity relation*, on subsets of $X$ such that the following axioms are satisfied:

(E1)   if $A\delta B$, then $A, B \neq \varnothing$,

(E2)   $A\delta(B \cup C)$ if and only if $A\delta B$ or $A\delta C$,

(E3)   if $A\delta B$, then $B\delta A$,

(E4)   if $A \cap B \neq \varnothing$, then $A\delta B$,

(E5)   if $A\bar{\delta}B$, then there exists $C$ such that $A\bar{\delta}C$ and $(X \smallsetminus C)\bar{\delta}B$.

A proximity space $(X, \delta)$ is said to be *separated* if it satisfies the following condition:

$$\text{if } x, y \in X, \text{ then } \{x\}\delta\{y\} \text{ implies } x = y$$

Spaces satisfying only axioms (E1)–(E4) were considered by Čech [9]. Other generalizations of E-proximity spaces can be found in [42].

The relation $\ll$ in a Čech proximity space is defined as

$$A \ll B \text{ if and only if } A\bar{\delta}(X \smallsetminus B).$$

If $A \ll B$, then $B$ is called a $\delta$-*neighborhood* of $A$. It is obvious that the relations $\delta$ and $\ll$ are interdefinable and the axioms of Čhech proximity space can be expressed in terms of $\ll$ as follows:

($\ll 1$)  $X \ll X$,

($\ll 2$)  if $A \ll B$, then $A \leqslant B$,

($\ll 3$)  if $A \leqslant B \ll C \leqslant D$ then $A \ll D$,

($\ll 4$)  if $A \ll B$, then $(X \smallsetminus B) \ll (X \smallsetminus A)$,

($\ll 5$)  if $A \ll B$ and $A \ll C$, then $A \ll B \cap C$.

In terms of $\ll$, axiom (E5) takes the form

($\ll 6$)  if $A \ll B$, then for some $C$: $A \ll C \ll B$.

Note that axioms (E1)–(E4) are the same as axioms of contact algebras (C1)–(C4); moreover, axiom (E5) or ($\ll 6$) is the same as axiom (Nor). Owing to this fact, it is possible to use proximity spaces for constructing models of contact algebras.

A standard example of E-proximity space comes from metric spaces $(X, d)$. Using the distance $d(A, B) = \inf\{d(a, b) : a \in A, b \in B\}$ between two sets $A$ and $B$ of a metric space, we define the *proximity relation*

$$A\delta B \text{ if and only if } d(A, B) = 0.$$

In this case, all the axioms of E-proximity space are satisfied.

A relational kind of proximity spaces is considered in [**60**]: for a given relational system $(X, R)$, where $X \neq \varnothing$ and $R$ is a binary relation in $X$, the relation $\delta_R$ on subsets of $X$ is defined as

$$A\delta_R B \text{ if and only if } (\exists x \in A)(\exists y \in B)(xRy).$$

In this case, axiom (E1) and the right and left implications in axiom (E2) are satisfied by any $R$, axiom (E3) is satisfied if $R$

is symmetric, axiom (E4) is satisfied if $R$ is reflexive, and axiom (E5) is satisfied if $R$ is transitive relation.

For an example of a Čech proximity space we can consider a system $(X, R)$ with a reflexive symmetric relation $R$, and for an example of an E-proximity space we can take a system $(X, R)$, where $R$ is an equivalence relation. These examples will be used for presenting discrete models of contact algebras in the following section.

Now, we use E-proximity spaces to construct topological models of contact algebras with a nonstandard proximity model of contact relation.

Every Čech proximity space $(X, \delta)$ defines a topology in $X$ in the following way. Let $Cl(A) = \{x \in X : \{x\}\delta A\}$. Then $Cl$ is a Kuratowski closure operator defining a topology in $X$. The following assertion shows how $Cl$ and $Int$ are connected with the relation $\ll$.

**Lemma 2.5.1** ([42]). *The following assertions hold:*

(1) $A \ll B$ *implies* $Cl(A) \ll B$,

(2) $A \ll B$ *implies* $A \ll Int(B)$.

Having a topology in a proximity space $X$, we can consider the set of regular closed subsets of $X$ with respect to this topology. Consider the Boolean algebra $RC(X, \delta)$ of regular closed sets with respect to the introduced topology in $(X, \delta)$. Since axioms (C1)–(C4) are the same as the axioms of proximity space, we conclude that $RC(X, \delta)$ is a contact algebra. We show that axioms (Nor) and (Ext) are also satisfied. Axiom (Nor) follows from the following stronger version of axiom ($\ll$ 6):

($\ll 6'$)     if $A \ll B$, then $A \ll C \ll B$ for some regular
          closed set $C$.

Indeed, let $A \ll B$. By axiom ($\ll$ 6), there is a subset $D$ (not necessarily regular and closed) such that $A \ll D \ll B$. By Lemma 2.5.1,

$$A \ll \text{Int}(D) \subseteq \text{Cl}(\text{Int}(D)) \subseteq \text{Cl}(D) \ll B.$$

Hence $A \ll \text{Cl}(\text{Int}(D)) \ll B$. Then $C = \text{Cl}(\text{Int}(D))$ is the required regular closed subset.

To verify axiom (Ext), assume that $A \neq \varnothing$ is a regular closed set. Then there is a point $x \in \text{Int}(A)$ and, consequently, $\{x\} \ll A$. By ($\ll 6'$), we get a regular closed set $B$ such that $\{x\} \ll B \ll A$ and, consequently, $B \neq \varnothing$ and $B \ll A$. Thus, axiom (Ext) is satisfied.

The above arguments lead to the following assertion.

**Theorem 2.5.2** ([61]). *Let $(X, \delta)$ be an E-proximity space, and let $\text{RC}(X, \delta)$ be the Boolean algebra of regular closed sets in $(X, \delta)$. Then $(\text{RC}(X, \delta), \delta)$ is a contact algebra with contact $\delta$ satisfying axioms* (Nor) *and* (Ext).

Note that the proximity contact defined in Theorem 2.5.2 is not necessarily the standard topological contact for regular closed sets. For example, consider the metric space of rational numbers and the corresponding proximity space. For the regular closed sets $A = \{x : 0 \leqslant x^2 \leqslant 2\}$ and $B = \{x : 2 \leqslant x^2 \leqslant 4\}$ we have $d(A, B) = 0$, which implies $A \delta B$. But these sets are not in the relation of the topological contact because $A \cap B = \varnothing$. If we consider the same sets over the real numbers, then both proximity and topological contacts hold. In the further consideration, we have exactly $A \cap B = \{\sqrt{2}\}$. The reason is that the space of the rational numbers does not have points (in this case, the point $\sqrt{2}$) enough to describe the standard contact, but this can be done with the help for the proximity contact. Thus, the proximity contact is more suitable for describing the real picture between regular closed sets. Generalizing the notion of an Efremovič proximity space in different ways, we can obtain models with proximity-like contacts for other contact algebras (some examples are contained in [16]).

## 2.6. Contact algebras with predicate of boundedness

In this section, we extend the language of contact algebras by introducing the predicate of boundedness. To explain this notion at the intuitive level, we consider the real line $R$. A regular closed set $a$ in $R$ is bounded if it is contained in a closed interval $[x, y]$ of $R$. A generalization to the space $R^n$ is obvious: closed spheres should be taken instead of $[x, y]$.

The notion of boundedness was used in topology by Hu [35] and in proximity spaces by Leader [37] (cf. also [42]).

The *boundedness* is defined as a class $\mathcal{B}$ of subsets of a space $X$ such that

(B1)   $\varnothing \in \mathcal{B}$,

(B2)   if $B \in \mathcal{B}$ and $A \subseteq B$, then $A \in \mathcal{B}$,

(B3)   if $A, B \in \mathcal{B}$, then $A \cup B \in \mathcal{B}$.

From the formal point of view, it is a fixed ideal of sets in $X$. The boundedness predicates related to the topology of $X$ are of great interest. For example, in $R^n$, the set of bounded regular closed regions coincides with the set of compact regular closed regions. Note that $R^n$ is a locally compact Hausdorff space (recall that a topological space $X$ is *locally compact* if for every point $x \in X$ there is a compact regular closed set $a$ such that $x \in \text{Int}(a)$). Therefore, the above definition can be taken for a topological definition of boundedness in locally compact spaces. Based on this definition of boundedness, Leader [37] introduced *local proximity spaces* by adding the following axioms to axioms (B1)–(B3) in the definition of Čech proximity spaces:

(B4)   if $A\delta B$, then $\exists C \in \mathcal{B}$ such that $C \subseteq B$ and $A\delta C$,

(C)   if $A \in \mathcal{B}$ and $A \ll C$, then $\exists C \in \mathcal{B}$ such that $A \ll C \ll B$
         $(A \ll B \Leftrightarrow A\bar{\delta} - B)$.

Note that axiom (C) is equivalent to the conjunction of the axiom

(B5)  if $A \in \mathcal{B}$, then $\exists B \in \mathcal{B}$ such that $A \ll B$

and the Efremovič axiom

(E5)  if $A \ll B$, then $\exists C$ such that $A \ll C \ll B$.

A typical example of a local proximity space is any locally compact Hausdorff spaces $X$ with $A\delta B$ defined as $\mathrm{Cl}(A) \cap \mathrm{Cl}(B) \neq \varnothing$ and $A \in B$ if and only if $\mathrm{Cl}(A)$ is a compact subset of $X$. Leader used this example to develop the local compactification theory for local proximity spaces.

Using an analogy between contact algebras and proximity spaces (cf. Section 2.5), we introduce a *local contact algebra* as a system $\underline{B} = (B, 0, 1, +, ., *, C, \mathcal{B})$, where $(B, 0, 1, +, ., *, C)$ is a contact algebra and $\mathcal{B}$ is a subset of $B$ satisfying axioms similar to (B1)–(B5) and denoted in the same way:

(B1)  $0 \in \mathcal{B}$,

(B2)  if $b \in \mathcal{B}$ and $a \leqslant b$, then $a \in \mathcal{B}$,

(B3)  if $a, b \in \mathcal{B}$, then $a + b \in \mathcal{B}$,

(B4)  if $aCb$, then $\exists c \in \mathcal{B}$ such that $c \leqslant b$ and $aCc$,

(B5)  if $a \in \mathcal{B}$, then $\exists b \in \mathcal{B}$ such that $a \ll b$ $(a \ll b \Leftrightarrow a\overline{C}b^*)$.

We say that $\underline{B}$ is *connected* (*extensional* or *normal*) if it satisfies axiom (Con) ((Ext) or (Nor)).

Standard examples of local contact algebras can be obtained from a locally compact space $X$: the contact algebra $RC(X)$ and the set of bounded regions $\mathcal{B}(X)$ coinciding with the compact regular closed sets in $X$.

In mereotopology, the notion of boundedness was first used by Roeper in [49], where it was referred to as the *limitedness*. The *region-based topology* introduced by Roeger is equivalent to the local contact algebras satisfying axioms (Ext) and (Nor). The axioms of Roeper are (C1)–(C4), (B1)–(B4), and the following:

(R)  if $a \in \mathcal{B}$, $b \neq 0$ and $a \ll b$, then $\exists c \in \mathcal{B}$ such that $c \neq 0$ and

$a \ll c \lll b.$

Note that axiom (R) is close to Leader's axiom (C) and de Vries' axiom (P6). Roeper did not make any reference to their works, and it is quite impressive that he independently worked out some ideas and methods of proximity theory. For example, his definition of a point as a *coincidence set* is the same as a bounded cluster introduced by Leader. Roeper gave an elegant proof of the fact that every complete contact algebra satisfying axioms (Ext) and (Nor) (the *complete region-based topology* in the terminology of Roeper) is isomorphic to the local contact algebra $\mathrm{RC}(X)$ of regular closed sets of a Hausdorff locally compact space $X$ and that there is a one-to-one correspondence between region-based topologies (up to an isomorphism) and Hausdorff locally compact spaces (up to a homeomorphism). Another proof of the Roeper theorem is contained in [61], where the Leader compactification theorem is generalized to local proximity spaces.

The goal of this section is to expand the Roeper embedding theorem to the case of local contact algebras under additional axioms (Con), (Ext), and (Nor).

**Theorem 2.6.1** (representation of local contact algebras). *Let $\underline{B} = (B, C, \mathcal{B})$ be a local contact algebra. Then there exists a locally compact semiregular $\mathrm{T}_0$-space $(X, \tau)$ and a dense $C$-separable embedding $h$ of $\underline{B}$ in the local contact algebra of regular closed sets $\mathrm{RC}(X)$. Moreover,*

(1) $\underline{B}$ *satisfies* (Con) *if and only if $X$ is connected,*

(2) $\underline{B}$ *satisfies* (Ext) *if and only if $X$ is weakly regular,*

(3) $\underline{B}$ *satisfies* (Nor) *if and only if $X$ is $\varkappa$-normal,*

(4) *if $\underline{B}$ is a complete algebra, then $h$ is an isomorphism between $\underline{B}$ and the complete local contact algebra $\mathrm{RC}(X)$.*

The proof of this theorem is similar to that of Theorem 2.3.9, but, instead of ultrafilters and clans, *bounded ultrafilters and bounded clans* (ultrafilters and clans possessing bounded regions) and the constructions from Section 2.3 are used.

A stronger result (similar to Theorem 2.4.1) for locally compact weakly regular $T_1$-spaces can be obtained under the assumption that local algebras satisfy axiom (Ext). In this case, the role of points is played by *bounded maximal clans*. If both axioms (Ext) and (Nor) are assumed, locally compact Hausdorff spaces are obtained (as was established by Roeper). If, in addition, axiom (Con) is satisfied, we obtain a connected space.

As was noted by Roeper [49], if axiom (R) (equivalently, axioms (Ext) and (Nor)) is not assumed, it is impossible to establish a one-to-one correspondence between local contact algebras and locally compact spaces.

## 2.7. Algebras of regions based on non-Boolean lattices

It is reasonable to weaken the Boolean part of a contact algebra since it constitutes the mereological basis for the contact algebra, but the basic mereological relations (part-of, overlap, and underlap) admit equivalent definitions in terms of the lattice operations. Another reason is to examine how much the lattice properties affect properties of the mereological relations. We give two examples. The relations overlap and underlap are extensional in the following sense:

(Ext-O)     $a = b$ if and only if $(\forall c)(aOc \leftrightarrow bOc)$,

(Ext-U)     $a = b$ if and only if $(\forall c)(aUc \leftrightarrow bUc)$.

If we restrict the Boolean part to a distributive lattice with 0 and 1, then (Ext-O) and (Ext-U) are not necessarily valid. In the case of a distributive lattice, (Ext-O) is equivalent to the following stronger condition:

(Ext-O')     $a \leqslant b$ if and only if $(\forall c)(aOc \rightarrow bOc)$.

Similarly, (Ext-U) is equivalent to the following:

(Ext-U')     $a \leqslant b$ if and only if $(\forall c)(bUc \rightarrow aUc)$.

For more details we refer to [22] and [18]).

The following lemma illustrates the importance of the extensionality principles for the representability results in Boolean contact algebras of regular closed sets.

**Lemma 2.7.1** ([18]). *Suppose that $X$ is a topological space, $L = (L, 0, 1, +, .)$ is a lattice, and $h$ is an embedding of the upper semi-lattice $(L, 0, 1, +)$ in the lattice $C(X)$ of closed sets of $X$. Let $\mathbf{B} = \{h(a) : a \in L\}$ be the closed base of a topology for $X$.*

(1) *The following conditions are equivalent*:
   (a) *$L$ is $U$-extensional,*
   (b) *$\mathbf{B} \subseteq \mathrm{RC}(X)$,*
   (c) *$h(a.b) = \mathrm{Cl}(\mathrm{Int}(h(a) \cap h(b)))$ for all $a, b \in L$,*
   (d) *$h$ is the dual dense embedding of $L$ in $\mathrm{RC}(X)$.*
(2) *If some of conditions (a)–(d) in (1) are satisfied, then*
   (a) *$L$ is an $U$-extensional distributive lattice,*
   (b) *$X$ is a semiregular space.*

Lemma 2.7.1 shows that in order for a lattice $L$ to be embedded in $\mathrm{RC}(X)$ so that the image of $L$ to form a basis of closed sets for $X$, the lattice $L$ should be distributive and U-extensional with semiregular topology.

Theorem 2.7.2 below shows that such representability results can also take place for distributive U-extensional lattices with axioms (C1)–(C4).

**Theorem 2.7.2** (topological representation of $U$-extensional distributive contact lattices, [18]). *Let $D = (D, 0, 1, +, ., C)$ be an $U$-extensional distributive contact lattice. Then there exists a semiregular $T_0-$space $X$ and the dual dense embedding $h$ of $D$ in $\mathrm{RC}(X)$ such that $\{h(a) : a \in D\}$ is a basis of closed sets for $X$.*

Note that the embedding of a distributive contact lattice in $\mathrm{RC}(X)$ is possible even if the lattice is not necessarily U-extensional provided that we omit the condition that $\{h(a) : a \in D\}$ generates the topology of $X$. This shows that this assumption has a lattice equivalent in the form of the U-extensionality of the lattice.

We can further weaken the mereological part of contact algebras. In particular, examples with nondistributive lattices were considered in [22]. The question is: What can be regarded as a nice point-based representation? For a candidate for topological modelling of nondistributive contact lattices we can consider regions in some bi-topological spaces. Let $(X, \tau_1, \tau_2)$ be a space with two different topologies $\tau_1$ and $\tau_2$. A set $a \subseteq X$ is called a *mixed regular closed set* if $a = \mathrm{Cl}_1(\mathrm{Int}_2(a))$. We set $0 = \varnothing$, $1 = X$, $a + b = a \cup b$, $a.b = \mathrm{Cl}_1(\mathrm{Int}_2(a \cap b))$, and $aCb$ if and only if $a \cap b \neq \varnothing$. Then such mixed regions form a (not necessarily distributive) lattice and $C$ satisfies axioms (C1)– (C4). By a result of Urquhart [59], any lattice can be embedded in such a special lattice. The problem is that such a representation does not hold for the contact relation. The question of finding a satisfactory model and representation theory for nondistributive contact lattices remains still open.

The further generalization is to drop the entire lattice part and consider only some mereotopological relations with suitable axioms that are valid in the standard Boolean model. For example, as we can see for contact algebras, all the RCC-8 relations are definable and their definition uses only $O$, $C$, $\leqslant$. and $\ll$. Thus, it is of interest to find the complete set of axioms for $O$, $C$, $\leqslant$, and $\ll$. The author does not know whether there are results in this direction.

## 2.8. Precontact algebras and discrete spaces

In this section, we describe discrete nontopological models of contact algebras. We begin with a general definition.

A *precontact algebra* is a system $\underline{B} = (B, C) = (B, 0, 1, +, ., *, C)$, where $(B, 0, 1, +, ., *)$ is a Boolean algebra and $C$ is a binary relation, called a *precontact*, satisfying the following axioms:

(C1)  if $aCb$, then $a, b \neq 0$,

(C2′)  $aC(b + c)$ if and only if $aCb$ or $aCc$,

(C2″)    $(a + b)Cc$ if and only if $aCc$ or $bCc$.

Note that $\underline{B}$ is a contact algebra if it satisfies axioms (C3) (i.e., $aCb \to bCa$) and (C4) (i.e., $a.b \neq 0 \to aCb$).

Precontact algebras were considered in [14] and in [19], where they are called *proximity algebras*.

We give nontopological examples of precontact and contact algebras using the notion of an *adjacency space* introduced by Galton [29, 30]. An *adjacency space* is a relational system $(X, R)$, where $X \neq \varnothing$ is a set whose elements are called *cells* and $R$ is a binary *adjacency relation* on cells. By a *region* in $(X, R)$ we mean any subset of $X$. We say that two regions $a, b \subseteq X$ are in the *adjacency contact* $C_R$ and write $aC_Rb$ if $(\exists x \in a)(\exists y \in b)(xRy)$. Such a binary relation was used in [60] for defining relational proximity spaces, but its interpretation there was different from Galton's one. Galton assumed that $R$ is reflexive symmetric, whereas $R$ is arbitrary in [19]. An intuitive example of an adjacency space, adopted from Galton, is a chess-board table with cells – squares such that two squares are adjacent if they have a common point. This is an example of a reflexive symmetric adjacency relation. However, there are also several nonreflexive nonsymmetric adjacency relations, for example: "$a$ to be next on the left of $b$" or "$a$ to be on the top of $b$," etc., which motivates the choice of an arbitrary binary relation of $R$ in [19].

It is easy to prove the following assertion.

**Lemma 2.8.1** ([19]). *Let $(X, R)$ be an adjacency space, and let $\underline{B}(X) = (B(X), C_R)$, where $B(X)$ is the Boolean algebra of subsets of $X$ and $C_R$ is the adjacency contact. Then the following assertions hold:*

(1) $\underline{B}(X) = (B(X), C_R)$ *is a precontact algebra,*

(2) $\underline{B}$ *satisfies* (C3) *if and only if $R$ is symmetric,*

(3) $\underline{B}$ *satisfies* (C4) *if and only if $R$ is reflexive,*

(4) $\underline{B}$ *is a contact algebra if $R$ is reflexive and symmetric,*

(5) $\underline{B}$ *satisfies* (Nor) *if and only if $R$ is transitive,*

(6) $\underline{B}$ *satisfies* (Con) *if and only if $R$ is connected in the sense of graphs, i.e. if $x \neq y$ then there is an $R$-path from $x$ to $y$.*

With every precontact algebra we can associate a *canonical adjacency space* $X(\underline{B}) = (X(B), R_B)$ taking the set of all ultrafilters of $\underline{B}$ ("points" of $\underline{B}$) for $X(B)$ and setting for two ultrafilters $F$ and $G$

$$FR_BG \Leftrightarrow F \times G \subseteq C \Leftrightarrow F\rho G,$$

where the relation $\rho$ was introduced in Section 2.3. The mapping $h(a) = \{F \in X(B) : a \in F\}$ is the Stone embedding of $\underline{B}$ in the Boolean algebra of subsets of $X(B)$. By Lemma 2.3.1, $h$ preserves the relation of precontact. Thus, the following representation result holds.

**Theorem 2.8.2** (representation of precontact algebras in adjacency spaces, [**19**]). *Suppose that $\underline{B}$ is a precontact algebra, $(X(B), R_B)$ is the canonical adjacency space, and $B(X(B))$ is the precontact algebra over the canonical space. Then the following assertions hold:*

(1) *$h$ is an embedding of $\underline{B}$ in $B(X(B))$,*

(2) *$\underline{B}$ satisfies (C3) if and only if $B(X(B))$ satisfies (C3),*

(3) *$\underline{B}$ satisfies (C4) if and only if $B(X(B))$ satisfies (C4),*

(4) *$\underline{b}$ is a precontact algebra if and only if $B(X(B))$ is a contact algebra,*

(5) *$\underline{B}$ satisfies (Nor) if and only if $B(X(B))$ satisfies (Nor).*

As was noted in [**19**], the canonical adjacency space of a connected contact algebra is not in general a connected adjacency space. That is why Theorem 2.8.2 does not cover the case of connected contact algebras, unlike topological representation theorems.

Theorem 2.8.2 gives examples of nontopological discrete representations of contact algebras and normal contact algebras. This fact is remarkable because this means that a contact algebra has two essentially different representations: a discrete representation in a reflexive symmetric adjacency space and the other in a topological space. The points of the discrete representation are ultrafilters and the contact is realized by a binary adjacency relation between ultrafilters, whereas points in the topological representation are clans, i.e., special collections of ultrafilters. Considering both representations in the same space, we see that every region $a$ have two representations: the first, $h_{ultrafilters}(a)$ containing only ultrafilters and the second, $h_{clans}(a)$ containing $h_{ultrafilters}(a)$ and including some clans. Note that $h_{clans}(a)$ is a regular closed set with a boundary containing only clans and the ultrafilters are included only in $\text{Int}(h_{clans}(a)$. These representations reminiscent to consider ultrafilter-points as analogs of *atoms*, and clan-points can be regarded as analogs of *molecules*. Respectively, the representation theory is, in a sense, some kind of establishing certain atomistic micro-structure of the space, in which different kinds of points constitute the microlevels of the regions. Note that this interpretation is quite disputable and arise serious philosophical questions about the atomicity of space. More about this discussion in the realm of the top-level ontology and mereology can be found, for example, in [53].

At the first glance, topological modelling of precontact algebras is not possible because the standard topological contact satisfies additional axioms (C3) and (C4). However, for an arbitrary precontact algebra we can define an additional relation of contact $C^{\#}$ as follows: $aC^{\#}b$ if and only if $aCb$ or $bCa$ or $a.b \neq 0$. It is obvious that $C^{\#}$ satisfies all the axioms of contact algebra. Hence we can look for topological models of precontact algebras withe lements represented by regular closed sets of a topological space $X$ and the contact $C^{\#}$ is represented as the standard topological contact. This is possible to be done but in topological structures of a more complicated nature, containing a binary relation $R$ between some points of the space. We refer to [14] for definitions and topological representation theorems of precontact algebras in more detail.

# 3. Region-Based Propositional Modal Logics of Space

In this section, we present a language for propositional, quantifier-free logics of the region-based theory of space. The consideration of a quantifier-free language is mainly motivated by the necessity to obtain decidable fragments of some well-known systems of region-based theory of space related to RCC. We present three kinds of semantics:

- *algebraic semantics* based on algebras of regions,
- *topological semantics* based on contact algebras of some classes of topological spaces,
- *Kripke-type semantics* based on Kripke structures regarded as adjacency spaces.

The main tools in the proof of completeness theorems are the representation theorems for contact and precontact algebras from Section 2. We use a language similar to that of *relative modal logic* introduced by von Wright [66], which motivates us to call the considered logics *region-based propositional modal logics of space* (RPMLS). Another motivation is that Kripke-type semantics is very closed to the Kripke semantics in modal logic. Moreover, almost all known techniques of modal logic (in particular, modal definability, filtration, canonical-model constructions, etc.) used for proving completeness theorems can be transferred to our case with slight modifications. In addition, the language has a direct translation into the minimal modal logic K + universal modality, which also motivates our choice. However, the "modal" qualification of our logical language is not obligatory and it can be considered as a quantifier-free version of some first-order language. Note that the introduced language is a simplified version of the language of RCC-8 with Boolean terms, used by Wolter and Zakharyaschev [65].

The material of this section is mainly based on [3].

## 3.1. Syntax and semantics of RPMLS

### Syntax

The *language* $\mathbf{L}(\mathbf{C}, \leqslant)$ of *region-based propositional modal logics of space* (RPMLS) consists of

- a denumerable set Var of *Boolean variables*,
- *Boolean operations*:: . (Boolean meet), + (Boolean join), * (Boolean complement), and 0, 1 (Boolean constants),
- *propositional connectives*: $\neg, \wedge, \vee, \Rightarrow, \Leftrightarrow$, and propositional constants $\top$ and $\bot$,
- *modal connectives*: $\leqslant$ (part-of) and $C$ (contact).

The set of *Boolean terms* $\mathbf{B}$ is defined in a standard way: from Boolean atoms and Boolean constants by means of Boolean operations.

*Atomic formulas* are formulas of the form $a \leqslant b$ and $aCb$, where $a$ and $b$ are Boolean terms.

*Complex formulas* (or simply *formulas*) are defined in a standard way from atomic formulas and propositional constants $\bot$ and $\top$ by means of propositional connectives.

*Abbreviations*:

$a = b \overset{\text{def}}{=} (a \leqslant b) \wedge (b \leqslant a)$,

$a \neq b \overset{\text{def}}{=} \neg(a = b)$,

$a\overline{C}b \overset{\text{def}}{=} \neg(aCb)$,

$aOb \overset{\text{def}}{=} a.b \neq 0$ (overlap),

$a \ll b \overset{\text{def}}{=} a\overline{C}b^*$ (nontangential inclusion).

*Substitution.* Let $\alpha$ be a Boolean term or a formula, and let $p_1, \ldots, p_n$ be a list of different Boolean variables. We write $\alpha(p_1, \ldots, p_n)$ to indicate that $p_1, \ldots, p_n$ can occur in $\alpha$.

If $b_1, \ldots, b_n$ are Boolean terms, then $\alpha(b_1, \ldots, b_n)$ or, more precisely $\alpha(p_1/b_1, \ldots, p_n/b_n)$ means the simultaneous substitution of $b_1, \ldots, b_n$ for $p_1, \ldots, p_n$. The formula $\alpha(b_1, \ldots, b_n)$ is called a *substitutional instance of* $\alpha$. If we consider $p_1, \ldots, p_n$ as meta variables for Boolean terms, then $\alpha(p_1, \ldots, p_n)$ is called a "schema." Schemes are usually understood as schemes of axioms of some axiomatic systems.

Let $A = A(q_1, \ldots, q_n)$ be a formula of the propositional calculus built up by different propositional variables $q_1, \ldots, q_n$ and the propositional connectives $\neg, \wedge, \vee, \Rightarrow, \Leftrightarrow, \bot$, and $\top$. Let $\alpha_1, \ldots, \alpha_n$ be formulas of our language. Then $A(\alpha_1, \ldots, \alpha_n)$ or, more precisely, $A(q_1/\alpha_1, \ldots, q_n/\alpha_n)$ is called the *substitutional instance of the propositional formula A.*

## Semantics

First of all, we introduce an algebraic semantics of the language $\mathbf{L}(\leqslant, C)$. Let $\underline{B} = (B, 0, 1, ., +, *, C)$ be a precontact algebra. A mapping $v$ from Var into $B$ is called a *valuation*. It is extended to arbitrary Boolean terms by induction in a standard way: $v(a.b) = v(a).v(b)$, $v(a + b) = v(a) + v(b)$, $v(a^*) = v(a)^*$, $v(0) = 0$, and $v(1) = 1$.

A pair $M = (\underline{B}, v)$, where $\underline{B}$ is a precontact algebra and $v$ is a valuation in $B$, is called an *algebraic model* or an *interpretation* in $\underline{B}$. The truth of a formula $\alpha$ in $(\underline{B}, v)$, in symbols $(\underline{B}, v) \models \alpha$, is defined inductively as follows:

$(\underline{B}, v) \models a \leqslant b$ if and only if $v(a) \leqslant v(b)$,

$(\underline{B}, v) \models aCb$ if and only if $v(a)Cv(b)$,

$(\underline{B}, v) \models \alpha \wedge \beta$ if and only if $(\underline{B}, v) \models \alpha$ and $(\underline{B}, v) \models \beta$,

$(\underline{B}, v) \models \alpha \vee \beta$ if and only if $(\underline{B}, v) \models \alpha$ or $(\underline{B}, v) \models \beta$,

$(\underline{B}, v) \models \neg\alpha$ if and only if $(\underline{B}, v) \not\models \alpha$.

We say that $M$ is a *model of a formula* $\alpha$ if $M \models \alpha$ and $M$ is a *model of the set of formulas* $A$ if $M$ is a model of all members of $A$.

We say that $\alpha$ is *true in a precontact algebra* $\underline{B}$ if $\alpha$ is true in all interpretations in $\underline{B}$. If $\Sigma$ is a class of precontact algebras, $\alpha$ is said to be *true in* $\Sigma$ if $\alpha$ is true in all members of $\Sigma$. The set of all formulas true in $\Sigma$ is called the *logic* of $\Sigma$ and is denoted by $\mathcal{L}(\Sigma)$. This is a semantic definition of logic.

Let $\Sigma$ be a class of topological spaces. The *topological semantics* of $\mathbf{L}(\mathbf{C}, \leqslant)$ in $\Sigma$ consists of interpretations in contact algebras $RC(X)$ of regular closed sets of topological spaces $X \in \Sigma$. Pairs $(X, v)$, where $X$ is a topological space and $v$ is a valuation in $RC(X)$, are referred to as *topological model* or *topological interpretation*. If $\alpha$ is true in $RC(X)$, we write "$\alpha$ is true in $X$" for brevity.

Let $\Sigma$ be a class of relational systems $(X, R)$ considered as *adjacency spaces* (cf. Section 2.8). The *Kripke semantics* of $\mathbf{L}(C, \leqslant)$ in $\Sigma$ consists of interpretations in precontact algebras over structures $(W, R) \in \Sigma$. As in modal logic, structures of the form $(W, R)$ are called *frames* (*Kripke frames* or *Kripke structures*) and the Kripke semantics is called *relational semantics*. Triples $(X, R, v)$, where $v$ is a valuation in the precontact algebra over $(W, R)$, is called a *Kripke model* or a *Kripke interpretation*. If $\alpha$ is true in the precontact algebra $(B(X), C_R)$ over the frame $(X, R)$, we write $\alpha$ is true in $(X, R)$ for brevity. The class of all frames is denoted by $\Sigma_{all}$. Note that the truth of a formula $aCb$ in the Kripke model $(X, R, v)$ can be expressed in the equivalent way as follows:

$(X, R, v) \models aCb$ if and only if $(\exists x, y \in X)(xRy$ and $x \in v(a)$ and $y \in v(b))$ if and only if $v(a)C_Rv(b)$ if and only if $v(a) \cap \langle R \rangle v(b) \neq \varnothing$, where $\langle R \rangle v(b) = \{x \in X : (\exists y \in X)(xRy$ and $y \in v(b))\}$.

## A translation into modal logic $K$
## with universal modality

Owing to the relational semantics, we can define a translation $\tau$ of our language into the modal logic $K_U$ with standard modalities, denoted by $[R]A$ or $\langle R \rangle A$, and universal modalities, denoted by $[U]A$ or $\langle U \rangle A$. The modalities $[R]A$ and $\langle R \rangle A$ are interpreted by the relation $R$ in the modal frames, whereas $[U]A$ and $\langle U \rangle A$ are interpreted by the universal relation $U = W \times W$ in the frames $(W, R)$. The formal definition of $\tau$ is the following.

- *For Boolean terms*: If $p$ is a Boolean variable, then $\tau p = p$ considered as a propositional variable in $K_U$, $\tau a^* = \neg \tau a$, $\tau(a + b) = \tau a \vee \tau b$, $\tau(a.b) = \tau a \wedge \tau b$, $\tau 0 = \bot$, and $\tau 1 = \top$.
- *For atomic formulas*: $\tau(a \leqslant b) = [U](\tau a \Rightarrow \tau b)$, and $\tau(aCb) = \langle U \rangle (\tau a \wedge \langle R \rangle \tau b)$.
- *For compound formulas*: $\tau \neg A = \neg \tau A$, $\tau(A \wedge B) = \tau A \wedge \tau B$, $\tau(A \vee B) = \tau A \vee \tau B$, $\tau(A \Rightarrow B) = \tau A \Rightarrow \tau B$, and $\tau(A \Leftrightarrow B) = \tau A \Leftrightarrow \tau B$.

The following assertion is easily proved by induction.

**Lemma 3.1.1** (on translation, [3]). *Let $F = (W, R)$ be a frame. Then for any formula $A$ the following is true: $F \models A$ in the sense of* RPMLS *if and only if $F \models \tau A$ in the sense of the modal logic $K_U$.*

If we consider only reflexive symmetric frames corresponding to the adjacency representation of contact algebras, then the above-introduced translation is in the logic KTB + universal modality (T is the code of the reflexivity axiom $[R]p \Rightarrow p$ and B is the code of the symmetry axiom $p \Rightarrow [R]\langle R \rangle p$ from modal logic).

## 3.2. Modal definability and undefinability in Kripke semantics

### Modal definability

The modal definability of a class of frames by a formula is defined in the same way as the global modal definability in modal logic. Namely, we say that a class $\Sigma$ of frames is *modally definable by a formula* $\alpha$ if for every frame $\mathcal{F} = (X, R)$

$$\mathcal{F} \in \Sigma \text{ if and only if } \mathcal{F} \models \alpha.$$

If $\Sigma$ is defined by a first-order formula $F$, then we say that $F$ is *modally definable by* $\alpha$ or $F$ *is a first-order equivalent of* $\alpha$.

**Lemma 3.2.1** (modal definability: first-order examples, [3]). *Let $\mathcal{F} = (W, R)$ be a frame, and let $p$, $q$ be Boolean variables. Then the following equivalencies hold:*

(1) [nonemptiness of $R$]
$R \neq \emptyset \Leftrightarrow \mathcal{F} \models 1C1$,

(2) [right seriality of $R$]
$(\forall x \in W)(\exists y \in W)(xRy) \Leftrightarrow \mathcal{F} \models (p \neq 0 \Rightarrow pC1)$,

(3) [left seriality of $R$]
$(\forall y \in W)(\exists x \in W)(xRy) \Leftrightarrow \mathcal{F} \models (p \neq 0 \Rightarrow 1Cp)$,

(4) [weak seriality of $R$]
$(\forall x \in W)(\exists y \in W)(xRy \vee yRx) \Leftrightarrow \mathcal{F} \models (p \neq 0 \Rightarrow 1Cp \vee pC1)$,

(5) [reflexivity of $R$]
$(\forall x \in W)(xRx) \Leftrightarrow \mathcal{F} \models (p \neq 0 \Rightarrow pCp)$,

(6) [symmetry of $R$]
$(\forall x, y \in W)(xRy \rightarrow yRx) \Leftrightarrow \mathcal{F} \models (pCq \Rightarrow qCp)$,

(7) [definability of overlap]
$(\forall x, y \in W)(xRy \leftrightarrow x = y) \Leftrightarrow \mathcal{F} \models (pCq \Leftrightarrow p.q \neq 0)$,

(8) [universality of $R$]
$(\forall x, y \in W)(xRy) \Leftrightarrow \mathcal{F} \models (a \neq 0 \wedge b \neq 0 \Rightarrow aCb)$

Note that the first-order conditions in (1), (3), (4), and (8) are not modally definable in the classical modal language. Below we will show that there are examples of definable first-order conditions in modal logic that are not modally definable in our language. For an example the transitivity condition $R$ can be considered.

Since the reflexive symmetric frames are important for our purposes, we denote $\Sigma_{\mathrm{ref}}$ ($\Sigma_{\mathrm{sym}}$ or $\Sigma_{\mathrm{ref,sym}}$) for the class of all reflexive (symmetric or reflexive and symmetric) frames and $\Sigma_e$ for the class of all equivalence relations. For the corresponding formulas which modally define these properties we use the notation

(Ref)   $p \neq 0 \Rightarrow pCp$,

(Sym)   $pCq \Rightarrow qCp$.

A relation $R$ (or a frame $(W, R)$) is said to be *connected* if for all $x \neq y \in W$ there exists an $R$-path from $x$ to $y$.

Let $n > 0$ be a natural number. A relation $R$ (or $\mathcal{F}$), regarded as a graph, is said to be $n$-colorable if it is an $n$-colorable graph (i.e. all points can be colored by the colors from a given set of $n$ colors in such a way that any two points connected by $R$ have different colors).

**Lemma 3.2.2** (modal definability: second-order examples, [3]). *The following assertions hold for a frame $\mathcal{F} = (W, R)$ :*

(1) [connectedness of $R$]
$\mathcal{F}$ *is connected if and only if*
$\mathcal{F} \models (p \neq 0 \wedge p \neq 1 \Rightarrow pCp^*)$.

(2) [non-$n$-colorability of $R$]
$\mathcal{F}$ *is not $n$-colorable if and only if*
$\mathcal{F} \models (\bigvee_{i=1,\ldots,n} p_i = 1 \wedge \bigwedge_{i \neq j, i, j = 1, \ldots, n} p_i \sqcap p_j = 0$
$\Rightarrow \bigvee_{i=1,\ldots,n} (p_i C p_i))$.

## Modal undefinability

For obtaining examples of the *modal undefinability* results, the following simple assertion is very useful.

**Lemma 3.2.3** (modal undefinability criterion). *Let $\Sigma$ and $\Sigma'$ be two classes of frames such that $\Sigma \subseteq \Sigma'$, $\Sigma \neq \Sigma'$ and they determine the same logics, $\mathcal{L}(\Sigma) = \mathcal{L}(\Sigma')$. Then the class $\Sigma$ is not modally definable.*

To use this criterion, we need to show that different classes of frames can determine the same logics. In the case of the classical modal language, the notion of a $p$-morphism was used for such a purpose. We introduce a similar notion adapted for the language of RPMLS.

Let $\mathcal{F} = (W, R)$ and $\mathcal{F}' = (W', R')$ be two frames. A surjective function $f$ from $W$ to $W'$ is called a *p-morphism* from $\mathcal{F}$ to $\mathcal{F}'$ if for any $x, y \in W$ and $x', y' \in W'$ the following conditions are satisfied:

(P1)    if $xRy$, then $f(x)R'f(y)$,

(P2)    if $x'R'y'$, then $(\exists x, y \in W)(x' = f(x), y' = f(y), xRy)$.

If $v$ is a valuation in $W$ and $v'$ is a valuation in $W'$, then $f$ is a $p$-morphism from $(W, R, v)$ to $(W', R', v')$ provided that for any Boolean variable $p$ and $x \in W$ we have

$$x \in v(p) \text{ if and only if } f(x) \in v'(p).$$

The following assertion can be proved in the same way as the corresponding analog in modal logic.

**Lemma 3.2.4** (*p*-morphism, [3]). *Let $f$ be a p-morphism from a model $\mathcal{M}$ to a model $\mathcal{M}'$. Then for any formula $\varphi$*

$$\mathcal{M} \models \varphi \text{ if and only if } \mathcal{M}' \models \varphi.$$

**Lemma 3.2.5** ([3]). *The following assertions hold.*

(1) *The logic $\mathcal{L}(\Sigma_{\text{ref,sym}})$ of all reflexive symmetric frames coincides with the logic $\mathcal{L}(\Sigma_e)$ of all equivalence relations.*

(2) *The class $\Sigma_e$ is not modally definable.*

IDEA OF THE PROOF. (1) Let $\mathcal{F} = (W, R)$ be a reflexive symmetric frame, and let $R_0 = \{\{x, y\} : xRy\}$. Define $W' = \{(x, \alpha) : x \in \alpha \text{ and } \alpha \in R_0\}$, $(x, \alpha)R'(y, \beta)$ if and only if $\alpha = \beta$. Let $f(x, \alpha) = x$. It is obvious that $R'$ is an equivalence relation in $W'$ and $f$ is a $p$-morphism from the frame $(W', R')$ to the frame $(W, R)$. Consequently, the logics $\mathcal{L}(\Sigma_e)$ and $\mathcal{L}(\Sigma_{\text{ref,sym}})$ coincide.

(2) By the criterion of modal undefinability (Lemma 3.2.3), the class $\Sigma_e$ is not modally definable.                                    □

Similarly, it is possible to prove that the first-order condition of transitivity alone is not modally definable.

Lemmas 3.2.1 and 3.2.5 show that RPMLS and the classical modal language are essentially different from the point of view of the modal definability.

## 3.3. Axiomatizations and completeness theorems

### Axiomatization

We first introduce the axiomatic system $\mathbb{L}_{\text{min}}^{\text{precont}}$ for the minimal logic of all precontact algebras. It is a Hilbert-type axiomatic system consisting of axioms and inference rules.

*Axioms of* $\mathbb{L}_{\text{min}}^{\text{precont}}$

I. The complete set of axiom schemes of classical propositional logic (or all formulas which are substitution instances of tautologies of classical propositional logic),

II. The set of axiom schemes for Boolean algebra in terms of the part-of $\leqslant$ ($a$, $b$, and $c$ are arbitrary Boolean terms):

$$a \leqslant a, \ (a \leqslant b) \wedge (b \leqslant c) \Rightarrow (a \leqslant c), \ 0 \leqslant a, \ a \leqslant 1,$$

$$(c \leqslant a.b) \Leftrightarrow (c \leqslant a) \wedge (c \leqslant b), \ (a + b \leqslant c) \Leftrightarrow (a \leqslant c) \wedge (b \leqslant c),$$

$$(a.(b + c)) \leqslant (a.b) + (a.c),$$

$$(c.a \leqslant 0) \Leftrightarrow (c \leqslant a^*), \ a^{**} \leqslant a.$$

III. The set of axiom schemes for the precontact $C$:

(C1)  $(aCb) \Rightarrow (a \neq 0) \wedge (b \neq 0),$

(C2)  $(aC(b + c)) \Leftrightarrow (aCb) \vee (aCc), \ ((b + c)Ca) \Leftrightarrow (bCa) \vee (cCa).$

*Inference rule of* $\mathbb{L}_{\text{min}}^{\text{precont}}$. Modus ponens: $A$ and $A \Rightarrow B$ imply $B$.

The notion of a *proof* in $\mathbb{L}_{\text{min}}^{\text{precont}}$ is standard. All provable formulas are called *theorems* of $\mathbb{L}_{\text{min}}^{\text{precont}}$. It is easy to see that the set of theorems of $\mathbb{L}_{\text{min}}^{\text{precont}}$ is closed under the *substitution rule*:

if $\alpha(p_1 \ldots, p_n)$ is a theorem of $\mathbb{L}_{\text{min}}^{\text{precont}}$ and $p_1, \ldots, p_n$ is a sequence of different Boolean variables, then for any Boolean terms $b_1, \ldots, b_n$, the formula $\alpha(b_1, \ldots, b_n)$ is a theorem of $\mathbb{L}_{\text{min}}^{\text{precont}}$.

We consider extensions of $\mathbb{L}_{\min}^{\text{precont}}$ by new axioms, for example, by some of the formulas from Lemma 3.2.1 considered as modal schemes (the variables $p$ and $q$ are arbitrary modal terms). The minimal logic of all contact algebras $\mathbb{L}_{\min}^{\text{cont}}$ is an extension of $\mathbb{L}_{\min}^{\text{precont}}$ by the axiom schemes

(C3)  $aCb \Rightarrow bCa,$

(C4)  $a.b \neq 0 \Rightarrow aCb.$

Let $\mathbb{L}$ be an extension of $\mathbb{L}_{\min}^{\text{precont}}$ by a set of arbitrary axiom schemes $Ax$. Denote it by $\mathbb{L}_{\min}^{\text{precont}} + Ax$ and call the *axiomatic extension* of $\mathbb{L}_{\min}^{\text{precont}}$. Similar notions are introduced for extensions of $\mathbb{L}_{\min}^{\text{cont}}$.

We also consider extensions $\mathbb{L}_{\min}^{\text{precont}} + R$ of $\mathbb{L}_{\min}^{\text{precont}}$ by an additional inference rule $R$. In this paper, we are interested only in some special rules, so a general definition of an inference rule is omitted. On the other hand, we assume that any set of rules determines proofs and theorems in the standard sense. We identify $\mathbb{L}$ with the set of its theorems and call it also a *logic*. Hereinafter, $\mathbb{L}$ is an arbitrary logic considered as an extension of $\mathbb{L}_{\min}^{\text{precont}}$.

## Canonical models

Let $\mathbb{L}$ be an arbitrary extension of $\mathbb{L}_{\min}^{\text{precont}}$. A set $\Gamma$ of formulas is called an $\mathbb{L}$-*theory* or a *theory* if it contains all theorems of $\mathbb{L}$ and is closed under the rule

(MP)  if $A$ and $A \Rightarrow B$ are in $\Gamma$, then $B$ in $\Gamma$.

For example, the set of all theorems of $\mathbb{L}$ is a theory; moreover, it is the smallest theory. A theory $\Gamma$ is said to be *consistent* if $\bot \notin \Gamma$ and *maximal* if it is consistent and $\Gamma \subseteq \Delta$ implies $\Gamma = \Delta$ for any consistent theory $\Delta$. Maximal theories are also referred to as *maximal consistent sets*.

Some well-known properties of theories are listed in the following assertion.

**Lemma 3.3.1.** *The following assertions hold.*

(1) *Let $\Gamma$ be a theory, and let $\alpha$ be a formula. Then the set $\Gamma + \alpha = \{\beta : \alpha \Rightarrow \beta \in \Gamma\}$ is the smallest theory containing $\Gamma$ and $\alpha$. The set $\Gamma + \alpha$ is inconsistent if and only if $\neg\alpha \in \Gamma$.*

(2) *The following conditions are equivalent for any theory $\Gamma$ :*
   (a) *$\Gamma$ is maximal,*
   (b) *for any formula $\alpha$, $\neg\alpha \in \Gamma$ if and only if $\alpha \notin \Gamma$,*
   (c) *for any formulas $\alpha$ and $\beta$, $\alpha \vee \beta \in \Gamma$ if and only if $\alpha \in \Gamma$ or $\beta \in \Gamma$.*

(3) *Any consistent theory can be extended to a maximal theory (the Lindenbaum lemma).*

The following assertion presents a semantical construction of maximal theories in $\mathbb{L}_{\min}^{\text{precont}}$.

**Lemma 3.3.2.** *Let $\mathcal{M}$ be a model. Then the set of formulas $\Gamma = \{\alpha : \mathcal{M} \models \alpha\}$ is a maximal $\mathbb{L}_{\min}^{\text{precont}}$-theory. If $\mathcal{M}$ is a model over contact algebra, then $\Gamma$ is a maximal $\mathbb{L}_{\min}^{\text{cont}}$-theory.*

A set of formulas $A$ is *consistent* in $\mathbb{L}$ if $A$ is contained in an $\mathbb{L}$-consistent theory and, consequently, $A$ is contained in a maximal $\mathbb{L}$-theory in view of the Lindenbaum lemma.

Let $S$ be a maximal theory in $\mathbb{L}$. Based on the Lindenbaum-algebra construction, we construct in a canonical way a precontact algebra associated with $S$. In the set of Boolean terms $\mathbf{B}$, we introduce the *equivalence relation*: $a \equiv b$ if and only if $a = b \in S$. Since $\equiv$ is a congruence relation depending on $S$, it is possible to consider equivalence classes of Boolean terms $|a| = \{b : a \equiv b\}$ and to define the canonical precontact algebra $\underline{B}_S$ over $S$ by setting $|a|.|b| = |a.b|$, $|a| + |b| = |a + b|$, $|a|^* = |a^*|$, $|a| \leqslant |b|$ if and only if $a \leqslant b \in S$, and $|a|C|b|$ if and only if $aCb \in S$.

Using the axioms of logic, we can prove that $\underline{B}_S$ is a precontact algebra and, if $\mathbb{L}$ is an extension of $\mathbb{L}_{\min}^{\text{cont}}$, $\underline{B}(S)$ is a contact algebra.

We define a canonical valuation for Boolean variables putting $v_S(p) = |p|$. Then the pair $\mathcal{M}_S = (\underline{B}_S, v_S)$ is called a *canonical model over $S$*. We have $v_S(a) = |a|$ for any Boolean term $a$. With $S$ we can canonically associate the *canonical frame $F_S =$*

$(W_S, R_S)$ of $S$ by taking for $F_S$ the canonical adjacency space of the canonical precontact algebra $\underline{B}_S$ (cf. Section 2.8). If $\mathbb{L}$ is an extension of $\mathbb{L}_{\min}^{\text{cont}}$, with $S$ we can associate the *canonical topological space* $X_S$ by taking for $X_S$ the canonical topological space corresponding to the contact algebra $\underline{B}_S$ (cf. Section 2.3).

The following assertion is proved in a standard way.

**Lemma 3.3.3.** *Let $\mathbb{L}$ be a logic. Then the following two conditions are satisfied by any formula $\alpha$:*

(1) *$\alpha$ is a theorem of $\mathbb{L}$,*

(2) *$\alpha$ is true in all canonical models $M_S$ of $\mathbb{L}$.*

Now, we can state a completeness theorem for the minimal logics $\mathbb{L}_{\min}^{\text{precont}}$ and $\mathbb{L}_{\min}^{\text{cont}}$.

**Theorem 3.3.4** (completeness of $\mathbb{L}_{\min}^{\text{precont}}$, **[3]**). *The following conditions are equivalent for any formula $\alpha$:*

(1) *$\alpha$ is a theorem of $\mathbb{L}_{\min}^{\text{precont}}$,*

(2) *$\alpha$ is true in all precontact algebras,*

(3) *$\alpha$ is true in all Kripke frames.*

PROOF. The implications (1) $\rightarrow$ (2) $\rightarrow$ (3) are obvious.

(2) $\rightarrow$ (1) Let $\alpha$ be true in all precontact algebras. Then $\alpha$ is true in all canonical models of $\mathbb{L}_{\min}^{\text{precont}}$ and $\alpha$ is a theorem of $\mathbb{L}_{\min}^{\text{precont}}$ in view of Lemma 3.3.3.

(3) $\rightarrow$ (1) Suppose that $\alpha$ is not a theorem of $\mathbb{L}_{\min}^{\text{precont}}$. Then there is a canonical model $M_S = (\underline{B}_S, v_S)$ such that $M_S \not\models \alpha$. By the representation theorem for precontact algebras in adjacency spaces (cf. Theorem 2.8.2), there exists a frame $(X, R)$ and an embedding $h$ of the canonical precontact algebra $\underline{B}_S$ in the precontact algebra $B(X)$ over the frame $(X, R)$. Define the valuation $v(p) = v_S(h(|p|))$. Then $(X, R, v) \not\models \alpha$, which means that $\alpha$ is not true in the Kripke frame $(X, R)$. The proof is complete. $\square$

**Theorem 3.3.5** (completeness of $\mathbb{L}_{\min}^{\text{cont}}$, **[3]**). *The following conditions are equivalent for any formula $\alpha$:*

(1) *$\alpha$ is a theorem of $\mathbb{L}_{\min}^{\text{cont}}$,*

(2) $\alpha$ *is true in all contact algebras,*

(3) $\alpha$ *is true in all reflexive symmetric Kripke frames,*

(4) $\alpha$ *is true in all topological spaces,*

(5) $\alpha$ *is true in all compact and semiregular* $T_0$-*spaces.*

PROOF. The implications $(1) \to (2) \to (3)$ and $(2) \to (4) \to (5)$ are obvious. The implications $(2) \to (1)$ and $(3) \to (1)$ are proved in the same way as in Theorem 3.3.4 with the help of Theorem 2.8.2. The implication $(5) \to (1)$ is proved in the same way as the implication $(3) \to (1)$ in Theorem 3.3.4 with the help of Theorem 2.3.9.                                                                           $\Box$

**Theorem 3.3.6** (completeness of $\mathbb{L}_{min}^{cont} + (Con)$, [3]). *The following conditions are equivalent for any formula* $\alpha$ :

(1) $\alpha$ *is a theorem of* $\mathbb{L}_{min}^{cont} + (Con)$,

(2) $\alpha$ *is true in all connected contact algebras,*

(3) $\alpha$ *is true in all connected topological spaces.*

Note that Theorem 3.3.6 does not assert the completeness with respect to Kripke semantics. This fact will be proved by another method in the following section.

Theorems 3.3.5 and 3.3.6 present weak completeness statements. The strong statements are also valid (cf. [3]). For the logic $\mathbb{L}_{min}^{cont}$ if can be formulated as follows.

**Theorem 3.3.7** (strong completeness of $\mathbb{L}_{min}^{cont}$). *The following conditions are equivalent for any set* $A$ *of formulas:*

(1) $A$ *is consistent in* $\mathbb{L}_{min}^{cont}$,

(2) $A$ *has an algebraic model,*

(3) $A$ *has a Kripke model,*

(4) $A$ *has a topological model.*

Theorem 3.3.5 asserts that the logic $\mathbb{L}_{min}^{cont}$ is complete with respect to both topological and discrete semantics (semantics with respect to Kripke frames).

## 3.4. Filtration with respect to Kripke semantics and small canonical models

### Filtration

Let $\Phi$ be a finite set of formulas closed under subformulas. Denote by $\Gamma_\Phi$ the smallest set of Boolean terms satisfying the following conditions:

- if $aCb \in \Phi$, then $a, b \in \Gamma_\Phi$,
- if $a \leqslant b \in \Phi$, then $a, b \in \Gamma_\Phi$,
- $\Gamma_\Phi$ is closed under subterms of its members,
- $\Gamma_\Phi$ contains 0, 1 and is closed under Boolean combinations of its members.

Note that $\Gamma_\Phi$ is infinite, but it is logically finite in the sense that there is a finite subset $\Gamma_\Phi^0$ of $\Gamma_\Phi$ such that any term in $\Gamma_\Phi$ is Boolean equivalent to an element of $\Gamma_\Phi^0$. If $n$ is the number of Boolean variables occurring in the formulas from $\Phi$, then the cardinality of $\Gamma_\Phi^0$ is equal to $2^{2^n}$. Denote by $\Phi'$ the set of all formulas containing only Boolean terms in $\Gamma_\Phi$. It is obvious that $\Phi \subseteq \Phi'$ and $\Phi'$ is infinite, but logically finite.

Let $\mathcal{M} = (W, R, v)$ be a model. We define the *equivalence relation* $\equiv$ in $W$ (depending on $\mathcal{M}$ and $\Phi$) in the same way as in the definition of a filtration in modal logic:

- $x \equiv y$ if and only if $(\forall a \in \Gamma_\Phi)(x \in v(a) \leftrightarrow y \in v(a))$,
- for $x \in W$ define $|x| = \{y \in W : x \equiv y\}$ and set $W' = \{|x| : x \in W\}$,
- for $|x|, |y| \in W'$ define: $|x|R'|y|$ if and only if $(\exists x' \equiv x)(\exists y' \equiv y)(x'Ry')$,
- for any Boolean variable $p \in \Gamma_\Phi$ define $v'(p) = \{|x| : x \in v(p)\}$.

The model $\mathcal{M}' = (W', R', v')$ is called a *filtration of the model* $\mathcal{M}$ through $\Phi$. Similarly, the frame $(W', R')$ is called a *filtration of the frame* $(W, R)$. The valuation $v'$ is called the *canonical valuation* of the filtration.

Note that the above defined filtration coincides with the so-called minimal filtration in the classical modal language.

**Lemma 3.4.1** (filtration, [3]).

(1) *The set $W'$ is finite and the cardinality of $W'$ is less than or equal to $n$, where $n$ is the cardinality of $\Gamma_\Phi^0$.*
(2) *For any $x, y \in W$, $xRy$ implies $|x|R'|y|$.*
(3) *For any Boolean term $a \in \Gamma_\Phi$ and $x \in W$, $x \in v(a)$ if and only if $|x| \in v'(a)$.*
(4) *For every formula $\psi \in \Phi'$, $\mathcal{M} \models \psi$ if and only if $\mathcal{M}' \models \psi$.*

Now, we describe a construction which is also used in filtration theory in modal logic. Let $(W', R', v')$ be a filtration of $(W, R, v)$ through $\Phi$, and let $w$ be the new valuation in the filtrated frame $(W', R')$. Then for each Boolean variable $p$ in $\Gamma_\Phi$ we define a Boolean term $b_{w(p)}$ obtained as a Boolean combination of terms in $\Gamma_\Phi$ as follows. For every $y \in W$ we set $b_{|y|} = \bigwedge \{b : b \in \Gamma_\Phi^0$ and $|y| \in v'(b)\}$. Then define $b_{w(p)} = \bigvee \{b_{|y|} : |y| \in w(p)\}$.

Using $b_{w(p)}$, we define a new valuation $w'$ in $(W, R)$ for variables from $\Gamma_\Phi$ as follows: $w'(p) = \{x \in W : x \in v(b_{w(p)})\}$.

The valuation $w$ defines also the substitution $\mathrm{Sub}_w$ for variables from $\Gamma_\Phi$ as follows: $\mathrm{Sub}_w(p) = b_{w(p)}$ and then extended inductively for Boolean terms from $\Gamma_\Phi$ and formulas from $\Phi$.

Then the following stronger version of the filtration lemma holds. There are no analog of this lemma in the classical modal logic.

**Lemma 3.4.2** (strong filtration, [3]). *Let $w$ be a valuation in $F' = (W', R')$, and let $w'$ be the corresponding valuation in $(W, R)$ defined by $b_{w(p)}$. Then for a term $a \in \Gamma_\Phi$ and a formula $\psi \in \Phi'$ the following assertions hold for any $x \in W$ :*

(1) $x \in v(\mathrm{Sub}_w(a))$ *if and only if* $x \in w'(a)$ *if and only if* $|x| \in w(a)$,
(2) $F, v \models \mathrm{Sub}_w(\psi)$ *if and only if* $F, w' \models \psi$ *if and only if* $F', w \models \psi$.

*If $w$ coincides with the canonical valuation $v'$, then $w'$ acts as $v$, i.e., for all $a \in \Gamma_\Phi$ and $\psi \in \Phi'$*

(3) $x \in w'(a)$ *if and only if* $x \in v(a)$,
(4) $F, w' \models \mathrm{Sub}_{v'}(\psi)$ *if and only if* $F, v \models \psi$.

Let $\Sigma$ be a class of frames. We say that $\Sigma$ (or the logic $\mathcal{L}(\Sigma)$) *admits a filtration* if for any formula $\varphi$ there is a finite set of formulas $\Phi$, closed under subformulas and containing $\varphi$, such that the filtrated frame $F'$ through $\Phi$ of any frame $F$ in $\Sigma$ belongs to $\Sigma$.

**Remark 3.4.3.** Suppose that $\Sigma$ admits a filtration and $\Sigma^{fin}$ is the class of all finite frames in $\Sigma$. Then the logics $\mathcal{L}(\Sigma)$ and $\mathcal{L}(\Sigma^{fin})$ coincide. Thus, $\mathcal{L}(\Sigma)$ possesses the finite model property.

We say that a class of frames $\Sigma$ is *determined* if there exists a set of formulas $A$ such that $\Sigma$ coincides with the class of all frames in which the formulas from $A$ are true. In this case, $\Sigma$ will denoted by $\Sigma_A$.

**Theorem 3.4.4** ([3]). *Every determined class of frames admits a filtration.*

PROOF. Let $\Sigma_A$ be a class of frames determined by a set of formulas $A$. Suppose that $F = (W, R) \in \Sigma_A$, $\varphi$ is a formula, $\Phi$ is the set of all subformulas of $\varphi$, and $F' = (W', R', v')$ is a filtration of the model $(F, v)$ through $\Phi$. We show that $F'$ belongs to $\Sigma_A$, i.e., all formulas in $A$ are true in $F'$. Assume the opposite, i.e. there exist a formula $\psi \in A$ and a valuation $w$ in $F'$ such that $F', w \not\models \psi$. Let $w'$ be the valuation determined by $w$ in $(W, R)$. By Lemma 3.4.2, $F, w' \not\models \psi$. Consequently $\psi$ is not true in $F$ and, consequently, $\psi$ is not true in $\Sigma_A$. We arrive at a contradiction. $\square$

## Small canonical models

Let $\mathbb{L}$ be a consistent extension of $\mathbb{L}_{\min}^{\mathrm{precont}}$, and let $S$ be a maximal theory in $\mathbb{L}$. Consider the canonical frame $F_S$ and the canonical model $M_S = (F_S, v_S)$ for $S$. Let $M'_S = (F'_S, v'_S)$ be any filtration of $M$. Then $M'_S$ is called a *small canonical model* of $\mathbb{L}$ and $F'_S$ is called a *small canonical frame* of $\mathbb{L}$.

**Lemma 3.4.5** (small canonical frame, [3]). *Let $\mathbb{L}$ be a consistent extension of $\mathbb{L}_{min}^{precont}$, let $A = \text{Th}(\mathbb{L})$ be the set of all theorems of $\mathbb{L}$, and let $\Sigma_{\mathbb{L}} = \Sigma_A$ be the set of frames determined by $A$. Then $\Sigma_{\mathbb{L}}$ contains all small canonical frames of $\mathbb{L}$.*

PROOF. Let $M'_S = (W'_S, R'_S, v'_S)$ be a small canonical model related to the maximal consistent theory $S$. It suffices to prove that all formulas in $A$ are true in the small canonical frame $F'_S = (W'_S, R'_S)$. Assume the opposite, i.e., for some $\psi \in A$ and valuation $w$ in $(W'_S, R'_S)$ we have $(W'_S, R'_S, w) \not\models \psi$. By Lemma 3.4.2, $(W_S, R_S, v_S) \not\models \text{Sub}_w(\psi)$. Hence $\text{Sub}_w(\psi) \notin S$. However, $\text{Sub}_w(\psi)$ is a substitution instance of a theorem $\psi$ of $\mathbb{L}$. Therefore, it belongs to the maximal theory $S$. We arrive at a contradiction. $\qquad\qquad\qquad\qquad\qquad\qquad\qquad\qquad\qquad\qquad\square$

## Weak completeness theorems
## for extensions of $\mathbb{L}_{min}^{precont}$

**Theorem 3.4.6** (weak completeness and the finite model property of all consistent extensions of $\mathbb{L}_{min}^{precont}$, [3]). *Let $\mathbb{L}$ be a consistent extension of $\mathbb{L}_{min}$, and let $\Sigma_{\mathbb{L}}$ be the class of frames determined by the set $\text{Th}(\mathbb{L})$ of all theorems of $\mathbb{L}$. Then the following conditions are equivalent for any formula $\varphi$ :*

(1) *$\varphi$ is a theorem of $\mathbb{L}$,*
(2) *$\varphi$ is true in $\Sigma_{\mathbb{L}}$,*
(3) *$\varphi$ is true in $\Sigma_{\mathbb{L}}^{fin}$.*

The proof is based on Lemmas 3.4.1 and 3.4.5.

Although Theorem 3.4.6 is not too informative concerning the frames of $\mathbb{L}$, but it asserts that there are no incomplete logics for the relational semantics under consideration and that all consistent logics are characterized by their finite frames. These facts have no analogs in the classical modal logic. The following assertion gives more information about axiomatic extensions of $\mathbb{L}_{\min}^{\text{precont}}$.

**Theorem 3.4.7** (weak completeness and the finite model property of all axiomatic extensions of $\mathbb{L}_{\min}$, [**3**]). *Suppose that $A$ is a set of formulas, $\Sigma_A$ is the class of all frames determined by $A$, and $\Sigma_A^{fin}$ is the class of all finite frames in $\Sigma_A$. Let $\mathbb{L}$ be an extension of $\mathbb{L}_{\min}$ with formulas from $A$ for additional axiom schemes. Then for any formula $\varphi$ the following conditions are equivalent:*

(1) *$\varphi$ is a theorem of $\mathbb{L}$,*

(2) *$\varphi$ is true in $\Sigma_A^{fin}$,*

(3) *$\varphi$ is true $\Sigma_A$.*

*Hence $\mathbb{L}$ possesses the finite model property and is decidable if $A$ is finite.*

**Corollary 3.4.8** ([**3**]). *The logics $\mathbb{L}_{\min}^{\text{cont}}$ and $\mathbb{L}_{\min}^{\text{cont}} + (\text{Con})$ are complete in the class of their finite frames and. consequently, are decidable.*

## 3.5. Logics related to RCC

According to Stell's formulation, the RCC system is equivalent to the contact algebras satisfying axioms (Ext) and (Con). In a sense, all extensions of the notion of contact algebras with axioms (Ext), (Con), and (Nor) are related to RCC as follows: RCC+(Nor) is an extension with good properties, whereas any other extension is a subsystem of RCC+(Nor). Considering all these eight types of contact algebras as first-order region-based theories of space, we introduce the following abbreviations:

WRCC – weak RCC based on axioms (C1)–(C4),

WRCC$_{\text{Con}}$ – weak connected RCC=WRCC+(Con),

WRCC$_{\text{Ext}}$ – weak extensional RCC=WRCC+(Ext),

WRCC$_{\text{Nor}}$ – weak normal RCC=WRCC+(Nor),

$\text{WRCC}_{\text{Con,Nor}}$ – weak connected normal
$$\text{RCC}=\text{WRCC}+(\text{Con})+(\text{Nor}),$$

$\text{WRCC}_{\text{Ext,Nor}}$ – weak extensional normal
$$\text{RCC}=\text{WRCC}+(\text{Ext})+(\text{Nor}),$$

$\text{RCC}$ – $\text{WRCC}+(\text{Ext})+(\text{Con})$,

$\text{RCC}_{\text{Nor}}$ – normal $\text{RCC}=\text{RCC}+(\text{Nor})$.

The goal of this section is to introduce propositional logics based on the language $\mathbf{L}(\leqslant, \mathbf{C})$ corresponding to each of these first-order systems. For propositional systems we put the letter "P" before the abbreviation of the corresponding first-order system. Two propositional systems were already introduced: PWRCC – $\mathbb{L}_{\min}^{\text{cont}}$ and $\text{PWRCC}_{\text{Con}}$ – $\mathbb{L}_{\min}^{\text{cont}} + (\text{Con})$. It is obvious that these systems are propositional (quantifier-free) analogs of WRCC and $\text{WRCC}_{\text{Con}}$ because all the axioms of the first-order systems WRCC and $\text{WRCC}_{\text{Con}}$ are universal formulas of the same form as quantifier-free axioms in PWRCC and $\text{PWRCC}_{\text{Con}}$. However, there are no analogs of axioms (Ext) and (Nor) in our language because they are not universal sentences. We imitate them by some inference rules analogous to the quantifier rules in the first-order logic. The rules have the same impact on the canonical contact algebras as the corresponding first-order axioms and will be used in the proof of the strong completeness theorem of the required logic with respect to the topological semantics which are suggested by the topological representation theorems for the corresponding contact algebras.

For an analog of axiom (Nor) we introduce the following *rule of normality*:

NOR $\dfrac{\alpha \Rightarrow (aCp \vee p^*Cb)}{\alpha \Rightarrow aCb}$, where $p$ is a Boolean variable that does not occur in $a$, $b$, and $\alpha$.

For an analog of the first-order axiom (Ext) we introduce the *rule of extensionality*

EXT   $\dfrac{\alpha \Rightarrow (p = 0 \vee aCp)}{\alpha \Rightarrow (a = 1)}$, where $p$ is a Boolean variable that

does not occur in $a$ and $\alpha$.

These rules are similar to the irreflexivity rule introduced by Gabbay [**27**] in the context of the classical modal logic. In the language under consideration, these rules were introduced in [**3**].

Taking into account the correspondences between the first-order RCC-like systems and the propositional systems, we present the diagram of extensions of $\mathbb{L}_{\min}^{\mathrm{cont}}$, where the logics are identified with the sets of additional axioms and rules.

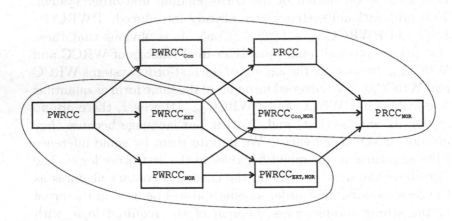

FIGURE 2

Consider the logics in this diagram. All logics satisfy the assumptions of Theorem 3.4.6 and, consequently, they are complete in certain classes of finite frames and possess the finite model property. Moreover, they are decidable because each of them has a finite set of axioms. The logics are strongly complete with respect to their intended topological semantics (cf. the following section). Concerning the weak completeness theorem, we see that the additional rules can be eliminated and thereby these rules do not affect the sets of the theorems. Thus, these eight logics collapse to the following two logics: PWRCC and PWRCC$_{\mathrm{Con}}$. Below we give some information about these two logics and their relationships with other systems in the literature. Then we discuss the admissibility of the introduced rules to PWRCC and PWRCC$_{\text{con}}$.

# PWRCC

As was already mentioned, PWRCC, propositional weak RCC, is $\mathbb{L}_{\min}^{\text{cont}}$. By Theorems 3.3.5 and 3.3.7, PWRCC is weakly and strongly complete in the class of all topological spaces (and in the smaller class of all semiregular and compact $T_0$-spaces) and in the class of all reflexive symmetric Kripke frames considered as adjacency spaces. Thus, PWRCC, is complete with respect to both topological and discrete semantics. By Corollary 3.4.8, PWRCC has the finite model property and, consequently, is decidable. As was noted in Section 2.1, all the RCC-8 relations are definable in our language by means of quantifier-free definitions. Therefore, we can use the same definitions in the language of propositional logics. This fact, together with the topological part of the completeness theorem, shows that the system PWRCC is equivalent to the system BRCC-8 (RCC-8 with Boolean terms), introduced by Wolter and Zakharyaschev [65], which can be interpreted in all topological spaces. Thus, PWRCC can be considered as an axiomatization of BRCC-8 with several completeness theorems. Wolter and Zakharyaschev [65] proved that the satisfiability problem for BRCC-8 is NP-complete. Respectively, the same assertion holds for PWRCC.

# PWRCC$_{\text{Con}}$

PWRCC$_{\text{Con}}$, propositional weak connected RCC, is an extension of PWRCC by the connectedness axiom

(Con)    $a \neq 0 \wedge a \neq 1 \Rightarrow aCa^*$

which defines the second-order connectedness property in frames. By Theorem 3.3.6, PWRCC$_{\text{Con}}$ is weakly and strongly complete in the class of all connected spaces and in the smaller class of all connected semiregular compact $T_0$-spaces. By Corollary 3.4.8,

PWRCC$_{\text{Con}}$ is weakly complete in the class of all finite connected reflexive symmetric frames and, consequently, has the finite model property and is decidable. Thus, PWRCC$_{\text{Con}}$ is weakly complete with respect to both topological and discrete semantics.

As in the case of PWRCC, we can conclude that PWRCC$_{\text{Con}}$ is equivalent to the logic BRCC-8 introduced by Wolter and Za-kharyaschev [65] who studied this logic in the class of all connected topological spaces. They proved that the satisfiability problem is PSPACE-complete. Their result implies a similar assertion for PWRCC$_{\text{Con}}$.

The completeness of PWRCC$_{\text{Con}}$ with respect to the class of connected reflexive symmetric adjacency spaces shows that the logic is equivalent to the logic GRCC (generalized region connection calculus) introduced semantically by Li and Ying [39] as a discrete version of RCC. Thus, PWRCC$_{\text{Con}}$ can be also understood as a complete axiomatization of GRCC.

BRCC-8-like systems were studied only with respect to their intended topological semantics. Since the modal logic S4 corresponds to the topological interpretation of the classical modal language, systems like BRCC-8 were also treated by means of a translation into the modal logic S4 + universal modality. Taking into account that PWRCC is complete in the class of all reflexive symmetric frames, we can conclude that the exact translation of these systems is in KTB + universal modality, which shows that the weaker modal logic KTB also has a spatial meaning. This translation is not used in this paper because for our purpose it is easier to exploit directly the relational semantics of the language of RPMLS.

## Admissibility of EXT and NOR

**Lemma 3.5.1** (admissibility of EXT in the logics PWRCC and PWRCC$_{\text{Con}}$, [3]). *The set of theorems of the logics* PWRCC *and* PWRCC$_{\text{Con}}$ *are closed with respect to the rule* EXT *and, consequently,* EXT *is an admissible rule in* PWRCC *and* PWRCC$_{\text{Con}}$.

PROOF. We use the completeness of the logics PWRCC and PWRCC$_{\text{Con}}$ with respect to the Kripke semantics and the $p$-morphism techniques in Section 3.2.

Denote by $L$ the logic PWRCC. We use the completeness of $L$ in the class $\Sigma_{\text{ref,sym}}$ of all reflexive symmetric frames. We prove that if $L \vdash \varphi \Rightarrow (p = 0 \vee aCp)$, where $p$ does not occur in $a$ and in $\varphi$, then $L \vdash \varphi \Rightarrow (a = 1)$. Assume the contrary: $L \nvdash \varphi \Rightarrow (a = 1)$. By the completeness theorem, we can choose a reflexive symmetric frame $\mathcal{F} = (W, R)$ and a valuation $V$ over $\mathcal{F}$ such that $\mathfrak{M} \vDash \varphi$ and $\mathfrak{M} \nvDash a = 1$, where $\mathfrak{M} = (\mathcal{F}, V)$. Thus, $W \setminus V(a) \neq \varnothing$. Consider two cases.

*Case* (a). $W \setminus \langle R \rangle(V(a)) \neq \varnothing$.

*Case* (b). $W \setminus \langle R \rangle(V(a)) = \varnothing$.

Here, $\langle R \rangle(V(a)) = \{x \in W : (\exists y \in V(a))(xRy)\}$.

In case (a), we choose a Boolean variable $p$ that does not occur in $a$ and $\varphi$ and a valuation $V'$ coinciding with $V$ for Boolean variables different from $p$: $V'(p) = W \setminus \langle R \rangle(V(a))$. It is clear that $\mathfrak{M}' \vDash \varphi$ and $\mathfrak{M}' \vDash \neg(p = 0 \vee aCp)$, where $\mathfrak{M}' = (\mathcal{F}, V')$. Hence $L \nvdash \varphi \Rightarrow (p = 0 \vee aCp)$.

In case (b), we choose $w_1 \in \langle R \rangle(V(a))$ and $w_0 \notin W$. We set $W_1 = W \cup \{w_0\}$, $R_1 = R \cup \{(w_0, w_0), (w_0, w_1), (w_1, w_0)\}$, $f(w) = w$ if $w \neq w_0$, $f(w_0) = w_1$, and $V_1(q) = f^{-1}(V(q))$. It is easy to verify that $\mathfrak{M}$ is the $p$-morphic image of $\mathfrak{M}_1 = ((W_1, R_1), V_1)$ under $f$, $(W_1, R_1)$ is a reflexive symmetric frame (consequently, it verifies the theorems of $L$). Thus, we can apply case (a) to $\mathfrak{M}_1$, which completes the proof for PWRCC.

For PWRCC$_{\text{Con}}$ we proceed in the same way. The only difference is that we start with a connected reflexive symmetric frame $(W, R)$. It is easy to see that the above construction of the $p$-morphic pre-image $(W', R')$ preserves the connectedness property. $\square$

**Lemma 3.5.2** (admissibility of NOR in the logics PWRCC and PWRCC$_{\text{Con}}$, [3]). *The set of theorems in the logics PWRCC* *and* PWRCC$_{\text{Con}}$ *are closed under the rule* NOR *and, consequently,* NOR *is an admissible rule in* PWRCC *and* PWRCC$_{\text{Con}}$.

The proof is similar to that of Lemma 3.5.1.

**Corollary 3.5.3** ([3]). *The following assertions hold.*

(1) PWRCC, PWRCC$_{\text{EXT}}$, PWRCC$_{\text{NOR}}$, *and* PWRCC$_{\text{EXT,NOR}}$ *have the same set of theorems.*

(2) PWRCC$_{\text{Con}}$, PRCC, PWRCC$_{\text{CON,NOR}}$, *and* PRCC$_{\text{NOR}}$ *have the same set of theorems.*

(3) *All the eight logics are weakly complete in the corresponding class of frames determined by their axioms, have the finite model property, and, consequently, are decidable. The satisfiability problem for the logics in* (1) *is* NP*-complete and for the logics in* (2) *is* PSPACE*-complete.*

## 3.6. Strong completeness theorems for RCC-like logics

The purpose of this section is to illustrate how to work with additional rules of type NOR and EXT and how to modify the canonical-model construction in the presence of such rules. For an example we consider the logic PWRCC$_{\text{NOR}}$ which is an extension of the logic PWRCC with the rule NOR. The material of this section mainly follows [3]. Some important results are supplied with sketches of proofs.

### PWRCC$_{\text{NOR}}$

We establish the strong completeness of topological semantics of PWRCC$_{\text{NOR}}$. The proof is a modification of the Henkin construction for the first-order logic.

For the canonical construction we must modify the notion of a theory in order to reflect the role of the rule NOR in the deduction. In this rule, the Boolean variable $p$ plays a special role like bounded variables in quantifier logics, and this will be incorporated in the new notion of a theory. First of all, we need some preparations.

If $A$ is a formula or a set of formulas, then Var $(A)$ denotes the set of Boolean variables occurring in the members of $A$. By Th (Var) we denote the set of all theorems of PWRCC$_{NOR}$ constructed form the set of all Boolean variables Var of our language. Sometimes, we need to extend the set of Boolean variables to a set Var$'$. In this case, Th (Var$'$) will denote the set of theorems constructed from Var$'$. Note that Th (Var) and Th (Var$'$) are not too different because theorems in Th (Var$'$) are versions of theorems in Th (Var). However, it is more convenient to consider them separately.

A pair $T = (\mathcal{V}, \Gamma)$ is called a NOR-*theory* if $\mathcal{V}$ is a set of Boolean variables and $\Gamma$ is a set of formulas satisfying the following conditions:

(1) all theorems of PWRCC$_{NOR}$ belong to $\Gamma$,

(2) if $\alpha, \alpha \Rightarrow \beta \in \Gamma$, then $\beta \in \Gamma$,

(3) if $\alpha \Rightarrow aCp \vee p^*Cb \in \Gamma$ for some Boolean variable $p \notin \mathcal{V} \cup$ Var $(\alpha \Rightarrow aCb)$, then $\alpha \Rightarrow aCb \in \Gamma$.

The variables in $\mathcal{V}$ are called the *free variables of $T$* and the members of $\Gamma$ are called *formulas* of $T$. We will also write $T = (T_1, T_2)$, where $T_1$ is the set of free variables of $T$ and $T_2$ is the set of formulas of $T$. We say that a formula $\alpha$ *belongs to $T$* and write $\alpha \in T$ if $\alpha \in T_2$. By (1) and (2), $T_2$ is a theory. We say that $T$ is *consistent* if $\perp \notin T_2$ or, equivalently, if $T_2$ is a consistent theory.

A set $A$ of formulas is said to be NOR-*consistent* if there is a consistent NOR-theory $T$ such that $A \subseteq T_2$.

A theory $T$ is called a *good* NOR-*theory* if out of $T_1$ there are infinitely many Boolean variables.

For example, $(\varnothing, \text{Th (Var)})$ is a good NOR-theory. If $\Gamma$ is a consistent theory, then the pair $T = (\text{Var}, \Gamma)$ is a consistent NOR-theory because $T$ is trivially closed under the rule NOR (out of Var there are no variables). But $T$ is not a good theory.

We say that $T$ is *included* in $T'$ and write $T \subseteq T'$ if $T_i \subseteq T_i'$, $i = 1, 2$. A theory $T$ is a *complete* NOR-*theory* if it is a consistent NOR-theory and for any formula $\alpha$ we have either $\alpha \in T_2$ or $\neg\alpha \in T_2$. A theory $T$ is called a *rich* NOR-*theory* if for any formula $\beta$ of the form $\alpha \Rightarrow aCb$

if $\beta \notin T_2$, then $\alpha \Rightarrow aCp \vee p^*Cb \notin T_2$ for some Boolean variable $p$.

Our next goal is to show that every consistent good NOR-theory can be extended to a complete rich NOR-theory. For this purpose, we formulate and prove several lemmas.

Let $\Gamma$ be a set of formulas, and let $\alpha$ be a formula. Denote $\Gamma + \alpha = \{\beta : \alpha \Rightarrow \beta \in \Gamma\}$. Let $T$ be an NOR-theory, and let $\alpha$ be a formula. Denote $T \oplus \alpha = (T_1 \cup \mathrm{Var}\,(\alpha), T_2 + \alpha)$.

The following preliminary assertion is used in the proof of the Lindenbaum lemma (Lemma 3.6.2 below).

**Lemma 3.6.1.** *Let $T$ be a good* NOR-*theory, and let $\alpha$ be a formula. Then*

(1) $T \oplus \alpha$ *is a good* NOR-*theory containing $T$, and $\alpha \in T_2$,*

(2) $T \oplus \alpha$ *is inconsistent if and only if $\neg\alpha \in T_2$,*

(3) *if for some $\beta$ of the form $\neg(\alpha \Rightarrow aCb)$ the theory $T \oplus \beta$ is consistent, then there is a Boolean variable $p \notin T_1 \cup \mathrm{Var}\,(\beta)$ such that $(T \oplus \beta) \oplus \neg(\alpha \Rightarrow aCp \vee p^*Cb)$ is consistent.*

**Lemma 3.6.2** (Lindenbaum lemma for NOR-theories). *Every good consistent* NOR-*theory $T = (\mathcal{V}, \Gamma)$ can be extended to a complete rich* NOR-*theory $T' = (\mathcal{V}', \Gamma')$.*

PROOF. Let $T = (\mathcal{V}, \Gamma)$ be a consistent good NOR-theory, and let $\alpha_1, \alpha_2 \ldots$ be an enumeration of all formulas. Introduce an increasing sequence of consistent good NOR-theories $T_n = (\mathcal{V}_n, \Gamma_n)$, $n = 1, 2, \ldots$, by induction. Let $T_1 = T$. Assume that $T_1, \ldots, T_n$ are already defined. To define $T_{n+1}$, we consider several cases.

*Case* 1. $T_n = (\mathcal{V}_n, \Gamma_n)$ is consistent.

(a) $\alpha_n$ is not of the form $\neg(\alpha \Rightarrow aCb)$. In this case, we put $T_{n+1} = T_n \oplus \alpha_n$. By Lemma 3.6.1, it is a good NOR-theory.

(b) $\alpha_n$ is of the form $\neg(\alpha \Rightarrow aCb)$. By Lemma 3.6.1(3), there exists a Boolean variable $p \notin V_n \cup \text{Var}(\alpha_n)$ such that $(T_n \oplus \alpha_n) \oplus \neg(\alpha \Rightarrow aCp \vee p^*Cb)$ is a consistent good NOR-theory. Let $p$ be the first NOR-theory possessing this property. In this case, we put $T_{n+1} = (T_n + \alpha_n) \oplus \neg(A \Rightarrow aCp \vee p^*Cb)$.

*Case* 2. $T_n + \alpha_n$ is not consistent. Then we put $T_{n+1} = T_n$. In this case, $\neg\alpha_n \in \Gamma_n$. Define $\Gamma' = \bigcup_{n=1}^\infty \Gamma_n$ and $\mathcal{V}' = \bigcup_{n=1}^\infty \mathcal{V}_n$. Then $T' = (\mathcal{V}', \Gamma')$ is the required NOR-theory. $\square$

Note that Lindenbaum lemma can be applied only to good NOR-theories, whereas the "goodness"' property is not essential for consistency because it depends on the amount of Boolean variables in our language. However, this fact is not too important, because the language can be extended. The following assertion clarifies this case.

**Lemma 3.6.3** (conservativeness). *Let $T = (V, \Gamma)$ be a consistent NOR-theory in the language based on the set $\text{Var}$ of Boolean variables, and let $\text{Var}'$ be an extension of $\text{Var}$ by a denumerable set of Boolean variables. Then there exists a consistent good NOR-theory $U = (\mathcal{W}, \Delta)$ in the language with $\text{Var}'$ such that $\Gamma \subseteq \Delta$.*

IDEA OF THE PROOF. Let $\Delta = \{\gamma : (\exists\beta \in \Gamma)(\beta \Rightarrow \delta \in \text{Th}(\text{Var}'))\}$, and let $U = (V, \Delta)$. Then $U$ is a consistent good theory in the language with $\text{Var}'$ such that $\Gamma \subseteq \Delta$. $\square$

**Corollary 3.6.4.** *The following assertions hold.*

(1) *A set of formulas is NOR-consistent if it is contained in a good consistent NOR-theory in an extension of the language by a countable infinite set of new Boolean variables.*

(2) *A set of formulas is NOR-consistent if it is contained in a complete rich NOR-theory in an extension of the language by a countable infinite set of new Boolean variables.*

The canonical construction uses complete rich theories $T = (\mathcal{V}, S)$. For canonical models we use the second component $S$. All other constructions are the same as in Section 3.3.

The following assertion shows the influence of the rule NOR on the canonical contact algebras: all of them are normal.

**Lemma 3.6.5.** *Let $T = (\mathcal{V}, S)$ be a complete rich NOR-theory, and let $(\underline{B}_S, v_S)$ be the canonical model corresponding to $S$. Then $\underline{B}_S$ satisfies axiom* (Nor).

PROOF. Let $T = (\mathcal{V}, S)$ be a complete rich theory, and let $(\underline{B}_S, v_S)$ be the corresponding canonical model. Suppose that $|a|\overline{C}|b|$. $aCb \notin S$. Since $S$ is rich, there exists a Boolean variable $p$ such that $aCp \vee p^*Cb \notin S$ and, consequently, $aCp \notin S$ and $p^*Cb \notin S$. Then $|a|\overline{C}|p|$ and $|p|^*\overline{C}|b|$, which proves that $\underline{B}_S$ is a normal contact algebra. $\qquad\square$

We are ready to prove the main result of this section.

**Theorem 3.6.6** (strong completeness of PWRCC$_{\text{NOR}}$, [3]). *Let $A$ be a set of formulas. Then the following conditions are equivalent:*

(1) *$A$ is NOR-consistent,*

(2) *$A$ has an algebraic model in the class of normal contact algebras,*

(3) *$A$ has a model in the class of all $\varkappa$-normal semiregular spaces,*

(4) *$A$ has a model in the class of compact semiregular $T_0$ $\varkappa$-normal spaces.*

PROOF. The implications $(4) \rightarrow (3) \rightarrow (2)$ are obvious. To prove the implication $(2) \rightarrow (1)$, we assume that $A$ has an algebraic model $\mathcal{M} = (\underline{B}, v)$ in the class of normal contact algebras. Let $\Gamma = \{\alpha : \mathcal{M} \models \alpha\}$. It is easy to show that $\Gamma$ is a consistent theory containing $A$. Thus, $T = (VAR, \Gamma)$ is a consistent (but not good) NOR-theory containing $A$. Hence $A$ is NOR-consistent.

$(1) \rightarrow (2)$ Assume that $A$ is NOR-consistent. By Corollary 3.6.4, $A$ is contained in some complete rich NOR-theory $T = (\mathcal{V}, S)$

in a possible extension of the language by a countable set of new Boolean variables. By Lemma 3.6.5, the canonical contact algebra $\underline{B}_S$ is normal. Hence the canonical model $(\underline{B}_S, v_S)$ is a model of $A$.

(2) → (4) This implication holds in view of the topological representation theorem (Theorem 2.3.9). □

**Corollary 3.6.7** (weak completeness of PWRCC$_{\text{NOR}}$ and PWRCC). *Let $\mathbb{L}$ be any of the logics PWRCC$_{\text{NOR}}$ and PWRCC. Then the following conditions are equivalent for any formula $\alpha$ :*

(1) *$\alpha$ is a theorem of $\mathbb{L}$,*

(2) *$\alpha$ is true in all normal contact algebras,*

(3) *$\alpha$ is true in all semiregular $\varkappa$-normal spaces,*

(4) *$\alpha$ is true in all compact semiregular $\varkappa$-normal $T_0$-spaces.*

PROOF. The required assertion is valid for PWRCC$_{\text{NOR}}$ by Theorem 3.6.6 and for PWRCC by the fact that NOR is an admissible rule in PWRCC. □

## Strong completeness theorem
## for PRCC-like logics

The proof of the completeness of PRCC$_{\text{NOR}}$ can be repeated for any logic in Fig. 2. The canonical construction depends on the choice of a rule, NOR or EXT. For example, if both rules are assumed, then the theories are closed under these rules. If these rules are not assumed either, the notion of a theory becomes standard. We formulate a completeness theorem in a uniform way for all the logics in Fig. 2. For the sake of simplicity, we consider only the algebraic semantics. Using representation theorems for contact algebras, the completeness theorem can be further generalized to some topological spaces.

**Theorem 3.6.8** (strong completeness of PWRCC-like logics). *Let $\mathbb{L}$ be any of the logics in Fig. 2, and let $A$ be a set of formulas. Then the following conditions are equivalent:*

(1) *$A$ is consistent in $\mathbb{L}$,*

(2) *A has an algebraic model in the class of contact algebras corresponding to* $\mathbb{L}$.

We formulate the strongest topological completeness theorem only for $\mathrm{PWRCC_{NOR,EXT}}$ and $\mathrm{PRCC_{NOR}}$.

**Theorem 3.6.9** (strong topological completeness of the logics $\mathrm{PWRCC_{NOR,EXT}}$ and $\mathrm{PRCC_{NOR}}$).

I. *The following conditions are equivalent for any set of formulas A :*
  (1) *A is consistent in* $\mathrm{PWRCC_{NOR,EXT}}$,
  (2) *A has a model in the class of all compact Hausdorff spaces,*
II. *If, in addition, axiom* (Con) *is satisfied, then the corresponding spaces are connected.*

**Corollary 3.6.10** (weak topological completeness theorem for PWRCC and $\mathrm{PWRCC_{NOR,EXT}}$). *Let* $\mathbb{L}$ *be any of the systems* PWRCC *or* $\mathrm{PWRCC_{NOR,EXT}}$. *Then the following conditions are equivalent for any formula* $\alpha$ :

(1) $\alpha$ *is a theorem of* $\mathbb{L}$,
(2) $\alpha$ *is true in all compact Hausdorff spaces.*

Note that Corollary 3.6.10 yields a stronger completeness result for PWRCC than Corollary 3.6.7. The following assertion states a similar result for $\mathrm{PWRCC_{Con}}$.

**Corollary 3.6.11** (weak topological completeness theorem for PWRCC and $\mathrm{PWRCC_{NOR,EXT}}$). *Let* $\mathbb{L}$ *be any of the systems* PWRCC *or* $\mathrm{PWRCC_{NOR,EXT}}$. *Then the following conditions are equivalent for any formula* $\alpha$ :

(1) $\alpha$ *is a theorem of* $\mathbb{L}$,
(2) $\alpha$ *is true in all compact connected Hausdorff spaces.*

## 3.7. Extending the language with new primitives

We consider some extensions of the language $\mathbf{L}(\leqslant, C)$ by new primitives: boundedness and connectedness of regions.

The *Boundedness* is a primitive one-place predicate in bounded contact algebras. Some of the boundedness axioms are not universal sentences, but, fortunately, they can be replaced with additional inference rules for the corresponding axiomatic system.

The *connectedness* is a definable one-place predicate in contact algebras with quantifiers. Thus, such a predicate must be taken for the connectedness predicate in a quantifier-free language; moreover, both sides of the equivalence in the definition must be imitated by suitable inference rules.

These examples show that some additional rules are very useful in the axiomatic characterizations of the predicates under consideration. Further we discuss the complete axiomatization of the predicates of connectedness and boundedness.

### Connectedness

The predicate of connectedness $\mathrm{Con}(a)$ was introduced in Section 2.1 as follows:

$(\#)$    $\mathrm{Con}(a)$ if and only if $(\forall b, c)(b \neq 0$ and $c \neq 0$ and
$\qquad b + c = a \to bCc)$.

We extend the language $\mathbf{L}(\leqslant, C)$ by the predicate Con. We can also extend the notion of an *atomic formula* by setting that $\mathrm{Con}(a)$ is an atomic formula for any Boolean term $a$. The desired topological semantics for $\mathrm{Con}(a)$ is as follows. If $(X, v)$ is a topological model, then

$(X, v) \models \mathrm{Con}(a)$ if and only if $v(a)$ is a connected regular closed set of $\mathrm{RC}(X)$.

We can also define the relational semantics in Kripke structures. We give a complete axiomatization of Con with respect to the topological semantics. The implication "$\Rightarrow$" in (#) suggests the following axiom:

(Connect)    $\text{Con}(a) \wedge p \neq 0 \wedge q \neq 0 \wedge a = p + q \Rightarrow pCq.$

The implication "$\Leftarrow$" in (#) suggests the following inference rule:

$$\text{CONNECT}\quad \frac{\alpha \wedge p \neq 0 \wedge q \neq 0 \wedge p + q = a \Rightarrow pCq}{\alpha \Rightarrow \text{Con}(a)},$$

where $p$ and $q$ are Boolean variables not occurring in $a$ and $\alpha$.

The *axiomatic system* PWRCC-Connect, for $\text{Con}(a)$ is an extension of the axiomatic system for PWRCC extended by axiom (Connect) and the inference rule CONNECT.

The following formula is an example of a nontrivial theorem of PWRCC-Connect:

$$\text{Con}(a) \wedge \text{Con}(b) \wedge aCb \Rightarrow \text{Con}(a + b).$$

The canonical-model-construction for PWRCC-Connect can be done in the same way as for $\text{PWRCC}_{\text{NOR}}$. The following lemma shows how axiom (Connect) and the inference rule CONNECT affect the canonical contact algebra.

**Lemma 3.7.1.** *Let $(\underline{B}_S, v_S)$ be a canonical model of* PWRCC-Connect. *Then for any $|a| \in B_S$*

$\text{Con}(|a|)$ *if and only if* $(\forall |p|, |q| \in B_S)(|p| \neq |0|$ *and* $|q| \neq |0|$ *and* $|a| = |p| + |q| \rightarrow |p|C|q|).$

Lemma 3.7.1 and the topological representation theorems for contact algebras lead to the following completeness result.

**Theorem 3.7.2** (topological strong completeness of PWR-CC-Connect). *The following conditions are equivalent for any set of formulas $A$ of* PWRCC-Connect:

(1) *$A$ is a consistent set in* PWRCC-Connect,

(2) *$A$ has an algebraic model in the class of all contact algebras with the definable predicate $\text{Con}(a)$,*

(3) *A has a model in the class of all topological spaces,*

(4) *A has a model in the class of all semiregular compact* $T_0$-*spaces.*

A similar completeness result can be established for the extensions of PWRCC-Connect by axiom (Con) and the rules EXT and NOR. However, the question about the completeness with respect to Kripke models and decidability is still open. It is of interest to clarify relationships between the contact $C$ and connectedness in special classes of contact algebras. Pratt-Hartmann [43] shows that in some natural contact algebras $C$ is definable by Con in some special sense. A natural candidate for $C$ in terms of Con can be contact algebras satisfying the condition

(C-connect)   if $aCb$, then $(\exists a', b')(a' \leqslant a$ and $b' \leqslant b$ and $\mathrm{Con}(a')$
and $\mathrm{Con}(b')$ and $a'Cb')$.

This condition asserts that a contact between two regions is realized between their connected parts. If (C-connect) is satisfied, we obtain the following equivalence defining $C$ in terms of Con:

$aCb$ if and only if $(\exists a' \neq 0, b' \neq 0)(a' \leqslant a$ and $b' \leqslant b$ and
$\mathrm{Con}(a')$ and $\mathrm{Con}(b')$ and $\mathrm{Con}(a + b))$.

## Boundedness

To define the quantifier-free logic of boundedness, we extend the language $\mathbf{L}(\leqslant, C)$ by a one-place predicate $B$ with the obvious extension of the notion of a formula. The algebraic semantics in local contact algebras (with the boundedness predicate $\mathcal{B}$) was introduced in Section 2.6. We recall the boundedness axioms:

(B1)   $0 \in \mathcal{B}$,

(B2)   if $b \in \mathcal{B}$ and $a \leqslant b$, then $a \in \mathcal{B}$,

(B3)   if $a, b \in \mathcal{B}$, then $a + b \in \mathcal{B}$,

(B4)   if $aCb$, then $\exists c \in \mathcal{B}$ such that $c \leqslant b$ and $aCc$,

(B5)    if $a \in \mathcal{B}$, then $\exists b \in \mathcal{B}$ such that $a \ll b$, $(a \ll b \Leftrightarrow a\overline{C}b^*)$.

Axioms (B1)–(B3) are universal sentences and have direct translation in the language $\mathbf{L}(\leqslant, C, B)$ by the following formulas (denoted by the same symbols):

(B1)    $B(0)$,

(B2)    $B(b) \wedge a \leqslant b \Rightarrow B(a)$,

(B3)    $B(a) \wedge B(b) \Rightarrow B(a + b)$.

Axioms (B4) and (B5) are not universal sentences and should be replaced with the following inference rules:

$$\text{BOUND-1} \quad \frac{\alpha \Rightarrow (B(p) \wedge p \leqslant b \Rightarrow a\overline{C}p)}{\alpha \Rightarrow a\overline{C}b},$$

where $p$ is a Boolean variable not occurring in $a, b, \alpha$,

$$\text{BOUND-2} \quad \frac{\alpha \Rightarrow a \ll p}{\alpha \Rightarrow \neg B(a)},$$

where $p$ is a Boolean variable not occurring in $a$ and $\alpha$.

Denote by PWRCC-Bound the extension of PWRCC in the language $\mathbf{L}(\leqslant, C, B)$ by axioms (B1)–(B3) and the inference rules BOUND-1 and BOUND-2. Then we can introduce canonical models for PWRCC-Bound.

**Lemma 3.7.3.** *Assume that* $(\underline{B}_S, v_S)$ *is a canonical model with* $B(|a|)$ *if and only if* $B(a) \in S$. *Then* $\underline{B}_S$ *is a local contact algebra.*

Lemma 3.7.3 and the corresponding topological representation theorems for local contact algebras lead to the following completeness result.

**Theorem 3.7.4** (strong topological completeness of PWR-CC-Bound). *The following assertions hold.*

(1) *A is a consistent set in* PWRCC-Bound,

(2) *A has an algebraic model in the class of all local contact algebras,*

(3) *A has a model in the class of all locally compact topological spaces,*

(4) *A has a model in the class of all locally compact semiregular $T_0$-spaces.*

Similar completeness theorems can be obtained for extensions of PWRCC-Bound by axioms (Con), (Ext), and (Nor). The question about the decidability of such systems is still open.

**Acknowledgements.** I want to express my thanks to Philippe Balbiani, Georgi Dimov, Ivo Düntsch, Tinko Tinchev and Michael Winter for the collaboration in the field and for many stimulating discussions. I am especially indebted to Georgi Dimov, who showed me how beautiful is topology, and to Tinko Tinchev, for our everyday talks on Logic and Mathematics. Tinko also prepared the figures included in this paper. Special thanks are due to Misha Zakharyaschev, who pointed me out the importance of the field as a boundary area between Logic, Classical mathematics and Computer science. The work is supported by the project NIP-1510 by the Bulgarian Ministry of Science and Education.

# References

1. Ph. Balbiani (Ed.), *Special Issue on Spatial Reasoning*, J. Appl. Non-Classical Logics **12** (2002), No. 3–4.

2. Ph. Balbiani, T. Tinchev, and D. Vakarelov, *Dynamic logic of region-based theory of discrete spaces*, J. Appl. Non-Classical Logics (2006). [To appear]

3. Ph. Balbiani, G. Dimov, T. Tinchev and D. Vakarelov, *Modal logics for region-based theory of space*, 2006. [Submitted]

4. B. Bennett, *A categorical axiomatization of region-based geometry*, Fundam. Inform. **46**, (2001), 145–158.

5. L. Biacino and G. Gerla, *Connection structures*, Notre Dame J. Formal Logic **32** (1991), 242–247.

6. L. Biacino and G. Gerla, *Connection structures: Grzegorczyk's and Whitehead's definition of point*, Notre Dame J. Formal Logic **37** (1996), 431–439.

7. P. Blackburn, M. de Rijke, and Y. Venema, *Modal Logic*, Cambridge Univ. Press, 2001.

8. A. Chagrov and M. Zakharyaschev. *Modal Logic*, Oxford Univ. Press, 1997.

9. E. Čech, *Topological Spaces*, Interscience, 1966.

10. B. L. Clarke, *A calculus of individuals based on 'connection*, Notre Dame J. Formal Logic **22** (1981), 204–218.

11. B. L. Clarke, *Individuals and points*, Notre Dame J. Formal Logic **26** (1985), 61–75.

12. A. Cohn and S. Hazarika, *Qualitative spatial representation and reasoning: An overview*, Fundam. Inform. **46** (2001), 1–29.

13. T. de Laguna, *Point, line and surface as sets of solids*, J. Philos. **19** (1922), 449–461.

14. G. Dimov and D. Vakarelov, *Topological representation of precontact algebras*, In: ReLMiCS'2005, St. Catharines, Canada, February 22-26, 2005, Proceedings, W. MacCaul, M. Winter, and I. Düntsch (Eds.), Lect. Notes Commp. Sci. **3929** Springer, 2006, pp. 1-16.

15. G. Dimov and D. Vakarelov, *Contact algebras and region-based theory of space. A proximity approach. I*, Fundam. Inform. (2006). [To appear]

16. G. Dimov and D. Vakarelov, *Contact algebras and region-based theory of space. A proximity approach. II*, Fundam. Inform. (2006). [To appear]

17. I. Düntsch (Ed.), Special issue on Qualitative Spatial Reasoning, Fundam. Inform. **46** (2001).

18. I. Düntsch, W. MacCaul, D. Vakarelov, and M. Winter, *Topological Representation of Contact Lattices*, Lect. Notes Comput. Sci., Springer, 2006. [To appear]

19. I. Düntsch and D. Vakarelov, *Region-based theory of discrete spaces: A proximity approach*, In: Proceedings of Fourth International Conference Journées de l'informatique Messine, Metz, France, 2003, Nadif, M., Napoli, A., SanJuan, E., and Sigayret, A. (Eds.), pp. 123-129,

20. I. Düntsch, H. Wang, and S. McCloskey, *A relational algebraic approach to Region Connection Calculus*, Theor. Comput. Sci. **255**, (2001), 63–83.

21. I. Düntsch and M. Winter, *A Representation theorem for Boolean Contact Algebras*, Theor. Comput. Sci. (B) **347** (2005), 498–512.

22. I. Düntsch and M. Winter, *Weak Contact Structures*, Lect. Notes Comput. Sci. **3929**, Springer, 2006. [To appear].

23. M. Egenhofer, R. Franzosa, *Point-set topological spatial relations*, Int. J. Geogr. Inform. Systems **5** (1991), 161–174.

24. R. Engelking, *General Topology*, PWN, Warszawa, 1977.

25. V. Efremovič, *Infinitesimal spaces*, Dokl. Akad. Nauk SSSR **76** (1951), 341–343.

26. V. V. Fedorčuk, *Boolean δ-algebras and quasi-open mappings*, Siberian Math. J. **14** (1973) 759–767.

27. D. Gabbay, *An irreflexivity lemma with applications to axiomatizations of conditions in tense frames*, In: Aspects of Philosophical Logic, U. Moenich (Ed.), Reidel-Dordrecht, 1981, pp. 67–89.

28. D. Gabelaia, R. Konchakov, A. Kurucz, F. Wolter and M. Zakharyaschev, *Combining spatial and temporal logics: expressiveness versus complexity*, J. Artif. Intell. Res. **23** (2005), 167–243.

29. A. Galton, *The mereotopology of discete spaces*, In: Spatial Information Theory, Proceedings of the International Conference COSIT'99, C. Freksa - D. M. Mark (Eds.), Lect. Notes Comput. Sci. **1661**, Springer-Verlag, 1999, pp. 251-266.

30. A. Galton, *Qualitative Spatial Change* Oxford Univ. Press, 2000.

31. G. Gerla, *Pointless geometries*, In: Handbook of Incidence Geometry, F. Buekenhout (Ed.), Elsevier, 1995, pp. 1015–1031.

32. A. Grzegorczyk, *Undecidability of some topological theories*, Fundam. Math. **38** (1951), 137–152.

33. A. Grzegorczyk, *The system of Leśnewski in relation to contemporary logical research*, Stud. Log. **3** (1955), 77–95.

34. A. Grzegorczyk, *Axiomatization of geometry without points*, Synthese textbf12 (1960), 228–235.

35. S. T. Hu, *Boundedness in a topological space*, J. Math. Pures Appl. **28**, (1949), 287–320.

36. R. Konchakov, A. Kurucz, F. Wolter and M. Zakharyaschev, *Spatial Logic + Temporal Logic = ?* In: Logic of Space, M. Aiello, I. Pratt, and J. van Benthem (Eds.), 2006. [To appear]

37. S. Leader, *Local proximity spaces*, Math. Ann. **169** (1967), 275–281.

38. *S. Leśnewski, 1927–1931. Collected works*, S. J. Surma, J. T. J. Srzednicki, and D. I. Barmett (Eds.), PWN-Kluwer, Vols 1 and 2, 1992.

39. S. Li and M. Ying, *Generalized Region Connection Calculus*, Artif. Intell. **160** (2004), no. 1–2, 1–34.

40. C. Lutz and F. Wolter, *Modal logics for topological relations*, Logical Meth. Computer Sci. (2006). [To appear]

41. T. Mormann, *Continuous lattices and Whiteheadian theory of space*, Log. Log. Philos. **6** (1998), 35–54.

42. S. A. Naimpally and B. D. Warrack, *Proximity Spaces*, Cambridge Univ. Press, 1970.

43. I. Pratt-Hartmann, *Empiricism and racionalism in region-based theories of space*, Funadam. Informa. **46**, (2001), 159–186.

44. I. Pratt-Hartmann, *A Topological constraint language with component counting*, J. Appl. Non-Classical Logics **12** (2002), no 3–4, 441–467.

45. I. Pratt-Hartmann, *First-order region-based theories of space*, In: Logic of Space, M. Aiello, I. Pratt and J. van Benthem (Eds.), 2006. [To appear]

46. I. Pratt and D. Schoop, *A complete axiom system for polygonal mereotopology of the real plane*, J. Philos. Logic **27** (1998), no. 6, 621–661.

47. I. Pratt and D. Schoop, *Expressivity in polygonal plane mereotopology*, J. Symb. Log. **65** (2000), 822–838.

48. D. A. Randell, Z. Cui and A. G. Cohn, *A spatial logic based on regions and connection*, In: Proceedings of the 3rd International Conference Knowledge Representation and Reasoning, B. Nebel, W. Swartout, and C. Rich (Eds.), Morgan Kaufmann, 1992, pp. 165–176.

49. P. Roeper, *Region-based topology*, J. Philos. Logic **26** (1997), 251–309.

50. D. J. Schoop, *Points in point-free mereotopology*, Fundam. Inform. **46**, (2001), 129–143.

51. E. Shchepin, *Real-valued functions and spaces close to normal*, Siberian Math. J. **13** (1972), 820–830.

52. R. Sikorski, *Boolean Algebras*, Springer, 1964.

53. P. Simons, *PARTS. A Study in Ontology*, Oxford, Clarendon Press, 1987.

54. J. Stell, *Boolean connection algebras: A new approach to the Region Connection Calculus*, Artif. Intell. **122** (2000), 111–136.

55. M. H. Stone, *The theory of representations for Boolean algebras*, Trans. Am. Math. Soc. **40**, (1937), 37–111

56. W. J. Thron, *Proximity structures and grills*, Math. Ann. **206** (1973), 35–62.

57. A. Tarski, *Les fondements de la gèomètrie des corps*, In: First Polish Mathematical Congress, Lwø'w, 1927; English transl.: *Foundations of the Geometry of Solids*, In: Logic, Semantics, Metamathematics, J. H. Woodger (Ed.), Clarendon Press, 1956, pp. 24–29.

58. A. Tarski, *On the foundations of Boolean Algebra*, [in German] Fundam. Inform. **24**, (1935), 177-198; English transl. in: Logic, Semantics, Metamathematics, J. H. Woodger (Ed.), Clarendon Press, 1956, pp. 320–341.

59. A. Urquhart, *A topological representation theory for lattices*, Algebra Univers. **8**, (1978), 45–58.

60. D. Vakarelov, *Proximity modal logic*, In: Proceedings of the 11th Amsterdam Colloquium, December 1997, pp. 301–308.

61. D. Vakarelov, G. Dimov, I. Düntsch, and B. Bennett, *A proximity approach to some region-based theories of space*, J. Appl. Non-Classical Logics **12** (2002), no. 3-4, 527-559.

62. D. Vakarelov, I. Düntsch, and B. Bennett, *A note on proximity spaces and connection based mereology*, In: Proceedings of the 2nd International Conference on Formal Ontology in Information Systems (FOIS'01), 2001, C. Welty - B. Smith (Eds.), pp. 139–150.

63. H. de Vries, *Compact Spaces and Compactifications*, Van Gorcum, 1962

64. A. N. Whitehead, *Process and Reality*, New York, MacMillan, 1929.

65. F. Wolter. and M. Zakharyaschev, *Spatial representation and reasoning in RCC-8 with Boolean region terms*, In: Proceedings of the 14th European Conference on Artificial Intelligence (ECAI 2000), Horn W. (Ed.), IOS Press, pp. 244–248.

66. G. H. von Wright, *A new system of modal logic*, In: Actes du XI
     Congrès International de Philosophie, Bruxelles 1953; Amsterdam
     1953, Vol. 5, pp. 59–63.

# Index